BRUNEL

IN LOVE WITH THE IMPOSSIBLE

BRUNEL

IN LOVE WITH THE IMPOSSIBLE

A celebration of the life, work and legacy
of Isambard Kingdom Brunel

Edited by Andrew Kelly and Melanie Kelly

Bristol Cultural Development Partnership

Published by:

Bristol Cultural Development Partnership

Leigh Court, Abbots Leigh, Bristol BS8 3RA

Design and typesetting by Qube Design Associates Ltd
Printed by HSW Print

2006

ISBN: Hardback 0955074207 Paperback 0955074215

We would like to thank all those who have provided illustrative material for this book. Their names are credited with the captions. In most instances those who have provided illustrations are also the copyright holders of the images. No image in this book should therefore be reproduced without first ascertaining who holds the reproduction rights and getting the appropriate permission. All cover images are used within the body of the book and credited accordingly there with the exception of the image on the front flap, *The Proposed Suspension Bridge from Rownham Ferry* by Samuel Jackson, c 1836 (Bristol's Museums, Galleries and Archives).

We are grateful for the support of the following organisations who have kindly sponsored this publication:

This collection of newly commissioned essays on the life, work and legacy of Isambard Kingdom Brunel is published as part of the Brunel 200 celebrations in 2006 marking the 200th anniversary of his birth. Brunel 200 is a partnership initiative led by the Bristol Cultural Development Partnership.

CONTENTS

ss Great Britain (*ss* Great Britain *Trust*), *photograph by Mandy Reynolds.*

FOREWORD

In 2006 the bicentenary of Isambard Kingdom Brunel's birth is being marked with a year of major exhibitions, educational programmes, publications, walks and trails, arts projects, competitions, conferences, debates and talks, and much more. This illustrated collection of essays, *Brunel: in love with the impossible*, forms part of the celebrations.

Brunel 200 builds on widespread interest in the achievements of those nineteenth-century innovators whose great works have lasted the test of time and remain relevant to today and tomorrow. As well as celebrating a man, his work and the achievements of the past, Brunel 200 looks to the future, drawing upon Brunel's vision and spirit to help create the new Brunels – individuals and teams able to meet the challenges and opportunities of our times.

This book shows how the impossible can sometimes be achieved and how nineteenth-century Britain was transformed. Brunel was not solely responsible for this transformation – and there was often a terrible price to pay – but his works built the modern world and have stood the test of time.

April 2006

Brunel's signature (University of Bristol).

Isambard Kingdom Brunel statuette
(private collection).

INTRODUCTION

Andrew Kelly and Melanie Kelly

He came second in the BBC's *Great Britons* contest.
When he died, his colleague and friend, Daniel Gooch,
said that 'the greatest of England's engineers was lost, the
man with the greatest originality of thought and power of
execution, bold in his plans but right'.[1] His funeral route
was lined with hundreds of professional colleagues and
members of the public. He was one of the most versatile,
audacious and inspirational engineers of the nineteenth
century. His astounding feats changed the world and
continue to shape the way we live today.

The son of a refugee from Revolutionary France, Isambard Kingdom Brunel created the Clifton Suspension Bridge, the Great Western Railway, Paddington Station, the Royal Albert Bridge, the ss *Great Britain* and the ss *Great Eastern*. He was an engineer and an artist, creating works both functional and beautiful. He was a polymath, embracing engineering, architecture, art and design. As L T C Rolt said in his classic biography, Brunel was the last of Europe's Renaissance men. His failures were glorious ones. And these are far outweighed by the extraordinary successes he has left us.

The great Victorian pioneers, exemplified by Brunel, challenged what was possible. There is much to dislike about the period: appalling working conditions, poverty and, for the vast majority of people (including Brunel himself), no semblance of work-life balance. But there is a great deal to admire, too, particularly in the case of Brunel: his bravery, his attention to detail while retaining the grand vision, his dedication, his partnership-building, his ability to deliver, the inspiration that he provided then and still provides.

Kenneth Clark said that Brunel 'remained all his life in love with the impossible'.[2] This book aims to be a tribute to Brunel, promoting his considerable achievements, assessing their impact and demonstrating why they continue to be relevant today. It is an honest account: Brunel's questionable management style is explored as much as his

View down the Avon with Hotwells and Clifton by Samuel Jackson, c 1825 (Bristol's Museums, Galleries and Archives).

View on the Avon at Hotwells, c 1836 by Samuel Jackson (Bristol's Museums, Galleries and Archives).

genius; and there is nothing to disguise the fact that people died working on his projects. The essays have been specially commissioned from experts in their fields, and the book includes illustrations that have never been published before.

In the opening section, Angus Buchanan provides an overview of Brunel's life and work and explains how he has come to represent the pinnacle of Victorian achievement. Andrew Nahum's chapter about Brunel's father, Sir Marc Brunel, a distinguished engineer in his own right, shows how his innovations in manufacturing ships' pulleys anticipated the later age of mass production. Christine MacLeod explores the reasons for engineers' rise to prominence in mid-nineteenth-century Britain as the professionals secured their place in the nation's pantheon alongside the traditional heroes of the battlefield and the political arena.

This is followed by a section on some of Brunel's major projects, all of which are linked to the city of Bristol. Adrian Andrews and Michael Pascoe tell for the first time the full, fascinating story of the Clifton Suspension Bridge, completed after Brunel's death as a tribute to him by his fellow engineers. Angus Buchanan looks at Brunel's work with the Bristol Dock Company, far less visible or romantic than his other projects in the city but an enterprise that nevertheless helped to ensure the survival of Bristol as a port in the 1840s at a time of great difficulty for the city. Steven Brindle's examination of the Great Western Railway (GWR) considers Brunel's engineering achievements and magnificent legacy in the light of the GWR main line's candidacy for the ultimate accolade of World Heritage Site status. Andrew Lambert tells how Brunel created the modern ship, and how that first prototype survived the vicissitudes and vagaries of the ocean to return to her birthplace over 120 years later and begin a long-term and ultimately triumphant restoration. The section concludes with a brief history by Nick Lee of the development of the Brunel archive held at the University of Bristol – an archive that has done much to ensure that Brunel's work remains alive and that has been of direct assistance to many of the authors in this book.

View of the Clifton Suspension Bridge from Hotwells (Bristol Ferry Boat Company).

The final section examines Brunel's work in a wider context. Michael Bailey paints a portrait of the new generation of young engineers of whom Brunel was a part – a generation remarkable for their innovatory engineering talents, managerial and communication skills, and sheer tenacity. He argues that Robert Stephenson is perhaps more important than Brunel as an engineer. Adrian Vaughan examines the contribution made by hundreds of manual labourers, surveyors, contractors, resident engineers, assistants and investors whose work on site and behind the scenes ensured Brunel's visions became a reality. Denis Smith investigates the dramatic and unprecedented social changes that came in the wake of the railway and other nineteenth-century technical developments. Marcus Waithe tells the story of Victorian opposition to the railways, looking particularly to the work of Thomas Carlyle and John Ruskin, men who expressed an ideological objection to the advent of mass transport. Finally, Claire O'Mahony explores the role of decorative detail in Victorian architecture with a particular focus on the railway station.

The chapters are accompanied by illustrated text boxes providing information on some of Brunel's other projects, people associated with his life and work, and related political, social and cultural events. A detailed timeline is included in the appendices.

Brunel's prolific and varied output and the far-reaching consequences of what he achieved mean we can only hope to provide a flavour of his life and work within this book. We have chosen to focus on Bristol as it was the scene of some of Brunel's major projects and the place where, in 2003, the plans for a massive celebration of his life and work started. It is also a city that has had an ambiguous relationship with Brunel, in that it tried to stop the ss *Great Britain* coming back but now celebrates her renewal with pride and enthusiasm.

Bristol was once England's second city, but at the time of Brunel's birth it was facing increasing competition, particularly from Birmingham, Manchester and Liverpool.

With the coming of the Great Western Railway (GWR) and Brunel's innovative steamships, the ss *Great Western* and *Great Britain*, the city looked set to be at the forefront of a new world of design, transport and travel. But, as through much of Bristol's history, isolated triumphs were offset by slow decline. Iron-hulled shipbuilding moved northwards, along with much of Britain's industrial economy, and the difficulty of gaining access to the docks, despite Brunel's efforts, defeated attempts to establish Bristol as a transatlantic terminus.[3] Already by the late 1830s, the majority of shareholders in the GWR were from London and a fifth of them were from Liverpool. Bristol's influence was diminished.[4]

The twentieth century continued to hold mixed fortunes for the city. In 1907 historian Stanley Hutton wrote optimistically:

> Notwithstanding… Bristol's failure in the past to take occasion by the hand, so magnificent is her roadstead, her nearness to America, and the volume of her ever-increasing trade, that it is not beyond the bounds of possibility she may yet take the place she has so long vacated, and be once more, as Burke proudly intimated in 1774, "the second city of the kingdom".[5]

But the Depression and two world wars took their toll, and unsympathetic post-war redevelopment saw the loss of the central historic fabric of the city and the destruction of traditional neighbourhoods. The city faced major problems in the late 1980s with recession, a downturn in manufacturing industry, the exodus of major companies to out-of-town locations, homelessness, poor housing conditions and school examination results, and traffic congestion that threatened both quality of life and economic prosperity.

However, Bristol came through. By 2000 Bristol had established partnership initiatives promoting and developing its visitor attractions, housing, sport and retail facilities, providing economic, social and community regeneration. Although their traditional trade is gone, the city-centre docks, which first brought Bristol its wealth and then stood as a symbol of its failure to adapt, have become key to its renewal, with thriving restaurants, walkways, visitor attractions and centres for arts, science, education, media and entertainment. Central to this are Brunel projects such as the ss *Great Britain*, and science and nature centres. Bristol is an international tourism destination and in May 2005 celebrated the start of a direct transatlantic service between Bristol International Airport and Newark, New Jersey. At long last, Brunel's vision of an integrated transport system from London to New York is possible.

We hope that *Brunel: in love with the impossible* not only celebrates that astounding little man in the stovepipe hat, but also shows that the impossible sometimes turns out to be achievable. We return at the end of this book to his legacy, but start with an overview of his captivating life.

ISAMBARD KINGDOM BRUNEL

Angus Buchanan

Brunel (detail) by D J Pound, after a photograph by John Jabez Edwin Mayall, engraving, published 1859 (private collection).

(Opposite) Portrait of Isambard Kingdom Brunel by John C Horsley, c 1843 (Bristol's Museums, Galleries and Archives).

In the autumn of 2002, BBC television conducted a competition to establish who, in the opinion of the viewing public, is the Greatest Briton. A series of immensely enjoyable programmes gave viewers an opportunity to vote from a long list of candidates, from which a shortlist of ten emerged. Isambard Kingdom Brunel figured prominently in this final selection, and he enjoyed the most support until the very last programme, when he was just pipped for the top position by Winston Churchill. He was certainly the only engineer to get into the top ten, which suggests that he is better known to the British general public than any other engineer, so it would not be unreasonable to regard him as 'the Greatest Engineer'.

The whole exercise, however, should not be taken too seriously, as the definition of what is meant by 'greatest' is infinitely variable, depending upon time and place and the personalities making the assessment. Nevertheless, some qualities are, by general agreement, attached to the notion of human greatness, and it is possible to make an assessment of whether or not a person possessed these qualities without pressing too far a comparative analysis with rival candidates. On such a reckoning, it can be determined without any doubt that Brunel possessed many of the qualities of greatness, and that he used them to powerful effect in a professional career which, although short and encompassing several near-disasters, was conspicuously successful.[1]

Isambard Kingdom Brunel was born on 9 April 1806 in Portsea, the third child and only son of Marc Isambard Brunel, who gave him the favoured family name of 'Isambard', and Sophia Brunel, whose maiden name 'Kingdom' became the boy's middle name. Marc was a Frenchman who had been the second son of a Norman farmer, and as such he had been destined for the church. But instead his practical skills persuaded his father to send him to be trained as an engineer. Even though he did not attend one of the great *écoles*, the young Marc benefited from exposure to French techniques of engineering training, which in time he was able to pass on to his son. Marc was a devoted Royalist and found the republicanism of the French Revolution repugnant, and it was while he was hiding from the Terror in 1792 that he met

Marc Isambard Brunel, mezzotint by James Carter after Samuel Drummond, published 1846, showing Thames Tunnel in background (Elton Collection: Ironbridge Gorge Museum Trust).

Brunel's birthplace (collection of Tony Triggs).

Sketch from Brunel's Thames Tunnel Journal, 29 June 1827, showing Brunel crawling to the head of the shield after an influx of debris (University of Bristol).

Sophia. She was the youngest child of a large family raised by a British naval contractor, and after he died she had been in the care of an elder brother who had sent her to France to improve her language skills. It was not a good time for foreigners to be in France, and she soon found herself being persecuted by officials. It was in these circumstances that she met Marc.

The two young people had fallen in love, only to be separated by the political crisis that was then embroiling the whole of Europe. Marc was able to get out of the country with a forged passport and went to the newly United States of America, where he established a promising engineering career for himself in New York. Sophia, meanwhile, had managed to secure her safe return to Britain, where she settled down to wait for Marc. After seven years, in 1799, an opportunity arose for Marc to present to the British Admiralty a scheme for mass-producing the wooden blocks used in thousands in the rigging of all the sailing ships of the day. He grasped the opportunity, sailed to Britain, married Sophia, and succeeded in getting a commission to install a battery of ingenious machines in the 'Block Mills' which he designed in Portsmouth Naval Dockyard. It was while he was engaged on this project, and living in Portsea with Sophia and their two daughters, that Isambard Kingdom Brunel was born.

The child was blessed from the outset with a loving and supportive family and Marc in particular devoted much time and thought to the education of his son, who proved to have quick and retentive powers of learning. There never seemed to be any doubt in the minds of father or son that the young man would follow his parent in becoming an engineer. In addition to his personal talents as an outstandingly good engineer, Marc was able to pass on to Isambard the French engineering skills that he himself had acquired in mathematics and engineering drawing – the practices which he regarded as 'the Engineer's Alphabet', essential to the formation of engineering ability. This was an advantage of enormous importance at a time when British academic engineering was non-existent, and when British engineers had to pick up whatever theoretical knowledge of their profession they could by learning from their own practical experience and that of their contemporaries. Marc also ensured that his son had the opportunity to spend a couple of years in France, when the Napoleonic Wars were over, to benefit directly from French theoretical instruction and from experience in some of the best French engineering workshops.

Much about Brunel's early life remains uncertain through lack of reliable documentary evidence, and any account must rely upon a mixture of indirect testimony, from family members, friends and servants, and subsequent biographers such as Richard Beamish; enlightened reconstruction of events, as attempted gracefully by Lady Noble; or straightforward guesswork, as by many popularisers.[2] But from the moment that he returned home in 1822 at the age of 16 and began to work as an informal apprentice in his father's office, the written records become more plentiful.

The Thames Tunnel, aquatint, hand-coloured, published by R H Laurie, 9 November 1830 (Elton Collection: Ironbridge Gorge Trust Museum). A fund-raising print issued during a period when work on the tunnel had been suspended.

The family had long since moved to London and had undergone various vicissitudes, but had settled fairly comfortably in Chelsea, where Marc was able to supervise his practice as an engineer and had access to the best engineering workshops in the country and opportunities for frequent deliberations with leading scientists and engineers. The young Brunel was able to share in these advantages of metropolitan life and there was plenty of work to keep him busy, involving a wide range of civil and mechanical engineering. His father sent him to visit sites for dock works and bridges in various parts of the country, and introduced him to lectures by Michael Faraday at the Royal Institution. One of these lectures, on the properties of newly discovered gasses, led the Brunels, father and son, to try to construct an alternative prime mover to the steam engine by harnessing the expansive power of carbon dioxide moving between a liquid and a gaseous state. They spent ten years working on this 'gaz engine' before finally abandoning the project.

Meanwhile, Marc had won the commission to use the tunnelling shield, for which he had taken out a patent, to drive a tunnel under the River Thames in the heart of London, between Rotherhithe and Wapping. Begun in 1825, this project took 18 years to complete, after an epic feat of engineering and protracted difficulties with funds and promoters. The young Brunel was involved from the outset and rapidly became his father's chief assistant on the project, spending three years practically living on the job and undertaking many hazardous operations to keep the shield moving forward. Several inundations from the river were overcome with a mixture of bravery and ingenuity, and the experience did much to mature in the young man, barely turned 20, his powers of leadership and engineering resourcefulness. But in January 1828 the Thames broke into the workings with devastating effect, causing several deaths and almost drowning Brunel, who suffered severe though unspecified internal injuries that kept him off work for many months. The project came to a halt and was not resumed for several years, by which time the young Brunel had acquired other commitments and did not resume his work on the tunnel.

At the age of 22, Brunel found his career in the doldrums. He recovered slowly from his accident and continued to help his father in negotiations about the Thames Tunnel, but for a young man of astonishing energy he found himself at a frustratingly loose end, at a time when some of his contemporaries were making great advances. His father had trained him well in the importance of recording progress on his enterprises, so we are fortunate to possess Brunel's diaries for this period, and these convey a clear impression of his vigorous search for a similar breakthrough. He was also encouraged to write a more personal account of his thoughts and aspirations, in a journal of a mere 36 small pages that gives a rare insight into his mind. It was here that he revealed his ambitions and pondered the wisdom of marriage, concluding with prophetic insight that 'My profession is after all my only fit wife'.[3]

While he was in this post-accident limbo, Brunel visited Bristol and became involved in the competition to design a bridge to span the Avon Gorge at Clifton, and he emerged from convoluted arguments about the best way of achieving this by winning the competition early in 1831. Hardly had the project started, however, before the Bristol Riots of that year halted operations, so that it did not really get going until five years later. By this time, Brunel's involvement with Bristol had produced other fruit, first in the shape of a commission to improve the Floating Harbour of Bristol Docks, and then, even more significantly, in the form of his appointment as engineer to what was to become the Great Western Railway (GWR) in March 1833. This gave him the opportunity for which he had been searching, and he grasped it with enormous gusto. He devoted his tremendous energy to surveying the ground between London and

'Brunel's Drawing No 3' (detail) submitted for the first Clifton Suspension Bridge competition (University of Bristol).

Artist's impression of Brunel's 'Giant's Hole' design (below right) (National Archives).

Bristol, choosing the best route for his railway, shepherding the necessary legislation through parliament, raising funds and conducting property negotiations to facilitate the project, creating an engineering team to build the railway, and overseeing every aspect of the work from the time it started, in 1835, until the line was opened throughout in 1841. And that was only the first step in what developed in Brunel's lifetime into one of the major railway networks in the country, fulfilling his vision of a high-speed passenger transport service that led the world.[4]

Lithograph after Joseph Walter of the launch of the Great Britain, *1843 (Bristol's Museum, Galleries and Archives).*

Great Britain *returning to Bristol in 1970 (Bristol United Press).*

The Great Britain *'floating' on its glass sea, 2005 (ss Great Britain Trust), photograph by Mandy Reynolds.*

The Bristol connection thus served Brunel well, but this was not the end of the relationship. The early promise of the GWR encouraged the promoters to launch a subsidiary company to establish a steamship link between Bristol and New York – an enterprise that was widely regarded as impracticable, but demonstrated by Brunel to represent the future of transoceanic transport. The Great Western Steamship Company was formed in 1835 to build the ss *Great Western*, a wooden-hulled paddle steamer powered by Maudslay side-lever engines and designed by Brunel to carry a commercially viable complement of passengers and cargo in addition to the fuel it would need for the two-week voyage. The ship quickly established itself as a great success, and when the company called for a sister-ship so that it could compete as a mail-carrying service, Brunel designed the ss *Great Britain* which, like its predecessor, was built in Bristol. It was, however, substantially larger and required its own dry dock in which it could be built. Brunel, moreover, complicated the operation by deciding, first, to make it an iron-hulled ship – the first large, iron ship – and, second, to drive it by a screw propeller instead of the conventional paddle wheels. These innovations made it necessary for Brunel to design his own steam engine for the ship, which he did with considerable ingenuity, developing an idea of his father's. The ship was launched in July 1843, but could not escape from the Bristol Floating Harbour until the end of 1844, when Brunel took advantage of a particularly high tide to squeeze it through the entrance lock. Once out into the river, it did not return to Bristol until its rusting hull was brought home in triumph from the Falkland Islands in the summer of 1970, to occupy the dry dock in which it had been built and to become the outstanding monument to the industrial age in the city of Bristol.[5]

Despite this close and creative relationship between Brunel and Bristol, he never made his home in the city. He remained a metropolitan man, with his home and office located close to the centre of government in Westminster, and it was from here that he supervised his increasingly intricate web of projects and commissions. Most of these were in south west Britain, but he had projects in many other parts of the country, and became involved in railway enterprises in Ireland, Italy, India and Australia. His last great project was his third ship, the giant ss *Great Eastern*, iron-

The Great Eastern *(private collection).*

Robert Howlett photograph of the Great Eastern, *1857 (Institution of Civil Engineers).*

View of paddle engine room of the Great Eastern, The Illustrated London News, *3 September 1859 (University of Bristol).*

hulled and with two sets of engines, one driving paddle wheels and the other a screw. It was the largest ship to be built until the last decade of the nineteenth century, and being far too big to be built in Bristol it was constructed on the banks of the Thames in Millwall. The ship presented novel and peculiar problems of construction, all of which were overcome to make it a considerable technical success, but for a variety of reasons it was never a commercial success, and there is little doubt that anxiety over its construction contributed to Brunel's early death in 1859, at the age of 53.

Brunel's first project as an engineer in his own right, the Clifton Bridge, an elegant suspension bridge with wrought-iron chains, ran out of funds and was not completed in Brunel's lifetime. But the Institution of Civil Engineers – of which Brunel had been a vice-president and a devoted member – made the happy decision to undertake its

Royal Yacht Britannia *at the opening of the Royal Albert Bridge, Saltash, oil on canvas by Thomas Valentine Robins, 1859 (Elton Collection: Ironbridge Gorge Museum Trust).*

100th anniversary poster for Royal Albert Bridge, original painting by Terence Cuneo, 1959 (National Archives).

Mary Horsley: sketch for a portrait by her brother John, c 1830s (private collection).

Photograph of Henry Marc Brunel (University of Bristol).

completion as a memorial to him, and as such it was opened in 1864. He had built many other bridges in his career, including the Hungerford Bridge across the Thames, where Charing Cross railway bridge now stands, and dozens of railway bridges of all shapes and sizes, with some particularly notable constructions in brick and wood. It was iron, however, that was the most promising material for the development of wide spans, and especially wrought iron, which was becoming available in bulk and in large plates for the first time during Brunel's career. He was quick to realise its potential in avoiding the dangers of cast iron in tension, and gave much thought to the evolution of girders appropriate to the circumstances for which they were required. With the railway bridge at Windsor and the road bridge at Balmoral, he came close to anticipating the sort of lattice girder bridge that became almost universal in railway practice. But for his largest spans he developed, in the Chepstow railway bridge and the Royal Albert Bridge over the Tamar at Plymouth, types of wrought-iron trusses that were his characteristic contribution to modern bridge technology. These were technically successful even though they were comparatively expensive and the design was not subsequently pursued.[6]

Over almost three decades, Brunel produced an astonishing range of successful railway, bridge and steamship works. He also took an active part in the organisation of the Great Exhibition in Hyde Park in 1851, and when the Crystal Palace was moved to Sydenham after the Exhibition, Sir Joseph Paxton commissioned him to build the distinctive water towers for the enlarged structure.[7] During the Crimean War, in 1854-56, Brunel provided suggestions for floating gun batteries, which were not taken up, and for a complete pre-fabricated hospital which was built at Renkioi on the Dardanelles.

As his business prospered, he acquired an estate at Watcombe, near Torquay in Devon, where he spent a lot of effort in planning the garden and in designing a house of which only the foundations had been laid at the time of his death. He had married Mary Horsley in 1836, and despite his hyper-active career he found time to spend with his family and to design the fittings of his lavishly furnished house in Duke Street, Westminster, for which he commissioned the top artists of the day to paint pictures. The family came to consist of two boys, Isambard and Henry Marc, and a girl, Florence. Both the sons died without issue, but the blood line was passed on by Florence, who married an Eton house master called Arthur James. The family has been careful of Brunel's memory, providing the first and official biography in 1870, and saving for posterity the bulk of the marvellous archive of documents that now resides in the University of Bristol Library.[8]

Against this background of an exceptionally creative career it is possible to assess Brunel's life and work as an engineer. He was, in the first place, a man of extraordinary engineering vision. He had the capacity to grasp broad concepts, while never failing to attend to the details of the operation. Although his Great Western Railway was definitely not the first railway, Brunel quickly perceived a potential for high-speed passenger transport that his predecessors had not appreciated. George and Robert Stephenson had pioneered the national system of railways, but they saw it as a network linking the main centres of population and allowing the smooth transit of goods at modest speeds. Brunel had travelled on the Liverpool & Manchester Railway in 1831, the year after it opened, and yearned for the chance to improve on it, as he noted hopefully in his diary: '… the time is not far off when we shall be able to take our coffee and write while going, noiseless and smoothly, at 45 miles per hour – let me try.'[9]

Working from first principles for a fast passenger service, he designed the GWR to be as smooth as possible, and to that end he committed it to the broad gauge of seven feet instead of the conventional four feet 8½ inches favoured by the Stephensons.

Sketches of station lampposts for Bristol (University of Bristol).

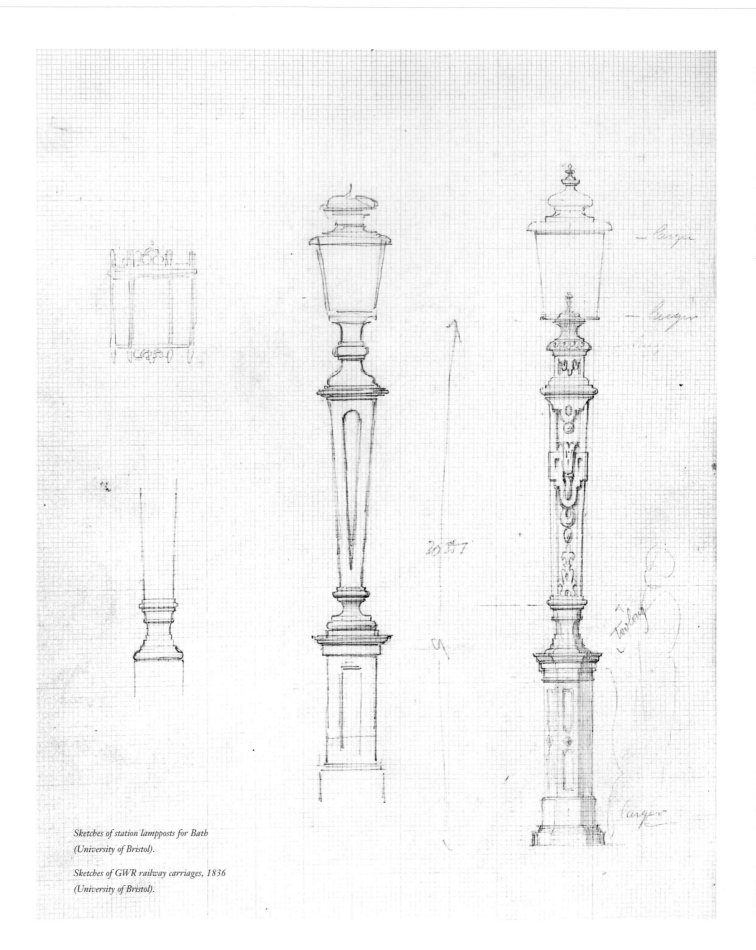

*Sketches of station lampposts for Bath
(University of Bristol).*

*Sketches of GWR railway carriages, 1836
(University of Bristol).*

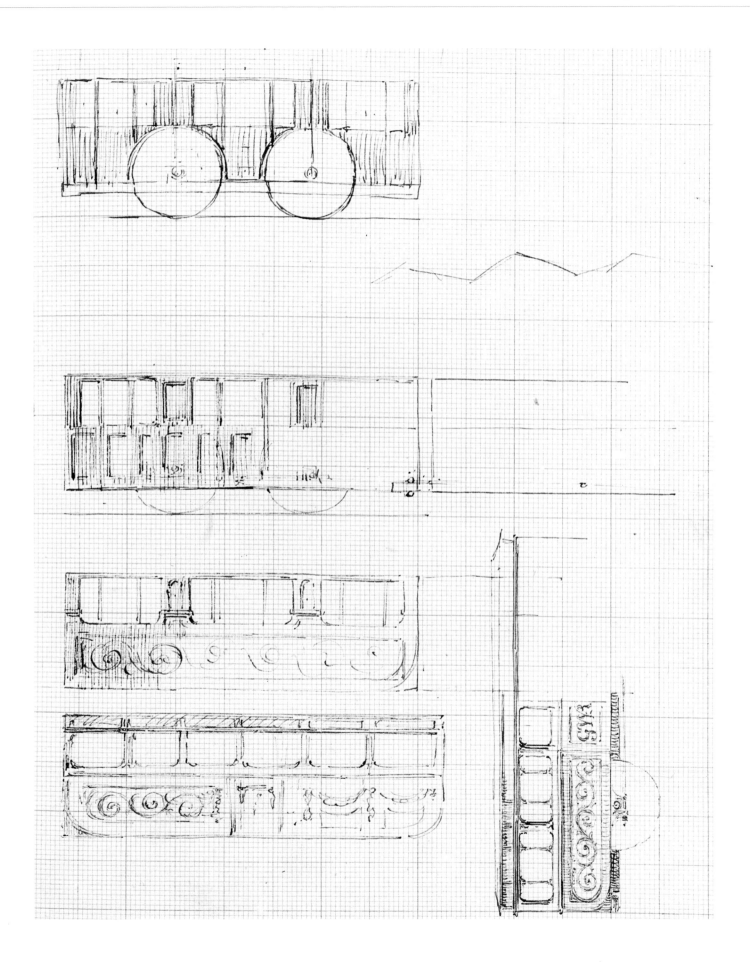

He also aimed at enhancing the stability of the track by laying it on continuous longitudinal sleepers. This distinctive track was ultimately a burden to the GWR because it made it different from all other railways and created formidable problems of transhipment, but technically the broad gauge was a success and enabled the GWR to seize the initiative in high-speed passenger traffic for two decades.

Brunel's vision of transoceanic steamship services was of similar character, and showed qualities of engineering genius. His scientific rival, Dr Lardner, had argued authoritatively that it would be impossible to build a workable steamship for long hauls because the bulk of fuel needed would take all the space needed for passengers and cargo. Brunel proved otherwise, showing in practice that the total requirements for fuel decline in proportion to increases in the size of a ship, so that larger ships become commercially viable. His first steamship converted the business of transoceanic transport to his view, and even though he failed to pursue his advantage in this respect, his second and third ships both introduced outstanding innovations and promoted the rapid transformation of ocean transport.

It does not follow from these instances of Brunel's vision that he always got it right: his failure to anticipate the long-term consequences of the introduction of the broad gauge is a case in point, and the enthusiasm with which he embraced the idea of atmospheric propulsion was even more obviously misguided. But with Brunel, even his failures and disasters tended to be monumental, so that he attracted admiration at least for his panache. The atmospheric system introduced by him on the South Devon Railway was not, after all, completely wrong in principle, and if it had worked well it is possible that it would have become a commercial success. But things did not work

View of South Devon Railway passing through tunnel near Dawlish after atmospheric system was abandoned, watercolour by W Dawson, c 1848 (private collection).

Pumping station for atmospheric railway, GWR Sketchbook (University of Bristol).

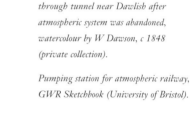

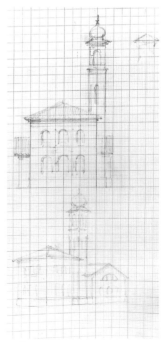

View of Clifton with artist's impression of proposed bridge, c early 1830s (National Trust).

well, and even though the decision to go straight from small-scale experiment to main-line implementation would be regarded as unacceptable today, he could not be blamed for some of the component failures such as the perishing of the leather longitudinal valve on the pressure pipe. The vision was impressive, but it proved to be defective.

Brunel's vision contributed substantially to his charisma. He had the ability to charm people with his ideas, and to inspire them to follow his leadership. His personal journal, intended for his own eyes alone, contains some curious reflections on his desire to impress people whom he passed in the street, and it seems likely that there was an element of compensation here for his small stature. He admitted as much to himself in an astonishing piece of self analysis:

> My self conceit and love of glory or rather approbation vie with each other which shall Govern me. The latter is so strong that even of a dark night riding home when I pass some unknown person who perhaps does not even look at me I catch myself trying to look big on my little pony.[10]

Like many people of fairly small build, he asserted himself to overcome a perceived limitation and in the process discovered powers of leadership which he employed very effectively. We have the eloquent testimony of Richard Beamish that even as a young man working on the Thames Tunnel project, Brunel displayed such qualities in a crisis: 'I saw that each man, with his eyes upon Isambard Brunel, stood firmly prepared to execute the orders he should receive with resolution and intrepidity.'[11]

One of the most impressive features of Brunel was his dynamism. He learnt how to drive himself to prodigious feats of sustained activity, and then drove others whom he expected to perform likewise. All the records indicate his enormous capacity for work, so that when his career became fully geared up in the mid-1830s he was able to run several major projects concurrently, with the assistance of a team which did his bidding and served him loyally. He became known as the 'Little Giant': a restless, highly mobile bundle of energy – admirable, but challenging to work with. Temperamentally, he seems to have had little time for holidays, although he always took the opportunity to visit things that interested him during his travels on business, and once he had acquired a family it became necessary for him to set up his wife and children in the summer months, usually in a seaside resort, where he would join them for a few days. Even his recreation was undertaken with vigour: a trip to Paris to see the Revolution in 1848 became a mission to acquire special furniture. And, famously, a party for his children in 1843 became a conjuring display that ended with him swallowing a sovereign and requiring intricate surgery. Leading surgeon Sir Benjamin Brodie performed a tracheotomy using a long-handled pair of forceps that Brunel may have designed himself. When this proved unsuccessful, Brunel devised a board,

Watercolour of Avon Gorge showing observatory and Clifton tower of the bridge, 1851 (private collection).

pivoted between two uprights, to which he was strapped and rapidly turned head over heels. The centrifugal force dislodged the coin, which dropped from his mouth. This was the dynamism of a driven man.

It is sometimes suggested that Brunel was not a great inventor, because he took out no patents. This, however, was a matter of principle with him: he objected to the patent system, seeing it as an unwarranted check on freedom of enterprise. But he was undoubtedly a great innovator, being exceptionally alert to new ideas, and quick to take advantage of any that were of use to him. He showed inventive talent in his work on the gaz engine, for which his father had taken out a patent, as he did also for the tunnel shield and other inventions: Marc did not share his son's aversion to the system. With his recognition of the value of the iron ship and screw propulsion, his perception was shrewd and rewarding. With the electric telegraph and creosote, he was prepared to try out good ideas. The atmospheric system, on the other hand, was not such a happy example of his readiness to adopt an untried novelty.

Brunel showed outstanding talents in organising large operations and establishing viable managerial procedures. These were manifest in his complex railway undertakings and ship-building activities, and in the elaborate detail with which he worked out methods of moving large bridge sections into position, or of launching ships. The floating of the great trusses of the Royal Albert Bridge, for example, required an operation that was almost military in its precision, and Brunel presided over it with complete authority. This helps to explain why he was so mortified when, having planned the novel sideways launch of the *Great Eastern* in the same way, he discovered on the day that the directors had sold tickets to the public for what they conceived to be a celebratory display. The result was a disaster, and the launch was long delayed. All Brunel's attention to detail in these operations is open to the criticism that he should have delegated more of his authority to his assistants, and there is some substance in these criticisms. But for Brunel, being in charge of an organisation or operation meant being responsible for it, and he wished to carry the full responsibility himself. He had the qualities of a one-man organiser rather than of an organisation man.

In his combination of talents, Brunel reflected more fully than most of his contemporaries the spirit of mid-Victorian Britain. He was imbued with the *zeitgeist* of entrepreneurial enthusiasm that made possible the construction of the railways and the transformation of oceanic navigation, the development of huge industrial enterprises and the confident assertion of imperial power, all characteristic of his period. He was outstandingly a man of his times, and it was because his talents matched his opportunities so closely that he was able to grasp those opportunities and create engineering works that were staggering in their scale, their elegance and their fitness for purpose. Earlier engineers, like his brilliant father, had less scope for their

Photograph of Brunel taken moments before his collapse on board the Great Eastern, *1859, from stereoview (above) (ss* Great Britain Trust*).*

Brunel with colleagues at unsuccessful launch of the Great Eastern, *1857 (left) (Institution of Civil Engineers).*

talents, while later engineers were, in a sense, victims of the success of their profession which, in response to its rapid expansion, also became increasingly specialised and reluctant to tolerate the easy transfer of skills from one area of engineering to another. Thus Brunel, together with a handful of his contemporaries, and particularly his great friend and rival Robert Stephenson, was able to flourish as an engineering polymath, undertaking several specialisms simultaneously. On any reckoning, therefore, it is this small group of mid-nineteenth-century British engineers that constitutes a sort of pantheon or Heroic Age of engineering, and Brunel was undoubtedly one of the most colourful figures among them.[12]

Brunel had one other advantage that deserves to be mentioned. In addition to his natural talents as an engineer and their assiduous cultivation by his devoted parents, and his brilliance in seizing the opportunities with which he was presented, it was fortunate for the image of Brunel in the mind of posterity that his career coincided with the birth of photography. W H Fox Talbot published his first portfolio of prints produced from his novel 'negative' process as *The Pencil of Nature* in 1844, and among the very first such images were two of works by Brunel – the Hungerford Bridge and the *Great Britain* in Bristol Harbour. Photography improved rapidly in the following decade, and by the time that the *Great Eastern* was being constructed in Millwall, several photographers were engaged to catch images of the work. One of these, a young man called Robert Howlett, was able to take pictures of Brunel as he stood in front of the huge chains installed to restrain the launch of the ship, top hat on his head and cigar in his mouth, hands thrust into his waistcoat pockets as he confidently faced the tasks ahead. These have become truly iconic images, establishing the face of Brunel in the collective memory in a manner achieved by no other engineer. Whether or not he is, in any meaningful sense, 'the greatest engineer', these images have certainly ensured that he is the best known.

JOHN CALLCOTT HORSLEY

On 5 July 1836 Brunel married Mary Horsley, eldest child of William Horsley, an organist and composer, and his wife Elizabeth Hutchins Callcott.

Brunel had been introduced to the Horsley family through his brother-in-law, Sir Benjamin Hawes, in the early 1830s, and he had often visited their home in High Row, London, where he enjoyed amateur theatricals, charades, music and oratories. After their marriage, Mary became a popular hostess at their elegant London home and 'showed every sign of being satisfied with the role of being an adornment in the social life of a successful professional man'.[1]

Mary's eldest brother was the artist and Royal Academician John Callcott Horsley. The inventory of property at Brunel's home in Duke Street compiled in 1858 includes Horsley's paintings *A Calm* and *Romeo and Juliet*, and two Horsley portraits. Horsley accompanied his brother-in-law on a visit to some of his railway projects in November 1834, sleeping rough, rising early and enduring Brunel's prodigious snoring. They also went to Italy together in 1842 and to witness the Paris Revolution in April 1848.

Horsley had travelled through England during the 1830s, sketching Elizabethan and Jacobean houses that would later feature in his paintings of period domestic scenes. He also illustrated books and created two frescoes for the new House of Parliament. In 1845, he took up the post of headmaster of the School of Design at Somerset House, remaining there for two years.

His first wife, Elvira, died in 1852. This was followed by the deaths of two of their sons in 1854 and of their third son in 1857. He married his second wife, Rosamund, in 1854 and they had seven children including a future surgeon, a writer, an architect and a painter. In the 1860s he moved with his family to Cranbrook in Kent, which had become an artists' colony, and began to paint some more contemporary subjects.

Horsley was closely associated with the Royal Academy, organising the old master exhibitions for nearly 30 years and being its treasurer from 1882 to 1897. A deeply religious man, he was often ridiculed in the press for his disapproval of female nudes and was nick-named 'clothes-Horsley' in the pages of *Punch*. He was described in his obituary in *The Times* as 'one of the most sweet-tempered of men, and one of the most innocent'. He was buried in Kensal Green Cemetery.

In 1870 Horsley wrote to Brunel's son, Isambard, who had requested some reminiscences of his father. He recalled:

I remember with singular distinctiveness the first time I ever saw him, when I was a lad of fourteen, and had just obtained my studentship at the Royal Academy. He criticised with great keenness and judgment a drawing which I had with me, and at the same time gave me a lesson on paper straining. From that time till his death he was my most intimate friend. Being naturally imbued with artistic taste and perception of a very high order, his critical remarks were always of great value, and were made with an amount of good humour which softened their occasionally somewhat trying pungency. He had a remarkably accurate eye for proportion, as well as taste for form. This evinced in every line to be found in his sketch books, and in all the architectural features of his various works.

So small an incident as the choice of colour in the original carriages of the Great Western Railway, and any decorative work called for on the line, gave public evidence of his taste in colour; but those who remember the gradual arrangement and fitting up of his house in Duke Street will want no assurance from me of your father's rare artistic feeling. He passed, I believe, the pleasantest of his leisure moments in decorating that house, and well do I remember our visits in search of rare furniture, china, bronzes, &c., with which he filled it, till it became one of the most remarkable and attractive houses in London.[2]

THE HOWLETT PHOTOGRAPH

Robert Howlett took a series of photographs of the *Great Eastern* during the ill-fated attempts to launch the ship in November and December 1857. Nine were turned into engravings used in *The Illustrated Times'* 'Leviathan Number' (16 January 1858) and a set of 15 were sold as stereoscopic slides.

They were also sold as prints and distributed as promotional cards for Howlett's photographic business, thereby demonstrating how early photographers had already spotted the commercial opportunities of mass-produced images.

It was during his assignment at the ship that Howlett took what has become the most famous image of Brunel. The engineer is standing, isolated and self-assured, before the chains of the checking drums used to control the launch. The photograph has been described as blending the real and the symbolic to show 'the fully-fledged national industrial hero, icon of the Industrial Age and exemplar of its achievements and dominant values'.[1] It is ironic that at the time Brunel was facing imminent disaster as the launch proved such a protracted and stressful enterprise.

In 1853 Howlett had joined the Photographic Institution, a business established by Joseph Cundall in New Bond Street, London. Regular clients included the Queen and Prince Albert. A series of portraits Howlett had taken of soldiers, entitled 'Crimean Heroes', was shown as part of the annual London exhibition of the Photographic Society – later the Royal Photographic Society – in 1857. Howlett also took a series of portraits

Cover of Watershed brochure for Brunel's Kingdom *exhibition, 1985 (Watershed Media Centre).*

of leading British artists, some of which were exhibited in the Art Treasures Exhibition in Manchester. These included one of Brunel's brother-in-law, John Callcott Horsley.

Robert Howlett died of a fever at the age of 27. His colleagues assumed this was due to overwork and his repeated exposure to the hazardous chemicals then used in making photographs.

Isambard Kingdom Brunel and the launching chains of the Great Eastern, photograph by Robert Howlett, 1857 (Getty Images).

MARC ISAMBARD BRUNEL

Andrew Nahum

Portrait of Marc Brunel, c 1802 (Institution of Civil Engineers).

The spectacular brilliance and energy of I K Brunel explored by Angus Buchanan has often tended to relegate his father, Marc Isambard Brunel, to the role of precursor – a kind of John the Baptist to his greater son. But Marc Brunel was also highly original and inventive. He played a crucial role as the educator of his son and was equally important to the development of both civil and mechanical engineering.

Marc Brunel is perhaps also less well known because he was most active in a period when engineering was beginning to be defined as a profession, and indeed helped himself to define it, but in advance of the great transformations to industry and transport resulting from the creation of the railway system and other public works, which made later engineers like Robert Stephenson, and Isambard, into the 'superstars' of their age.

Sir Isambard Brunel retiring from the opening celebrations at the Thames Tunnel, The Illustrated London News, *25 March 1843 (University of Bristol).*

But in Marc Brunel we can see many of the roots of his son Brunel's character. Both shared an extraordinary profligacy – almost a promiscuity – of ideas and talents, so that different schemes and enterprises jostled incessantly with each other. In Marc's case, almost any of his specialisms, if he had been content with one alone, would have been enough to secure a living, and perhaps also enduring fame. His bridge designs, his surveys in America for inland ship navigations, his machinery and wood machinery at Battersea sawmill and even, perhaps, his mechanisation of boot manufacture could have been career enough for one man. Tunnelling, too, could have been a career, although the Thames Tunnel came rather too late in life for him to capitalise on it. The tunnel, incidentally, seems always to have attracted a disproportionate amount of attention in the histories of Marc Brunel, particularly because of its incredible completion in the teeth of appalling problems, and also because the project initiated his son into the direction of a major civil engineering project, and perhaps also instilled in him a particular heroic and personally unstinting style of commitment and management.

The attention given to the tunnel obscures what is arguably Marc Brunel's most significant achievement, viewed from the perspective of the history of technology: the design and installation of suites of machines at Portsmouth for the manufacture of pulley blocks for the Royal Navy, the first effective purpose-designed and integrated system for mass production in the world. These 'block mills' were significant in the history of mass production, successful, well known and well regarded, yet they had little effect on the history of mass production or on the style of British manufacture in the nineteenth century.

Marc Brunel's prolific range of ventures must be seen in the light of an era when the engineer was not a mere consultant, as so often today, hired to verify or to realise other people's schemes and projects. Engineers like Marc and Isambard Brunel were, in themselves, representatives of a new species of entrepreneur. Like many of his contemporaries in the emerging field of engineering, they spread themselves over a range of projects from machine-tool design to civil engineering and architecture. Because they were gifted with an uncanny and often self-taught understanding of material things and their relationships, they could use this uncommon understanding to propose new schemes and new solutions to manufacturing, to building and to transport, which recommended themselves to investors because, if everything went right, they promised a cheaper way of performing some necessary service or offered a new facility that people would pay for.

For Marc Brunel, this fluency resulted in part from natural talent, but also from a fine mathematical and scientific training at a specific period in French history. It was also combined with practical experience in the French navy. To understand these factors it is necessary to have a concise account of Marc Brunel's earlier life, and this also helps us towards an understanding both of his own career and of the character and style of his son, Isambard.

Marc was born in 1769 at Hacqueville, near Rouen, in Normandy, to a long-established family with reasonable means. Marc's mother died when he was seven, and according to often-repeated family legend the intention was to educate the scholarly and obviously highly intelligent Marc for the priesthood. His elder brother showed an interest in running the farm, and though the farm was substantial, it was not the father's intention that it should be divided in an effort to support two families. Like many younger sons in those days, Marc would have to find a separate career and the church was seen as a sensible choice that would guarantee some social standing and a good living for a bright child.

Throughout boyhood, Marc made things with his hands and was fascinated by practical and mechanical problems: the wheelwright shrinking the red-hot iron tyre onto a cart wheel; the work of the shipwrights and the gear of the ships in the harbour at Rouen; the arrival, from England, of huge castings for a steam engine – then a great wonder and novelty – which stood on the quay. In spite of this evidence of vocation, he was sent to a seminary to prepare him for the church. However, the principal of this school, understanding his gifts, allowed him to study mathematics and drawing.

Eventually, Marc moved to live with his cousin and her husband, François Carpentier, who was, in fact, American consul in the city, and to study to enter the navy, tutored by Professor Jean-Noël Dulague, a notable astronomer at the Royal College in Rouen. It has been asserted by biographers that, in this period, Brunel was also taught by Gaspard Monge, the mathematician and designer who developed the new technique of orthographic projection for engineering drawing, by which three-dimensional designs could be fully represented on paper and passed to the machinists and engineers to guide construction. However, a difficulty with this theory is that Monge was, in this period, teaching in Paris and Mézières, and so seems unlikely to have met Brunel, except perhaps in his role of examiner of naval cadets. Another point is that Marc Brunel's drawings for his machines do not appear to use any convention of Monge's type for the projection of three dimensions. They show a beautiful sense of proportion and a good hand, but they are classic architectural views showing front and side elevations and a plan. Therefore, the suggestion that Brunel was a conduit for the passage of this element of engineering development into Britain seems to be wishful thinking.

It is hard to underestimate the importance of the navy in nurturing the new field of engineering during this period. Since the strategic importance of sea power was so great to both France and Britain, their navies were huge enterprises, and much of the latest scientific work in astronomy and hydrography was sponsored by them. Naval officers, responsible as they were for the navigation of ships, were required to become highly adept at mathematics and trigonometry.

The ships themselves were expressions of the highest constructional arts those societies could muster, and each fighting ship can be regarded as an ensemble of intricate mechanical devices. Furthermore, at sea, each was a self-contained and self-sufficient unit, carrying aboard a range of highly skilled craftsmen, including shipwright-carpenters, blacksmiths and riggers, who could effect almost any repair, anywhere in the world. Jobs in foreign parts with no facilities, such as rigging a new mast weighing several tons, using pulleys and shearlegs improvised from what was on board, were routinely undertaken. Sailors could therefore gain extraordinary insight, for the times, into the practical problems of manipulating heavy loads, and acquire an intuitive understanding of what structures and fastenings were appropriate and safe for different loads.

Brunel took up a commission at sea in 1786. When he returned six years later, the Revolution had occurred and France was in a very different state. Brunel spent about 18 months in France, during which he ran into danger for expressing Royalist opinions in Paris and escaped back to Normandy, which was itself a predominantly

Royalist area. As the Paris government tightened its grip, he became increasingly at risk of capture. With his cousin's help, he found a passage to America, though not before becoming deeply attracted to an English girl, Sophia Kingdom, whom the Carpentiers were sheltering.

Marc Brunel eventually gained US citizenship and became chief engineer of New York City. During his time in America, he undertook various works including surveys for navigations of inland waterways, and some architecture, designing a theatre – a modification of an award-winning but unrealised design for a Congress building in Washington – and a cannon foundry for New York. He decided to travel to England in 1799, partly, we surmise, to seek his fortune in the home of manufacture and industry, but also to meet again and marry his sweetheart, Sophia Kingdom, now back in England.

All this caused Samuel Smiles, the great recorder of the Industrial Revolution and its heroes, to observe, not altogether approvingly, that 'the career of Brunel was of a more romantic character than falls to the ordinary lot of mechanical engineers'.[1] However, romance was almost a defining characteristic of the careers of both of the Brunels. In England, Brunel hurried to see Sophia, then living in London with her brother, and the two soon announced their engagement. Meanwhile, he began promoting a scheme he had conceived in America – the idea of the manufacture of pulley blocks for the sailing ships of the Royal Navy.

The Distinguished Men of Science of Great Britain Living in the Year 1807-8 (with detail) (Institution of Civil Engineers). The picture depicts 51 of the leading British men in the field of science assembled in the library of the Royal Institution. The props (a globe, folio stand and an array of folio books) identify them as members of the intellectual elite. Marc Brunel (referred to as Sir M I Brunel) is seated at the desk between John Dalton, chemist and atomic theorist, and Matthew Boulton, mechanical engineer. Directly behind him stands Henry Maudslay with Samuel Bentham beside him.

British Imperial power in the eighteenth century was underpinned by the Royal Navy and a single fighting ship, such as Nelson's *Victory*, needed about a thousand pulley blocks. Indeed, during the Napoleonic wars, the navy purchased about 100,000 a year. The pulley block was therefore a natural subject for the early application of mass production.

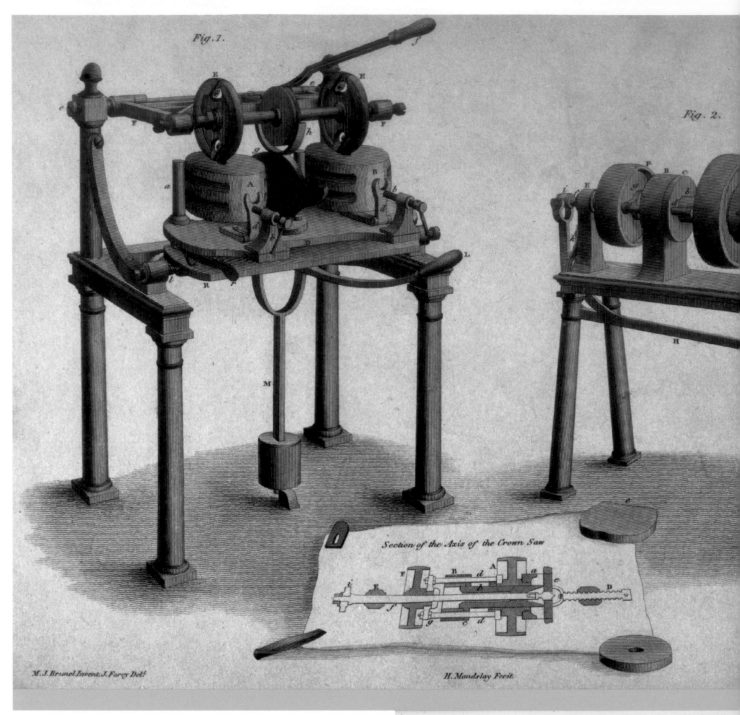

Fig. 1.

Fig. 2.

Section of the Axis of the Crown Saw

M. J. Brunel Invent. J. Farey Delt.

H. Maudslay Fecit.

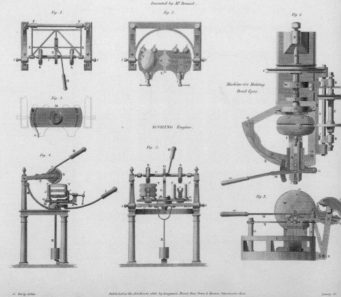

Invented by Mr Brunel.

Fig. 1. Fig. 2. Fig. 6.

Fig. 3.

Machine for Making
Dead Eyes.

SCORING Engine.

Fig. 4. Fig. 5.

Fig. 7.

J. Farey delin.

Published as the Act directs 1816, by Longman, Hurst, Rees, Orme & Brown, Paternoster Row.

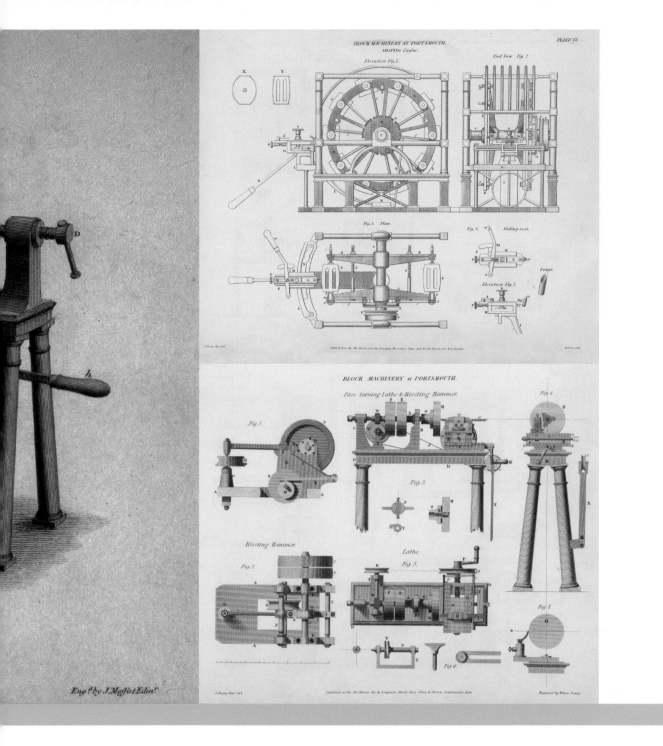

Engravings by J Moffat after drawings by J Farcy of Marc Brunel's machinery designs from Vol 2 of The Cyclopaedia; or, Universal Dictionary of Arts, Sciences, and Literature *by Abraham Rees, published 1820 (Science Museum/Science and Society Picture Library). These plates show a scoring engine, rounding saw, face-turning lathe, riveting hammer, shaping engine and machine for making dead eyes.*

Machines for cutting and shaping wood were not new, but the singular idea behind Brunel's system, apart from the excellence of the individual machines, was the design of an integrated suite of machines where each did a particular job, in sequence and, in his imagined scheme, the pieces of work were passed from one machine to the next.

Through a fellow French émigré, Brunel was introduced to the great mechanician Henry Maudslay, who then had a small workshop near Oxford Street. Maudslay's growing business was to be the cradle of British engineering, for he employed and trained there, at various periods, a range of people who were to become the key figures in the birth of the machine tool industry, including Joseph Whitworth, Richard Roberts, James Nasmyth, inventor of the steam hammer, and Joseph Clement, who was to build Charles Babbage's mechanical computer – the Differential Engine.

According to a perhaps apocryphal tale from Samuel Smiles, Brunel initially visited Maudslay three times, showing him drawings and discussing the manufacture of parts – but without revealing their purpose – when the great man exclaimed: 'Ah! Now I see what you are thinking of; you want machinery for making blocks.' Maudslay then produced a series of small-scale working models, to be used by Brunel in promoting and selling his ideas. These exquisite models are today in the National Maritime Museum at Greenwich.

Brunel meanwhile attempted to sell his scheme, first to Taylors of Southampton, one of the main contractors to the Admiralty for blocks, receiving the reply that they had already perfected their process and that 'I have no hope of anything ever better being discovered, and I am convinced there cannot… Inventions of this kind are always so different in a model and in actual work'.[2] In 1802 Brunel got direct access to the Admiralty and to Samuel Bentham, then Inspector-General of Naval Works, and the brother of the Utilitarian philosopher, Jeremy Bentham, and was able to demonstrate the working of the models created by Maudslay.

Samuel Bentham was the inventor of some wood-working machinery on his own account, an enthusiastic moderniser, and was then driving through an extensive programme of improvements, particularly at Portsmouth, including the installation of steam power. He quickly became convinced of the virtues of Brunel's proposed system.[3]

Brunel arrived with his scheme, therefore, at a most fortunate moment, and with the ideal patron in place at the Admiralty. Installation of the equipment then proceeded at Portsmouth, with various sets of Brunel's production machines produced to make different sizes of pulley blocks. The design of the machines was very particular to the age, since one was only intended to perform a single task. A pulley block consists of the outer case or 'shell' and, within it, one or more 'sheaves' (the pulley wheels, over which the ropes run) turning on a shaft or 'pin'. In Brunel's system, one series of machines prepares the shells, sawing rough blanks, cutting the mortices (the slots) in which the sheaves turn, boring the shells for the pin, and cutting the smooth, rounded, exterior surfaces on the shaping machine, ten blocks at a time. Another parallel series of machines cut the blanks of lignum vitae for the sheaves, turned the exterior diameter to size, put in the trough or 'score' for the rope to run, bored the centre and cut the housing for the bronze bearing or 'coak'. There were also lathes to turn and polish the pins.

The system was progressively developed from about 1805 and, once established, worked faultlessly with the machines staying serviceable for about 150 years – well after the age of sail, and only becoming redundant through the decline in demand for pulley blocks. When working in their heyday they were a popular sight and, in an age fascinated by automata and by mechanism, were visited and marvelled at by royal visitors from various nations and eminent people of all types. The writer Walter Scott and his daughters were one appreciative group. The mortising machines were a particular feature, for the chisels made about 400 strokes a minute, and so seemed almost invisible, while the chips flew and the morticed slots in the blocks seemed to grow and lengthen 'without any evident cause'. The blockmills may have launched that fascination with 'the rhetoric of mass production' which also drew industrial tourists much later to Ford's Highland Park plant or today to robotic factories in Japan, where the managers turn the lights on – which the robots do not need – in a pure piece of theatre for the visitors.

A representative set of the surviving machines eventually came to the Science Museum in London, where they today form a featured group within its gallery on the history of science and technology entitled 'Making the Modern World'. The machines are still notable for their architectural appearance and the great beauty of their proportions. This, to a great extent, answers the claims made over the years for a role for Samuel Bentham in their design – a claim never made by Bentham himself – and for Maudslay – a claim specifically refuted by Maudslay's partner, Joshua Field, though Maudslay must certainly be credited with detailed solutions for the execution of the designs, and for the excellence of his workmanship. This point was also addressed by the Science Museum curator K R Gilbert in 1965, when he noted that the drawings in Brunel's sketchbook (today held in the National Maritime Museum) tally exactly with the finished machines and that 'where a dimensioned drawing can be identified with an existing machine a check has been made by actual measurement' and for the 'second size mortising machine … all the dimensions given in the sketchbook agree closely with the machine, with a single unimportant exception'.[4] Other machines also tally closely.

Other significant episodes in Marc Brunel's life included his boot-making venture, his Battersea sawmill, destroyed by fire, the family's temporary imprisonment in a debtor's

(Images over) Further engravings from The Cyclopaedia; or, Universal Dictionary of Arts, Sciences, and Literature (Science Museum/Science and Society Picture Library). These plates show a sawing machine, crown saw, coaking engine, boring machine, cornering saw and mortising machine.

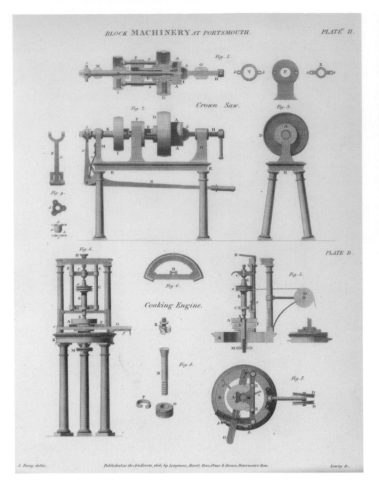

Fig. 1

Fig. 2

Crown Saw.

Fig. 3

Fig. 9

Fig. 4

Fig. 6

Coaking Engine.

Fig. 5

Fig. 8

Fig. 7

J. Farey delin. Published as the Act directs, 1806, by Longman, Hurst, Rees, Orme & Brown, Paternoster Row. Lowry Sc.

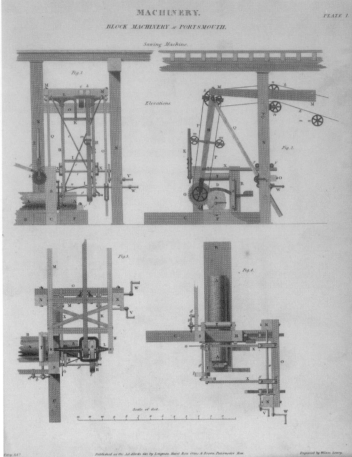

Sawing Machine.

Fig. 1

Elevations

Fig. 2

Fig. 3

Fig. 4

Scale of feet.

Farey delt. Published as the Act directs 1811 by Longman, Hurst, Rees, Orme & Brown, Paternoster Row. Engraved by Wilson, Lowry.

Second size Mortice machine

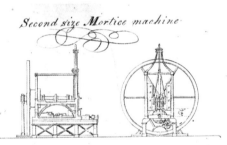

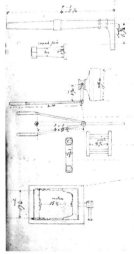

crank pin

Crank Spindle covered with thick steel

Cast Iron eccentric 7½ Inches Diameter

Drum 18 Inches Diameter

Fly Wheel 6 ft Diameter

a Moveable Cone at the joint 18 Inches Diameter

Cone thrown in and out by a common lever in sketch

Length of Wr Iron Connecting rod 2 ft 9 in sketch

Steel joint pin at top of connecting rod 1⅛ Diam

Chisel frame 8¾ Inches in the Clear in sketch

Chisel holder 1⅜ Wide

Block frame

Diam of tightening screw in block frame

1¾ Diameter two threads of sq ½ Inch

in a turn

Above is a page from Marc Brunel's notebook showing
second-size mortice machine (Science Museum).

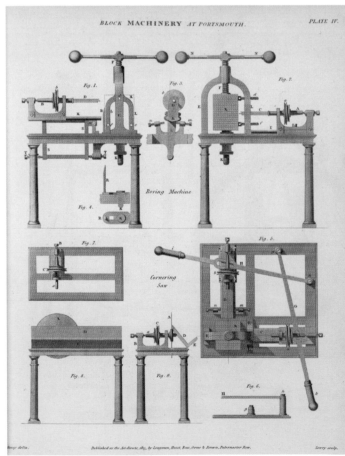

Fig. 1

Fig. 3

Fig. 2

Boring Machine

Fig. 4

Fig. 7

Fig. 5

Cornering Saw

Fig. 8

Fig. 9

Fig. 6

Farey delin. Published as the Act directs, 1813, by Longman, Hurst, Rees, Orme & Brown, Paternoster Row. Lowry sculp.

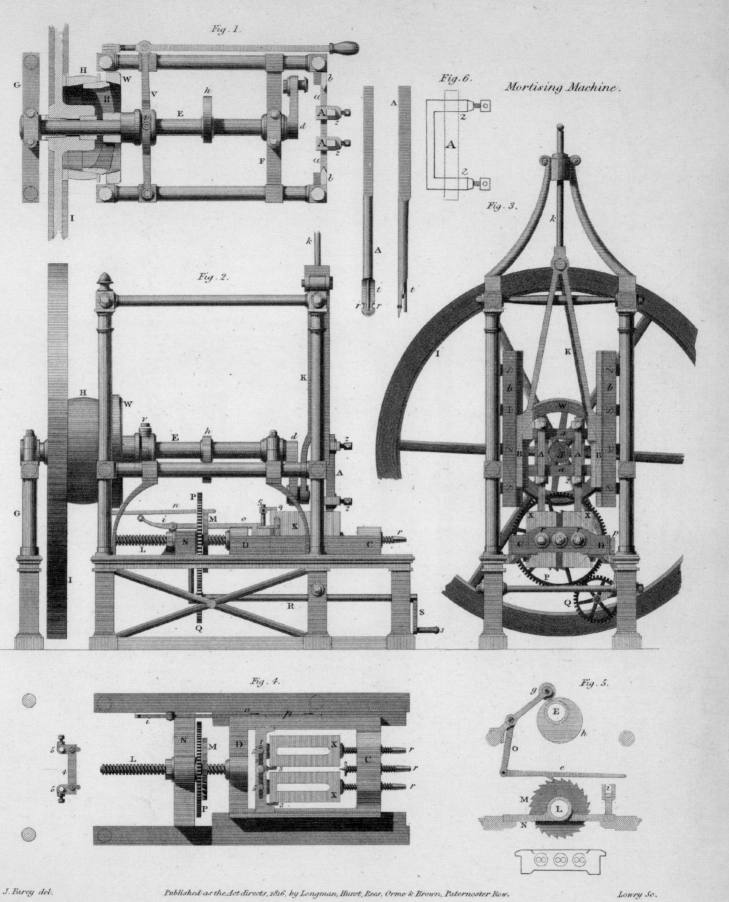

Mortising Machine.

Fig. 1.

Fig. 2.

Fig. 3.

Fig. 4.

Fig. 5.

Fig. 6.

J. Farey del.

Published as the Act directs, 1816, by Longman, Hurst, Rees, Orme & Brown, Paternoster Row.

Lowry Sc.

gaol and his celebrated but long-drawn-out Thames Tunnel project. Suffice it to say that Marc Brunel came through these to enjoy a relatively prosperous and respected position in later life. He was knighted for his services to British engineering, and in particular his achievements with the tunnel, in 1841.

The greatest puzzle in the life of Marc Brunel is why his feat of conceiving and designing the blockmills, which were so notable and so highly regarded, did not lead to similar techniques in other industries. In spite of Brunel's example, British machine tool makers concentrated on powerful and accurate 'universal' machines, like Whitworth's planing machines or Richard Roberts' lathes, and not on more easily minded special-purpose machines which were at the heart of the development of mass production. These 'general purpose' British machines, made to be guided by highly skilled and highly trained craftsmen, engineered the empire, making locomotives and looms and other products for Britain and for export – products which were often custom-built but incredibly durable and which endured for decade after decade in productive use.

The Royal Navy, as the largest industrial enterprise in Europe, and in a period of conflict and European instability, had offered a particular environment and uncommonly fertile ground for Brunel's new system. In addition, the idea of rational manufacture had fitted in well with Enlightenment ideals, and it is hard not to see a link between the Enlightenment spirit and Bentham's willingness to support an idealised scheme of manufacture. Thus it could be argued that Brunel's inspired, but costly and almost premature innovation, also gained support, in part, because the system resonated with an Enlightenment belief in the rational ordering of the universe and emerging hopes for rational systems of government, education and even punishment. But that had been a specific constellation of events and of spirit, quickly to be replaced by a more pragmatic Victorian mercantilism.

A long period was to elapse before another mass market suggested itself, and so British manufacturing and government largely ignored the principles Brunel had established until they awoke to the emerging power of what came to be known as 'the American system of mass production' some 50 years later. Ironically, shortly after the Great Exhibition of 1851, where British machine makers like Whitworth saw the orders for their new machines multiplied many fold by foreign interest, the industry was found to be deficient in an important area. Just three years later, in 1854, a British commission crossed the Atlantic to buy over 150 American gun-making machines to equip the Royal Small Arms Factory at Enfield. Similar techniques would eventually be applied to sewing machines, cars and computers. Arms, not pulley blocks, became the stepping stone into the new world of production and consumption.

Mortising machine (right), Yard No 1896, c 1800 (Science Museum/Science and Society Picture Library).

Brunel's coaking machine, c 1804 (Science Museum/Science and Society Picture Library).

THE THAMES TUNNEL

By the early nineteenth century, congestion in London, particularly around the busy dockyards in the east, was acute, exacerbated by the presence of the foul-smelling River Thames, which cut the city in two.

The construction of more bridges might have helped relieve the problem, but they would have needed to have been built high enough to allow the passage of ships. Looking for an alternative, the Thames Archway Company was formed in 1805 with the intention of constructing a tunnel to run beneath the river. Richard Trevithick, a Cornish miner and engineer, was appointed to supervise the initial stages, digging a pilot tunnel or driftaway. However, the traditional method of shoring up the tunnel sides and roof with timber proved unsuccessful in these difficult conditions and after a series of floods, the pilot was abandoned just 200 feet short of its target. The Thames Archway Company was dissolved in 1809.

In 1818 Marc Brunel patented a tunnelling shield made from iron, inspired by the head of the ship-worm which could bore through ships' timbers. Miners would work in a series of compartments inside the shield, excavating sections of earth held back by heavy wooden boards, which were removed and replaced one at a time to allow access to the face. Meanwhile, bricklayers, working closely behind the shield in the wake of the miners, would be constructing the tunnel lining. When all the earth within reach of the boards had been dug out, the shield could be moved forward to begin the process again.

Convinced that the shield made the scheme to build a tunnel beneath the Thames a feasible proposition, the Thames Tunnel Company was formed in 1824 with Marc Brunel appointed as chief engineer. Work began near the church of St Mary's Rotherhithe in March 1825 with the digging of the initial shaft. A huge cylinder of brickwork, 50 feet in diameter and 42 feet high, was built upon an iron ring. As the brickwork was completed, workmen dug away the ground inside and beneath the ring. The weight of the bricks caused the cylinder to slowly sink until its top reached ground level. The tunnelling shield was then lowered into the shaft to begin its laborious progress beneath the river.

It had been assumed that the miners would encounter firm clay throughout their passage, but the shield soon struck loose gravel and sand mixed with rotting sewage, making the work even more dangerous and difficult than expected. The foul air in the tunnel caused fevers and blindness. One victim was William Armstrong, the engineer-in-charge, who suffered serious illness. He resigned and Marc's son, Isambard, was promoted to replace him. Eager to prove himself, Brunel often spent days at a time underground, taking only short naps between shifts, lying on a bricklayer's stage beside the shield. Fearful for his health, Marc assigned him three assistants in a vain attempt to reduce his punishing workload: one died of fever while another lost the vision in his left eye.

Although the bricklayers worked fast, there was always a risk that the water would break through the

The Thames Tunnel Workings, a watercolour by George Yates, 1826 (Southwark Culture & Heritage Services, South London Gallery Trust Collection).

'The Tunnel!!! or another Bubble Burst!', coloured etching satirising the hazards of using the proposed tunnel, 1827 (Science Museum/Science and Society Picture Library).

unsupported gap beneath the riverbed that was exposed each time the shield was moved forward. On 18 May 1827 the river burst in and a great wave rushed through the tunnel. Fortunately, the men had time to reach the shaft and were able to climb the stairs to the surface before the tunnel flooded. Displaying his typical bravery, Brunel rescued an old man who was struggling in the rising water by sliding down an iron tie rod and securing a rope around the man's waist.

Within 24 hours Brunel was inside a diving bell, borrowed from the West India Dock Company, inspecting the hole in the tunnel roof that had caused the flood, one foot resting on the completed brickwork, the other on the shield. His companion lost his hold upon the diving bell and had to grab Brunel's extended foot to avoid disappearing down the hole to a certain death. Brunel dragged him back up and returned to the surface before descending once more to continue his inspection. Later, when the hole was filled with a combination of iron rods and bags of clay, Brunel was the first to reach the shield, initially by boat on the floodwaters and then by crawling over the bank of earth the river had washed in. In his diary he wrote:

What a dream it now appears to me! Going down in the diving bell, finding and examining the hole!... The novelty of the thing, the excitement of the occasional risk attending our submarine (aquatic) excursions, the crowd of boats to witness our works all amused – the anxious watching of the shaft – seeing it full of water, rising and falling with the tide with the most provoking regularity…[1]

Work resumed in November, an occasion marked by a promotional banquet held within the tunnel, but disaster struck again in January 1828 when water burst in with even greater ferocity. Brunel was standing in the shield at the time and was swept away. He was trapped beneath a fallen timber, which damaged his knee and caused internal injuries. Managing to free himself, he waited and called out for the men he thought had been lost beneath the collapsed staging. He later wrote:

While standing there the effect was – *grand* – the roar of the rushing water in a confined passage, and by its velocity rushing past the opening was grand, *very grand*. I cannot compare it to anything, cannon can be nothing to it. At last it came bursting through

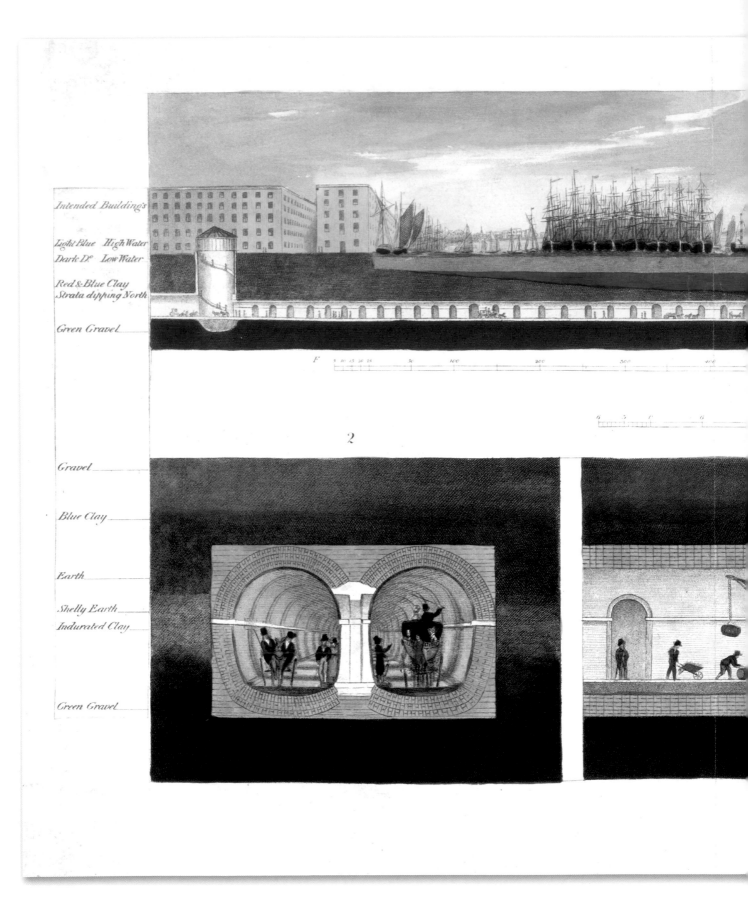

Engraving of the Thames Tunnel by T Blood, 1827, showing the tunnelling shield and the route beneath the river (Science Museum/Science and Society Picture Library).

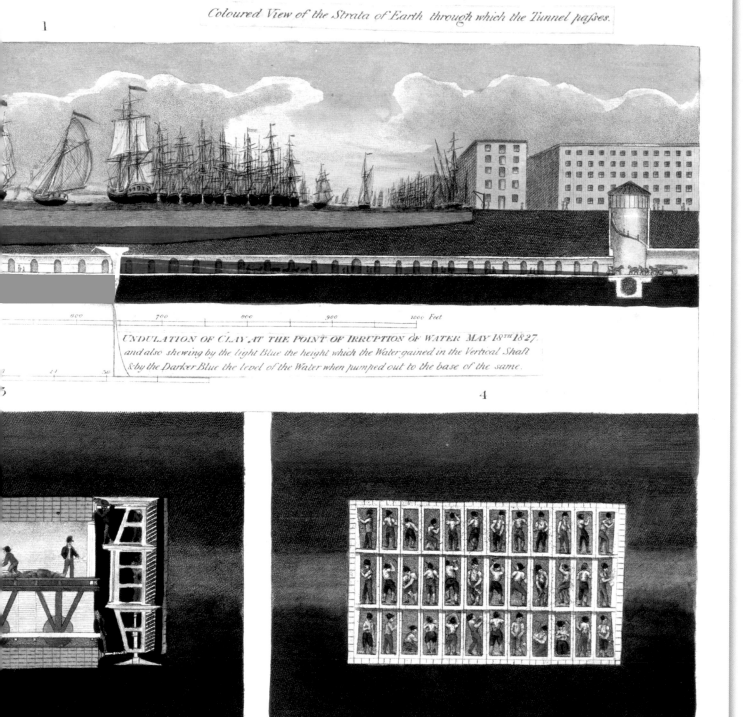

1

UNDULATION OF CLAY AT THE POINT OF IRRUPTION OF WATER MAY 18TH 1827.
and also shewing by the light Blue the height which the Water gained in the Vertical Shaft
& by the Darker Blue the level of the Water when pumped out to the base of the same.

4

T Blood Sculp.

the opening. I was then obliged to be off – but up to that moment, as far as my sensations were concerned, and distinct from the idea of the loss of the six poor fellows whose death I could not then foresee, kept there.

The sight and the whole affair was well worth the risk and I would willingly pay my share, £50 about, of the expenses of such a 'spectacle'.[2]

With his injuries, Brunel's association with the tunnel ended. Although Marc was again able to fill the breach, work came to a halt as the finances were exhausted and business confidence in the project was lost.

Marc persevered in drumming up interest again and work finally resumed in 1835. The Thames Tunnel was opened to foot traffic on 25 March 1843 and 50,000 people walked through it on that first day. It was the first tunnel to be built under a navigable river and hailed as the eighth wonder of the world. In 1869 it was converted to carry the East London Railway and became part of the London Underground in 1914. Today it is an International Landmark Site and around 14 million passengers travel through it each year on the East London Line.

The Royal Family arriving at the Tunnel, The Illustrated London News, *5 August 1843 (University of Bristol).*

The Thames Tunnel, aquatint, c 1830 (National Trust). Part of the fundraising material produced during the period work on the tunnel had been suspended.

The procession down the staircase at the opening of the Thames Tunnel (below), The Illustrated London News, *1 April 1843 (University of Bristol).*

Coloured lithograph (facing page) after an original drawing by B Dixie showing one of the proposed entrances to the Thames Tunnel and the spiral staircase leading to it, 1836 (Science Museum Pictorial/Science and Society Picture Library).

SOPHIA KINGDOM AND THE REIGN OF TERROR

The French Revolution began with the formation of the National Constituent Assembly and the storming of the Bastille in the summer of 1789 and ended with Napoleon Bonaparte's coup in November 1799. The period between the summers of 1793 and 1794 was marked by particularly brutal repression on the part of the central government and the suspension of many of the newly acquired civil rights. It became known as the Reign of Terror, or simply The Terror.

The two main revolutionary factions – the Girondins and the Jacobins – had become bitter rivals, and the Revolution was under threat from conspiracies hatched between members of the former French nobility, sympathetic clergy and agents from neighbouring countries, including Britain. In this climate of fear and suspicion, 31 Girondin leaders were arrested in June 1793 and the Jacobins installed a revolutionary dictatorship.

Outrage at the assassination of Jean-Paul Marat, a Jacobin who had masterminded the 1792 massacre of suspected counter-revolutionary prisoners, served to further the Jacobin cause. The moderate Georges Danton was removed from the Committee of Public Safety to make way for Maximilien de Robespierre. Under Robespierre's influence, the systematic repression of perceived enemies of the Revolution began.

(Top) Portrait of Sophia Kingdom, date and artist unknown (private collection).

(Above) Marc Brunel and Sophia Kingdom Brunel in old age (private collection).

Thousands were sent to the guillotine, condemned to death by the Revolutionary Tribunal in Paris. Those who died included not only members of the royal family, but also early supporters of the Revolution such as Madame Roland and Philippe Égalité as well as 'ordinary' citizens. By the end of 1793 provincial revolts had been crushed and the forces of Prussia, Britain, Spain and Austria pushed back from the French borders.

The Law of 22 Priarial, also known as the Law of the Reign of Terror, was enacted on 10 June 1794. It limited the opportunities for the accused to defend themselves and widened the scope of who could be brought before the Revolutionary Tribunal. Many thought that the centralising of power had gone too far and that the Republic would be damaged by this new ordinance. Robespierre was overthrown on 27 July. He was guillotined the following day along with his deputies and most of the members of the Commune. A more relaxed form of government in the form of the Directory was established and many of those imprisoned during The Terror were freed.

Marc Brunel would have been at risk during The Terror as he was not only a Royalist but had criticised Robespierre in public. When he fled France, he left behind him his future wife Sophia Kingdom, who was arrested as an enemy alien when Britain joined the alliance against France in October 1793. She was taken to a convent in the port of Gravelines, which accommodated the overflow from the town's prisons. A guillotine was set up in the courtyard and put to work whenever the cells became overcrowded. Sophia therefore lived in fear of being executed at any time. Families whose children had been taught by her during her time in France tried to secure her release but were rebuffed. With the fall of Robespierre, however, the Gravelines prisoners were finally freed. Sophia was reunited with Marc's friends the Carpentiers, who nursed her back to health before she returned to England in 1795.

Sophia endured a further three months in prison in 1821 when Marc Brunel's business failed and he was imprisoned for debt: Sophia remained with him throughout his ordeal. Marc's friends mounted a campaign to obtain a government grant of £5,000 to cover his debts and secure his release. The grant was partial compensation for the army-boots he had produced during the war but not been paid for. The Duke of Wellington intervened on his behalf after Marc threatened to emigrate to Russia and offer his services to Tsar Alexander unless the government paid him.

Sophia is widely thought to have been the daughter of a naval contractor who had links to Plymouth.[1] Sophia was married to Marc at Holborn on 1 November 1799. In her biography of the Brunels, Celia Brunel Noble characterises Marc as a romantic lover who even in old age and struck by the infirmities brought by his stroke was touchingly devoted to his wife. Visitors described the elderly couple sitting side by side with Marc occasionally bringing Sophia's hand to his lips to kiss. After Marc's death in 1849 Sophia spent her remaining years in seclusion, rarely venturing out and turning her room into a shrine to her late husband. The objects displayed there included a milk jug and tumbler she had kept from her time of imprisonment during The Terror.[2]

Execution of Marie Antoinette in the Place de la Revolution (now Place de la Concorde), October 1793 (Mary Evans/Explorer/ES Collection).

THE NINETEENTH-CENTURY ENGINEER AS CULTURAL HERO

Christine MacLeod

The Illustrated London News
memorial picture of Brunel, 24
September 1859 (University of Bristol).

In 1851, engineers were the toast of Great Britain. The catalogue for the Great Exhibition, published that year, said: 'These are the men of tubes and tunnels, they level hills, turn the course of streams, rear structures with a celerity and ease that shame the Pharaohs'.[1] The crowds that flocked to the exhibition probably marvelled no less at the railways on which they travelled to London than at the airy construction of the Crystal Palace and the powerful machinery on display inside it. Nearby, in Westminster Abbey, the recently inaugurated statues of Thomas Telford and James Watt towered over more delicate, aristocratic monuments; as astute observers recognised, their controversial presence in the Abbey was a cultural masterstroke, the symbol of a social revolution.[2]

It is perhaps unsurprising, therefore, that when the 'railway triumvirate', of Isambard Kingdom Brunel, Robert Stephenson and Joseph Locke, died within a year of one another in 1859-60, the shock was palpable. It was generally felt to be the end of an era.[3] For historians uncritically to accept this contemporary view, however, is to fall victim to 'the romance of the railways': the age of the heroic engineer had another half century to run. Not only was there a posthumous sequel to the Battle of the Gauges – won again by the narrow gauge, at least in the short term – but until the early twentieth century the rising status of the engineering profession was witnessed both by the commemoration of its luminaries in Westminster Abbey and in public spaces throughout Britain, and by a spate of honours from the state.[4] And, in recent years, there has been the extraordinary return of Brunel to the pinnacle of fame.

Engineering celebrities

The Brunels and the Stephensons became celebrities during the 1820s and 1830s – an unprecedented experience for a British engineer.[5] Marc and Isambard Kingdom Brunel first came to public notice with their ambitious project to tunnel beneath the Thames, which they 'puffed' in regularly published guidebooks and broadsheets.[6] An early publicity stunt in 1827, for the reassurance of shareholders, involved a banquet inside the tunnel workings, illuminated by gas-lit chandeliers, entertained by the band of the Coldstream Guards, and recorded in an oil painting.[7] Repeated inundations – one, which killed six men, being almost fatal for Isambard himself – maintained the press's attention and appeared to confirm the expectations of sceptics, but their taunts were silenced in 1843 by the tunnel's triumphant opening. Public curiosity was stoked by a souvenir industry, which portrayed the tunnel on prints, medallions, handkerchiefs and, most evocatively, in three-dimensional peepshows. Within four months, a million people had descended to view it, and for a decade its brightly lit roadways, lined with stalls, comprised London's major tourist attraction and 'premier gift shop'.[8]

Isambard Kingdom Brunel inherited his father's bravado, his 'spectacular feats of engineering, providing endless opportunities both for sardonic humour and respectful awe'.[9] Moreover, his dramatic successes as a railway and steamship pioneer allowed his occasional failures to be forgiven as the price of technological hubris. In 1845, for example, the *Railway Times* wrenched heroism out of the jaws of his defeat by the technical problems of the South Devon atmospheric railway:

> We do not take him for either a rogue or a fool but an enthusiast, blinded by the light of his own genius, an engineering knight-errant, always on the lookout for magic caves to be penetrated and enchanted rivers to be crossed, never so happy as when engaged "regardless of cost" in conquering some, to ordinary mortals, impossibility.[10]

As Adrian Vaughan comments, on the occasion in 1843 when Brunel's life was threatened by a gold coin lodged in his wind-pipe, the result of a conjuring trick that misfired, he 'had so far succeeded in becoming a household word that *The Times* reported regularly on the progress of his treatment'.[11]

It was claimed in 1842 'that in Germany and elsewhere abroad "nothing is asked of Englishmen, but about the Thames Tunnel and the Manchester Railroad"'.[12] As the tunnel had made the name of the Brunels, so the Liverpool & Manchester Railway, opened in 1830, had made that of George and Robert Stephenson. In his portrait by John Lucas, George Stephenson stands proudly in a landscape dominated by the railway as it crosses Chat Moss – the hardest technical challenge he had to confront in its construction.[13] (At his death in 1848 'whole-length' engravings of this portrait, by T L Atkinson, were offered for sale to subscribers at prices from two to six guineas.)[14] Stephenson evidently enjoyed relating his achievements. In 1844, for example, he was invited to speak about his life and career at an 'entertainment' in Newcastle upon Tyne, and Samuel Smiles often heard him talk at 'soirées' in the Leeds Mechanics Institute, where he was 'a great favourite'.[15] Stephenson was an obvious choice as first president of the Institution of Mechanical Engineers, founded in 1847; he was succeeded in 1849 by his son, Robert.[16]

Portrait of George Stephenson by John Lucas (Institution of Civil Engineers).

Banquet in the Thames Tunnel, *oil on board attributed to George Jones (Elton Collection: Ironbridge Gorge Museum Trust).*

This event was organised by Brunel in November 1827 to celebrate resumption of work following flooding.

The Tunnel Company's medallion of Marc Brunel supported by one of the tunnel excavators, The Illustrated London News, *1 April 1843 (University of Bristol).*

Medallion (below) commemorating the Tunnel opening, 1843 (Elton Collection: Ironbridge Gorge Museum Trust).

Selection of Thames Tunnel souvenirs (private collection).

The Thames Tunnel, a hand-coloured lithographed peepshow in six sections, c 1851 (Science Museum/Science and Society Picture Library). The cover scene shows the Thames Tunnel staircase with vignettes including the Crystal Palace and Parliament buildings.

Initially in his father's shadow, Robert Stephenson emerged during the 1840s to command both fame and admiration as a railway engineer and designer of bridges. In particular, his novel tubular designs for the Britannia and Conway Bridges, on the route to Holyhead, caught the public imagination. When the Britannia Bridge's gigantic iron tubes were floated out, preparatory to their being lifted into place a hundred feet above the water, cheering crowds lined the shore, accompanied by bands and cannons. Eight months later, in March 1850, Stephenson ceremoniously drove the first train through the first completed tube, across the Menai Straits.[17] August 1851 saw a banquet for 400 to celebrate the opening of the two bridges: Stephenson was lauded inordinately, and civil engineers were credited with the 'development of our national wealth and industry'.[18] The inclusion of the Britannia Bridge in several portraits of Stephenson by John Lucas bears witness to his pride in this enormously complex project. Lucas was also commissioned to paint the group portrait *Conference of Engineers at Britannia Bridge* (see page 223). This imaginary scene, with Stephenson at its centre and, incongruously, the completed bridge as backdrop, included Joseph Locke and Isambard Kingdom Brunel, who had advised on the procedure for floating the tubes.[19] For the next decade, Stephenson was rarely out of the public gaze, not only for his spectacular bridges, many of them overseas, but also as the commander of his racing yacht, *Titania*, and as an MP who enjoyed an extremely full social life.[20]

The railway triumvirate commemorated

Isambard Kingdom Brunel and Robert Stephenson died within a month of one another in the autumn of 1859. This accident of timing deepened the poignancy and sense of loss, and obituaries of Stephenson tended to pay tribute to both men. According to *The Times*, they were 'both men of rare genius'.[21] The tribute to Brunel

Brunel family headstone at Kensal Green (Barry Smith and The Friends of Kensal Green Cemetery).

Opening of the Clifton Suspension Bridge,
The Illustrated London News, *17*
December 1864 (University of Bristol).

in *The Illustrated London News* depicts him surrounded by a wreath of laurel intertwined with visual references to his achievements – iron chains, propeller blades, railway spikes, steaming funnels, and miniatures of his three great ships and the Saltash and Hungerford Bridges.[22] Brunel was buried beside his parents, under a simple headstone, in Kensal Green Cemetery, in north-west London.[23] Although many hundreds had lined the funeral route, this was essentially a quiet family affair. Stephenson's funeral, by contrast, was virtually a state occasion in Westminster Abbey, which prompted outpourings of national grief, echoed on Tyneside, where businesses closed, ships lowered their flags, and a thousand workmen from his locomotive workshops attended a special church service.[24]

The engineering profession was not content, however, to allow Brunel to go so modestly to his grave. Two schemes were soon in progress. His friends and colleagues, under the aegis of the Institution of Civil Engineers, determined to complete the Clifton Suspension Bridge, over the river Avon, at Bristol, which had languished for lack of funds since 1843.[25] A new company, established in 1861, raised capital of £30,000 and instructed Sir John Hawkshaw and W H Barlow to finish Brunel's project. While financial constraints unfortunately required Brunel's Egyptian detailing of the piers to be abandoned, they promoted the incorporation of the double chains from his Hungerford Bridge, which was being dismantled to make way for the Charing Cross railway bridge. The Clifton Suspension Bridge, festooned with flags, was ceremonially opened on 8 December 1864.[26] An inscription, high on the eastern pier, recorded the roles of Brunel and his two successors; it has since been joined by other commemorative plaques.

Secondly, it was proposed to erect a monument in the centre of London. A meeting in November 1859, attended principally by engineers and chaired by the Earl of Shelburne, chairman of the Great Western Railway (GWR), launched a public subscription, which raised over £2,300. Of the known 860 subscribers, approximately half were connected with the railways, in particular the GWR and other lines in the South West; 59 men identified themselves as civil engineers. Nearly all gave addresses in London, South and South West England, the South Midlands, or Wales – the regions which benefited most directly from Brunel's engineering talents. Collections were made in railway engineering workshops. Employees of the Cornwall Railway, for example, gave nearly £25, including over £4 donated by 30 groups comprising a named foreman and his team of three to 13 men; thus 187 men contributed a few hard-earned pence each.[27] When representatives of the major subscribers next met, in March 1860, they were informed that the Brunel family did not favour a memorial in St Paul's Cathedral but wished it to be 'a visible one', perhaps near to the statue of former prime minister George Canning in Westminster. Accordingly, they commissioned a bronze statue from Baron Carlo Marochetti, the Turin-born sculptor who had been based in Britain since 1848, to be placed in a 'public thoroughfare'. A

year later, they decided to unite Brunel's statue with that of Robert Stephenson, also being sculpted by Marochetti, on a site near to the statue of William Pitt, in the garden of St Margaret's, Westminster.[28] It would take more than another decade, however, to negotiate a permanent and separate site for each of them, neither as grand as St Paul's or the vicinity of the Houses of Parliament.

Meanwhile, the Brunel family commissioned a memorial window for Westminster Abbey.[29] Designed by the architect, R Norman Shaw, in close collaboration with the family and the Dean of Westminster, the window makes only oblique reference to Brunel's achievements or his profession.[30] Its principal theme, proposed by the Dean, is the history of the Temple in Jerusalem, depicting three subjects from both the Old and the New Testaments. A quatrefoil contains 'the Saviour in Glory, surrounded by angels' and, at Shaw's suggestion, above the legend bearing Brunel's name and dates, four allegorical figures represent his personal qualities: fortitude, justice, faith and charity.[31] When the window was installed, in the north aisle of the nave in 1869, it joined two that had been recently dedicated to Robert Stephenson and Joseph Locke respectively; both were brighter in colouring, with predominant reds and blues, while Brunel's was more 'sober'.[32]

The window erected to Stephenson's memory by his executors also differs markedly from Brunel's in its iconography: it explicitly commemorates Stephenson's work in particular and the engineering profession in general.[33] In a cinquefoil at the apex, the medallion portrait of Robert Stephenson is surrounded by portraits of five of his peers: George Stephenson, Thomas Telford, John Smeaton, James Watt and John

Rennie.[34] The window illustrates four of Stephenson's bridges and places them in a line of succession that stretches from the great engineering feats of biblical and ancient times. Linking these works are smaller medallion portraits of engineers and architects, from Tubal Cain and Noah to Sir Christopher Wren.[35] It is inscribed on the left, 'Robert Stephenson/MP DCL, Pres ICE/1803-1859' and, on the right, 'Son of George Stephenson/1781-1848/Father of Railways'. The executors also installed a commemorative brass, near to Stephenson's tomb in the nave, 'on which the engineer is represented in a standing position, with his arms folded across his breast'.[36] Only the inscription indicates that he had been an engineer, rather than a missionary or other saintly cleric.

Each member of the railway triumvirate was also commemorated with a biography; unfortunately, none served its subject well. The authorial services of Samuel Smiles – despite the success of his recently published biography of George Stephenson – were rejected by both the Brunel family and Robert Stephenson's executors. According to Smiles, the latter had 'made [unsuccessful] applications to several eminent literary men', before they approached the novelist and popular biographer, J C Jeaffreson.[37] Smiles subsequently produced 'a summary' of the younger Stephenson's life in tandem with his father's in a volume of his *Lives of the Engineers*.[38] A biography of Isambard Kingdom Brunel, written by his son, Isambard Brunel, appeared in 1870: in biographer L T C Rolt's words, it was 'heavy reading'.[39] The Brunels, Smiles later wrote, had been 'the only engineers I wished to add to my collection', but the family's preference was for tight control.[40]

Robert Stephenson memorial window, Westminster Abbey (with detail).

Brunel memorial window, Westminster Abbey.

(Both photographs copyright Dean and Chapter of Westminster).

As for Joseph Devey's biography of Joseph Locke, a reviewer astutely judged Locke's chances of fame in the long term:

> George Stephenson, his quondam master, with all his faults and shortcomings, was a hero of romance as compared with the steady, safe and cool-headed Locke. Yet Locke was an able engineer; indeed an abler, or at least a better educated, and hence more generally competent engineer, than his more celebrated master; but he was no original and no great inventor, and hence his biography would not stand comparison with that of George Stephenson in interest, even were it written by Mr Smiles himself.[41]

It is possible that the memory of Brunel and Robert Stephenson also would have burned more brightly in the late nineteenth century if they had enjoyed the benefit of Smiles' approach. There was a long hiatus before either found a good biographer.[42] Robert Stephenson's posthumous fame was largely a reflection of his father's. Brunel's persisted, it seems, largely thanks to one famous photograph, which was sold in the form of prints and *carte-de-visites* and made his stove-pipe hat and cigar an instantly recognisable visual reference.[43] A cartoonist in *Punch* was certainly exploiting that reference when he parodied the mourning of the broad gauge's demise in 1892.[44]

Name (or image) recognition, however, is not the same as heroic status. The railway triumvirate was, indeed, eclipsed by George Stephenson and James Watt. The first indication of this was the difficulty that the monument committees began to encounter in siting their statues in central London. In 1860 Brunel's monument was initially refused a site in Parliament Square by the Chief Commissioner of Works, William Cowper, on alleged aesthetic grounds. Those who objected to the presence in Trafalgar Square of Edward Jenner's statue would have been horrified to learn that Cowper was proposing that Brunel join him there, for 'now that Dr Jenner has been admitted, the place cannot be considered exclusively devoted to warriors and Kings'.[45]

In May 1861 Cowper reversed his earlier decision and conceded sites for both Brunel's and Robert Stephenson's statues on the north-west side of Parliament Square, which would be visible from the headquarters of the Institution of Civil Engineers.[46]

'The Ghost of Brunel Laments the Burial of the Broad Gauge', Punch *1892*
(Punch *Cartoon Library & Archive*).

Statue of Robert Stephenson, Euston Station, London, by Baron Carlo Marochetti (Conway Library, Courtauld Institute of Art).

Statue of Joseph Locke, Locke Park, Barnsley, Yorkshire by Baron Carlo Marochetti (Conway Library, Courtauld Institute of Art).

Shortly afterwards, however, he refused permission for Locke's statue to join them.[47] In June 1867 Marochetti informed Cowper that Brunel's and Stephenson's statues were ready; Cowper replied that Parliament Square should be considered only as a provisional site until somewhere 'more suitable' could be found.[48] There were further delays, until in December 1868 Cowper's successor withdrew the permission for even temporary sites in Parliament Square. His reasons were twofold: first, 'because the site should be reserved for statues of eminent statesmen'; second, 'because the two statues in question differ altogether in proportion from the statue of Canning'. On their respective pedestals, Canning stood over 26 feet high; the two engineers, only 15 feet.[49]

Henry Marc Brunel wrote despairingly to his brother that the Board of Works thought more about the size of their father's statue than its likeness, and conceded that it might be better placed in Bristol, on Clifton Down.[50] In 1871 Robert Stephenson's statue was erected in front of Euston railway station.[51] Although his supporters had had to resort to a privately owned site, it was an appropriate one, the more so because a statue of George Stephenson had stood in Euston's ticket hall since 1854. Less appropriately, a site on the Victoria Embankment was negotiated with the Metropolitan Board of Works for the statue of Brunel, where it was installed in 1877; surprisingly, Paddington station does not seem to have been considered.[52]

Meanwhile, Joseph Locke's statue went home to Barnsley in Yorkshire. Locke had been elected president of the Institution of Civil Engineers in 1857, a position that he held until his death on 18 September 1860. The Institution sponsored the appeal for his statue – also by Marochetti – which, having failed to find a metropolitan site, was unveiled in Locke Park, Barnsley, in 1866. The occasion was observed as a general holiday, with a procession of local dignitaries, accompanied by brass bands.[53] The park had been presented to the town of Barnsley by Locke's widow, Phoebe, in 1861 as a memorial to her husband, at a cost of £6,700. Locke had attended the grammar school, which simultaneously benefited from her generosity: with £2,000 she established the Locke Fund, which was to be used, in part, to provide scholarships for would-be colliery engineers. A further £1,000 was bestowed on the Roman Catholic school, to which Locke had been a 'liberal subscriber', despite having lapsed from the faith.[54] In front of the school, a 40-foot pinnacle, surmounted by a cross, was erected to witness the generosity of Joseph and Phoebe Locke.[55]

Statue of George Stephenson (marble) by John Gibson at St George's Hall, Liverpool (Conway Library, Courtauld Institute of Art).

Marble statue of George Stephenson by Edward Bailey, Euston Station, 10 September 1890 (National Railway Museum/Science and Society Picture Library).

Statue of James Watt by Alexander Munro, c 1860, Oxford University Museum (photograph by the author).

George Stephenson and James Watt

No doubt Barnsley was pleased with its windfall, but the outcome of the three commemorative appeals must have proved a serious disappointment to the subscribers and the engineering profession as a whole. Instead of occupying a combined, prestigious site at the heart of London, the three statues had been shuffled off to separate, relatively obscure places. Despite the aesthetic reasons offered, the implication remained that engineers (like physicians, such as Jenner), however successful or creative they might be, still ranked second to the traditional heroes of the battlefield and the political arena. This is not to say, however, that the reputation or social status of engineers declined in the years following the deaths of the railway triumvirate. Rather, it was the case that their standard bearers belonged to an older generation: none of the three, despite their mid-century celebrity, could approach the devotion to which either George Stephenson or James Watt still laid claim. Watt had been posthumously elevated to the status of a national hero in the 1820s, where he remained throughout the century; the Scots, in particular, continued to write heroic biographies and erect statues to his memory. Between 1850 and about 1908 Watt was commemorated with statues, busts or portrait medallions in Dundee, Edinburgh (three), Glasgow (five), Stirling and Greenock, but also in Manchester, Oxford, Northampton, Bradford, Birmingham (two) and Leeds, as well as on a set of bronze doors at the Victoria and Albert Museum in London.[56] Glasgow's engineers held an annual James Watt Anniversary Dinner, and the Greenock Philosophical Society sponsored an annual Watt Lecture, given by a succession of eminent scientists and engineers.[57] George Stephenson's reputation was adopted both by the engineering profession and his native Tyneside during the 1850s, and integrated into popular culture as the companion piece to Watt's.

Like Watt's, Stephenson's renown was multi-faceted: it drew on both local and professional support and enjoyed the patronage of members of the social elite. As yet there was no tradition of public memorials to men of science and engineering – Watt, and Newton before him, had been exceptional. It began in the 1850s and it began with George Stephenson.[58] During 1851 *The Times* announced that two commemorative schemes were in hand among Stephenson's former colleagues. The directors of the London & North Western Railway were installing a statue of 'heroic size' by John Gibson in Liverpool's new St George's Hall; Stephenson, seated, wears a toga, more in keeping with the building's classical architecture than with his persona as a representative of technological progress.[59] Meanwhile, the Institution of Mechanical Engineers had raised £3,000 to commemorate its first president with a statue at Euston station: 3,150 working men contributed approximately two shillings each, and 178 of his private friends, an average £14 each. Within a year, Edward Bailey completed a ten-foot high statue in marble, after the portrait by Lucas, which was judged to have 'skilfully preserved' Stephenson's 'form and... features'.[60]

Remarkably, when the statue was unveiled in April 1854, *The Times* complained that the occasion did not match the importance of its subject: the engineering profession had failed to capitalise on a major asset – Stephenson's reputation. 'Without much ceremonial observance,' it began, 'and in the presence only of the more active members of the committee', the statue had been inaugurated in Euston's grand hall. Yet here was 'a man who stands more nearly and intimately associated with the spirit of the nineteenth century than we are yet willing to recognize'; his 'life possesses an interest of the highest order'; his rise from 'a "hurrier" in a coal-pit shed, by the force of native genius... may well be regarded as proof that the days of romance are not yet over'. This, it seems, had not been sufficiently emphasised in the low-key ceremony, for the report concluded wistfully: 'Perhaps it is also to be viewed as a characteristic of the age that the fame of such a man is so quietly left to the good keeping of the works which he had achieved.'[61] It proved to be an unduly pessimistic conclusion.

When Stephenson had died in 1848, Samuel Smiles wrote a brief account of his life for *Eliza Cook's Journal*, which was reprinted by several newspapers. This having prompted the idea that he should write a full-length biography, Smiles secured an interview with Robert Stephenson.[62] Stephenson's response was initially cautious, because he doubted public interest, citing the recently published *Life* of Telford, which '"has fallen still-born from the press"'. Smiles was sure that the problem lay not with the subject matter but with its treatment, its lack of 'human interest'; if he wrote George Stephenson's biography, he 'would endeavour to treat of his character as a Man as well as an Engineer'.[63]

When *The Life of George Stephenson* appeared in 1857, Smiles' confidence was vindicated: it found an eager market, boosted by numerous, positive reviews, and in a little over a year, five editions (7,500 copies) were printed.[64] This was Smiles' first book, but notoriously it initiated a huge and influential output. The long delay between conception and publication may have been to his advantage: by 1857 public interest in inventors and engineers was growing strongly, undoubtedly boosted by the Great Exhibition, the feats of the railway triumvirate and the debates surrounding the future of the patent system.[65] Four years later the first edition of Smiles' *Lives of the Engineers* scarcely touched the booksellers' shelves: although priced at four guineas per four-volume set, approximately 6,000 copies were immediately sold. 'The *Saturday Review* expressed surprise "that the idea of handling the subject of engineering in this manner should not sooner have been seized"'.[66]

Detail from monument to George Stephenson, Newcastle upon Tyne, Tyne and Wear, by John Graham Lough, showing figure reclining on railway engine (Conway Library, Courtauld Institute of Art).

Trade Card of Robert Stephenson & Company, English engineer, mid-nineteenth century (National Railway Museum/Science and Society Picture Library).

The Stephensons' fame had a strong civic anchor, on Tyneside. In 1857 *The Builder* expressed outrage that the corporation of Newcastle upon Tyne proposed to demolish George Stephenson's cottage (Robert Stephenson's birthplace) in order to build a school. It drew parallels with the birthplaces of William Shakespeare and Sir Isaac Newton, which attracted thousands of visitors and which it would now be deemed 'a sort of sacrilege to destroy'. Newcastle should realise that 'as the years roll on, the fame of George Stephenson will increase'; if the school were built *next* to the cottage the teachers would be able, ever after, to use it as an object lesson, to inspire the emulation of 'a great man … who, by his perseverance and genius, benefited the world, and raised himself to a high condition'.[67] The Stephenson Memorial School at Willington Quay was ceremoniously opened in 1860; by then it had become a memorial to both Stephensons. The cottage had been demolished, but George Stephenson's birthplace at Wylam was expressly preserved.[68]

Newcastle's principal memorial to George Stephenson emerged from a public meeting in October 1858. Nearly £5,000 was subscribed in six months.[69] The monument, inaugurated in 1862, was sited in Neville Street, near the railway station – according to Smiles, 'a very thoroughfare of working-men, thousands of whom see it daily as they pass to and fro from their work'. Designed by John G Lough, Stephenson's bronze figure, in everyday costume and wearing his Northumberland plaid over one shoulder, stood 30 feet high on a stone pedestal, which was surrounded by four reclining figures emblematic of his inventive and engineering achievements: a miner holding a safety ('Geordie') lamp, an engine-driver leaning on a locomotive, a platelayer holding a 'fish-bellied' rail, and a blacksmith with hammer and anvil.[70] The festivities provided a stark contrast with the austere event at Euston eight years earlier. The day was declared a general holiday on Tyneside: a procession of 10,000, which contained many working men from Robert Stephenson & Co as well as other factories and friendly societies, took half an hour to pass by, and an estimated 70,000 people attended the ceremony, including 'almost every person of eminence connected with the trade of the north'.[71]

Other public and private memorials from around this time evince the high regard in which George Stephenson was held in the North East. His was one in a series of portrait heads painted in the spandrels of a newly converted courtyard at Wallington House, Northumberland, in 1855, by the young Pre-Raphaelite, William Bell Scott of Newcastle, who took as his theme the region's history from the Roman Wall to 'Iron and Coal'.[72] In 1858 copies of busts of both Stephensons were presented by the sculptor, Edward W Wyon, for display in Newcastle's town hall.[73] Across the Pennines, on the Victoria Monument in Lancaster's Dalton Square, Stephenson stands among other scientific and literary figures, holding a model locomotive.[74]

Indicative of Stephenson's growing national stature is his representation among the statues sculpted to decorate the great hall of the new University Museum at Oxford. Leaders in the fields of scientific knowledge which the museum was intended to promote, such as astronomy, geology and chemistry, were joined by others 'on special but very different grounds as benefactors to the human race, Bacon, Volta, Oersted, Watt, and Stephenson'.[75] He also became the subject of genre paintings. One of these canvases, *George Stephenson at Darlington* in 1823 by A Rankley, was exhibited at the Royal Academy and published, in 1862, by *The Illustrated London News*, which considered 'This little incident in the life of a great inventor and great public benefactor... of extreme interest'.[76]

The centenary of George Stephenson's birth, on 9 June 1881, was celebrated widely. Newcastle upon Tyne enthusiastically organised the grandest festivities.[77] A general holiday throughout Durham and Northumberland and special trains, arriving all morning from across northern England and southern Scotland, allowed many thousands of people to crowd the city's streets, watch the processions and buy centenary souvenirs from shops decorated with Stephenson memorabilia.[78] The showpiece was a procession of 16 modern locomotives, loaned by the chief railway companies, which steamed its way from Newcastle's central station to Stephenson's birthplace at Wylam and back again. The Newcastle Literary and Philosophical Society held an exhibition of Stephenson 'relics' and model locomotives, while a lecture on the *Rocket* and the history of early locomotive engineering attracted 'a large and interested audience'.[79] A banquet and a firework display completed the day's celebrations. In London, the Amalgamated Society of Railway Servants mounted an exhibition at the Crystal Palace, in aid of a new 'Stephenson wing' for its orphanage at

Commemorative plaque to George Stephenson, 1925, National Rail Museum, York (photograph by the author).

James Watt and George Stephenson on the membership certificate of the Associated Society of Locomotive Engineers and Firemen (Ironbridge Gorge Museum Trust).

Derby, and raised the £500 it needed to reach its £2,600 target. The exhibition demonstrated models of the latest safety devices adopted by the most progressive railway companies, with particular attention to those that would reduce the risks faced by employees, and hence the demand for orphanages.[80] Chesterfield, where Stephenson had spent his last years, held a gala, a dinner and a service in the church where Stephenson was buried.[81]

Celebrations also took place on the continent. In Rome, a grandiloquent commemorative tablet at the main station, paid for by railway employees, was unveiled in the presence of the British ambassador and a large crowd of spectators.[82] Already a plaque to mark the fiftieth anniversary of the *Rocket* had been placed on the façade of Turin station in November 1879.[83] In Holland, railway officials at Utrecht gave a concert to celebrate the anniversary, and it was reported that Stephenson's portrait had been enshrined at the centre of the Dutch railway map.[84] Budapest central station, opened in 1884, celebrates both Stephenson and Watt: their statues stand on either side of its main entrance.[85]

The two engineers also gaze at one another across the late nineteenth-century membership certificate of the Associated Society of Locomotive Engineers and Firemen, feature together or singly in a number of second-rate poems, and star in the grand narrative of the Industrial Revolution, constructed by Victorian historians since the 1830s – from which Brunel was normally absent.[86]

There are several reasons why it was George Stephenson, not Brunel, who came to share the laurels with Watt. First, Stephenson's reputation was regularly burnished by the Institution of Mechanical Engineers. As its first president, he enjoyed a particular status, which entailed kudos and commemoration. The Institution galvanised its members to subscribe to his monument, which from 1854 presided over Euston station; his image (and *Rocket*) appears on its bookplate, surrounded by the names of seven near-contemporaries;[87] naturally, his portraits grace the walls of its headquarters. Moreover, his reputation – more reliable than romantic and reckless – was probably the one preferred by most engineers.[88] Secondly, Stephenson had a better biographer: whatever Smiles' limitations, there is no doubting either the contemporary attractiveness or popularity of his pen-portraits. Other authors mined his works for information, which kept the story of Stephenson before the reading public. Thirdly, he enjoyed a strong local foundation: both George and Robert were born, lived and worked on Tyneside, and their locomotive workshops, Robert Stephenson & Co, provided a permanent source of employment there.[89] Brunel was born in Portsea, Portsmouth but had no other connection with the city; as an adult he lived in London, and therefore lacked a local identity; the locomotive workshops belonged to the GWR and other railway companies, dispersing his admirers across the south of England.

Brunel *redivivus*

Yet, Brunel is enjoying the latest laugh, though one hesitates to say 'last'. Not only is there a university, near London, named after him, but Bristol has adopted him – indeed, has virtually re-branded itself as 'Brunel City'.[90] This was not always the case. Testimony to the lack of a continuous association between Bristol and Brunel is his absence from an historical group portrait, *Some Who Have Made Bristol Famous* by Ernest Board, presented anonymously to the City Art Gallery in 1930.[91] Brunel's star began to rise again in 1957, in anticipation of the centenary of his death, with the publication of L T C Rolt's biography, described by Angus Buchanan as 'the outstanding work of engineering biography of the twentieth century'.[92] The centenary itself was marked by several exhibitions, a documentary broadcast on ITV and the unveiling of a memorial tablet on the Clifton Suspension Bridge.[93] It may have been Rolt's biography that prompted concern in both Britain and the United States to salvage the ss *Great Britain*, which had been beached in the Falkland Islands since 1937.

It was almost certainly the wreck's triumphant return to Bristol's Great Western dry dock in 1970, 127 years to the day since her launch, which sparked the city's rediscovery of Brunel. From its foundation in 1968, the Brunel Society had campaigned for her return, and continued for 20 years to raise awareness of Brunel's work through its activities both in Bristol and nationwide.[94] Thanks to the dedication and enthusiasm of volunteers, and the support of charitable donations, Brunel's great ship has been restored to become a major feature of Bristol's cityscape, visited by 100,000 people a year.[95]

In the course of this resurgence, she has revived interest in other major features which Bristol owes to Brunel – not least, the Clifton Suspension Bridge and Temple Meads

George Stephenson *by J R G Exley, print (© National Portrait Gallery, London). Design used as book plate for the Institution of Mechanical Engineers.*

(old) station. The seal was set on Bristol's identification with Brunel in 1982, when, in the rarest of twentieth-century tributes, two statues of him were unveiled on a single day – one in Bristol, the other at Paddington station, the London terminus of the Great Western Railway, which Brunel had built to join the two cities. The chairman of the Bristol and West Building Society, which had commissioned the statues, travelled by train between the two ceremonies.[96] Unlike Marochetti's statue of Brunel on the Embankment, neither of these statues by John Doubleday conveys the authority and self-confidence exuded by the great engineer. The Bristol statue, in particular, is a caricature, an icon to be recognised by anyone who has seen the photograph of Brunel standing in front of the *Great Eastern*'s anchor chains.

Some Who Have Made Bristol Famous by Ernest Board, 1930 (Bristol's Museums, Galleries and Archives). This symbolic gathering stands in front of Temple Gate, once part of the city's medieval defences, while Bristol biplanes fly overhead. Several members of the prominent local family of tobacco manufacturers, the Wills, are featured. The family had paid for the then recent extensions to the art gallery.

The past two decades have only added to Brunel's fame, with a succession of new biographies, television documentaries and second place in the *Great Britons* poll (Watt and George Stephenson, but no twentieth-century engineers, were in the top one hundred).[97] Perhaps Brunel has found his medium: his larger-than-life character and his ambitious projects make good television. It is hard for his successors not to pale by comparison – now, as much as in the later nineteenth century. The benefit to Bristol and the heritage industry may need to be set off against the damage possibly inflicted by this iconic image on recruitment into engineering careers and on the public engagement with science and technology. Do a stove-pipe hat and a cigar really convey the excitement of working with the cutting-edge technologies of the twenty-first century, or suggest the importance of the scientific and technological issues which urgently need to be debated?

L T C ROLT: ENGINEERING HISTORIAN

Lionel Thomas Caswall Rolt, known to friends and family as Tom, was born in Chester on 11 February 1910. He was distantly related to the mechanical engineer Kyrle Willans, who advised Rolt's parents to allow their son to leave Cheltenham School, where he was unhappy, and to let him be apprenticed.

In 1926 Rolt therefore joined the agricultural engineering firm of Bomford Bros near Evesham as a pupil. He began his full apprenticeship two years later at Kerr, Stuart & Co Ltd, locomotive engineers of Stoke-on-Trent. When the Kerr company collapsed, he finished his apprenticeship at R A Lister & Co Ltd of Dursley.

Having drifted through a succession of short-term engineering jobs, in 1934 Rolt set up a garage in partnership near Basingstoke, specialising in the servicing and restoration of vintage cars. He was also one of the founders of the Vintage Sports Car Club, which arose from associates with common interests.

Rolt had developed an interest in canals while holidaying with the Willans family on their narrow boat, *Cressy*. He bought *Cressy* in 1939 and it became home to himself and his first wife until 1951, when the marriage ended. His book, *Narrow Boat*, published in 1944, was inspired by his travels and is an affectionate account of the lives of those who lived and worked on the British canals. It is credited with leading to the formation of the Inland Waterways Association in 1946 and is still in print today. Rolt was the association's first secretary and in the immediate post-war years, he dedicated himself to preserving the canal system. He left

the Inland Waterways Association in 1951 as his desire that the canals should be restored for commercial use was at odds with the leisure interests of his fellow members. The previous year he had become involved in a successful preservation scheme to save the Talyllyn narrow-gauge slate quarry railway in north Wales. It was the first of many voluntary efforts. Today the Prince of Wales is the Talyllyn Railway's patron.

Rolt was a vice-president of the Newcomen Society, a member of the Science Museum's advisory council, helped to establish the National Railway Museum in York, advised the National Trust on industrial archaeology and was chairman of the Council for British Archaeology research committee on industrial archaeology and listing. He was appointed as the first president of the Association for Industrial Archaeology in 1973.

He lived with his second wife, Sonia, at Stanley Pontlarge, Gloucestershire where he wrote over 40 books on engineering history, making a significant contribution to the growing interest in industrial archaeology and the conservation of industrial buildings and machinery. His acclaimed biography of Brunel was published in 1957. Other works included *Railway Adventure* (1952), *Red for Danger* (1955), *Thomas Telford* (1958), *Victorian Engineering* (1970), *The Making of a Railway* (1971) and *From Sea to Sea* (1973). His three autobiographical books were published in a single volume entitled *The Landscape Trilogy* in 2001.

Rolt died at Stanley Pontlarge on 9 May 1974, and was buried in the churchyard of the small Norman Church on 14 May.

*Cover of first edition of
Rolt's biography of Brunel.*

THE BRUNEL STATUES

In March 2004 the correspondence columns of the Bristol *Evening Post* contained an exchange of letters about the future of the controversial Brunel statue outside the former Bristol & West head office. One proposal was simply to move the statue to a more appropriate location; Alan Elkan, chairman of Bristol Civic Society, called it an insult and wanted it melted down and recast in an anatomically more accurate form, while another correspondent favoured a radical resiting – at the bottom of the Bristol Channel.

John Doubleday's statue of Brunel at Paddington (photograph by David King).

The genesis of the two Brunel statues – both that in Bristol and a second version on the Paddington Railway Station concourse – was slightly bizarre. Bristol & West had earlier agreed to sponsor a statue of Charlie Chaplin to stand in London's Leicester Square: a project which caused some soul searching among the building society's executives. In the 1980s the societies were still conscious of their status as mutual organisations, and were anxious not to be seen 'wasting' savers' funds. Paying out good money to commemorate a comic actor might be considered flippant.

In the event, the ceremony in Leicester Square, enlivened by an urbane performance by guest-celebrity

Sir Ralph Richardson, attracted only modest media interest, and Bristol & West was barely mentioned in the handful of press notices.

Thus emboldened, the society next agreed to fund not one, but two statues of Isambard Kingdom Brunel. Both were to be sculpted by John Doubleday, creator of the Chaplin statue, and whose impressive list of commissions includes HRH Prince Philip, The Beatles in Liverpool, Sherlock Holmes and Nelson Mandela. There was then no commemoration in Bristol of the great engineer, and the proposed sponsorship was considered entirely appropriate.

The double-project now became a major public relations exercise, squeezing every ounce of publicity out of a unique day in May 1982 when the Paddington statue was unveiled to general acclaim in the morning, with musical entertainment and Bristol & West staff dressed up as Brunel look-alikes, the entourage then embarking on an Intercity 125 high-speed train for the second unveiling in Bristol barely two hours later.

John Sansom, who had been running the Bristol & West press office at the time, recalls the nearly disastrous unveiling:

> A crowd of the great-and-good stood around as Bristol's Lord Mayor tugged the cord to unveil the statue. As the figure slowly emerged, there was first a gasp, then tentative tittering… I feared the worst, but fortunately those who evidently found it all a bit comical pulled themselves together, and the moment passed.[1]

Those few moments were perhaps symbolic. The first note of disappointment with John Doubleday's Bristol sculpture was sounded minutes later when George Ferguson, a future president of the Royal Institute of British Architects, was heard to say to Andrew Breach, Bristol & West's chairman: 'Pity about the statue, Andrew, but at least you also sponsored a decent book about Brunel.'[2] Although some may think Brunel looks supremely confident and cocky, most informed opinion has been that the statue borders on the caricature, and that Brunel and his adopted city deserve something better.

Brunel statue in Bristol (photograph by Douglas Merritt).

BRISTOL

Booklet from City Information Bureau, The Centre, Bristol I.

TRAVEL BY TRAIN — BRITISH RAILWAYS

BRISTOL

ROMANTIC CENTRE FOR A DELIGHTFUL HOLIDAY

Booklet from Public Relations Officer, Council House, Bristol I.

TRAVEL BY TRAIN — WESTERN REGION

'SUSPENSA VIX VIA FIT'
THE SAGA OF THE BUILDING OF THE CLIFTON SUSPENSION BRIDGE

Michael Pascoe and Adrian Andrews

Railway promotional posters for Bristol: Windsor Terrace, the Clifton Suspension Bridge and part of the Hotwells, artwork by Claude Buckle, c 1952 and the Portway and Clifton Suspension Bridge, artwork by Leslie Wilcox, c 1957 (Bristol's Museums, Galleries and Archives).

The building of the Clifton Suspension Bridge, drawing by Samuel Jackson, 1836 (Bristol's Museums, Galleries and Archives).

A great gateway to the ancient port

The first proposal for a bridge at Clifton reflected the grandiose aspirations of the merchant classes of Bristol in the late eighteenth century. Several of the most prosperous were moving from living near their place of business in the old, crowded city to the lofty heights and clean air of Clifton to the west, overlooking the spectacular Avon Gorge. By 1790 a dozen grand terraces were being built there, including the longest crescent in Europe, Royal York Crescent. To match these grand buildings, in 1793 Bristol architect William Bridges designed a spectacular stone bridge with a massive central arch, 220 feet high spanning 180 feet with a 700-foot-long roadway. It was to be flanked by five, 40-foot-high terraced storeys, designated for commercial, social and industrial uses, ranging from a mill to a museum.

This imperial-scale triumphal gateway was intended to restore Bristol's status and help regain shipping business lost to the cheaper and more accessible ports of Liverpool and Southampton. The viability of Bridges' design assumed successful speculative development on both sides of the gorge.[1] Unfortunately, 1793 was also the year Britain declared war on Revolutionary France. The grand projects stopped, many builders and developers went bankrupt and the great bridge remained a dream.

Plan and elevation of a bridge across the Avon Gorge, Phenedus Daniel, after William Bridges, March 1793 (Bristol's Museums, Galleries and Archives).

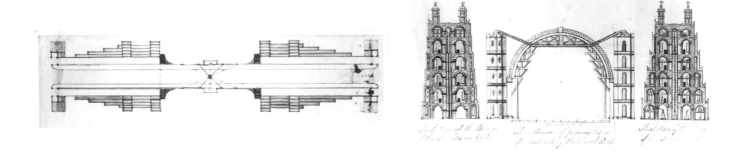

Bridges' proposal was designed for the site specified in a will written nearly 50 years earlier. William Vick, a wealthy Bristol wine merchant, died in 1754, leaving £1,000 to be invested until it reached £10,000 – the estimated cost of building a stone bridge 'of great public utility' across the Avon Gorge.[2] At that time, crossing the River Avon into Somerset entailed either ferry crossings in small boats – not ideal for transporting cattle or wheeled vehicles – or negotiating the steep hills and narrow city streets to cross Bristol Bridge. Vick also possibly foresaw the new bridge as linking roads to allow easier communications to the North, the West Country and South Wales. The location, although challenging in scale, given the need to provide clearance for tall-masted ships, was the last practical site close to the city for a bridge. A bequest was the only realistic means to fund a project likely to yield little profit in the short term.

Bridges' 1793 design (Bristol's Museums, Galleries and Archives).

In 1779, some years before Bristol was planning a great stone bridge, the world's first iron bridge had been built at Coalbrookdale in Shropshire. It drew admiring visitors from across the world as the first wonder of the Industrial Revolution. The second great cast-iron bridge was constructed at Sunderland in 1796, using 260 tons of iron (half that of Coalbrookdale), and was also flanked by multi-storey buildings. Its six great ribs took only ten days to erect over the busy River Wear. Within four years a new London Bridge in cast iron was proposed, with a magnificent single span of 600 feet. But there was a serious problem: cast iron is strong in compression and therefore ideal for arches, but brittle and prone to failure when subject to tension or excess load.

However, a recently developed type of iron was attracting the attention of engineers. Henry Cort of Hampshire had patented a technique in 1783 by which ferric oxide added to heated iron removed most of the carbon, making wrought iron a metal stronger and less brittle than cast iron. By Cort's process, wrought iron could be made 15 times quicker than cast and was therefore cheaper. Such new technology had produced a metal which was strong in tension, ideal for anchor chains and suspension bridges.

The first modern suspension bridge

In 1801 in Pennsylvania James Finley built the first modern suspension bridge with a modest span of 70 feet and hand-forged, wrought-iron loop chains with vertical suspenders supporting a level timber truss-braced deck. Finley patented this system in 1808 and, by the time of his death 20 years later, some 50 bridges had been built to his method, the longest with a span of 306 feet. Compared to stone, these bridges were cheap and quick to build and, despite several collapsing due to forging failures or extremes of climate, the concept spread rapidly.

The Wear Bridge, Sunderland, 1796, print (Adrian Andrews' collection).

Chain Bridge over the Potomac, Georgetown, Washington DC, USA, 1807, wood engraving published in The Family Magazine, *1839 (Adrian Andrews' collection).*

Essex-Merrimac chain bridge, Newburyport, Massachusetts, USA, 1810, from a stereoview (Adrian Andrews' collection)

By the 1800s British iron making led the world. The foremost engineers, Thomas Telford, Samuel Brown, Peter Barlow and Marc Brunel, were conducting experiments on the relative strengths and durability of iron in loops, bars, rods and wire. Wrought-iron bar chains were proved to require half the metal but possessed twice the strength of iron links. In 1818 retired Royal Navy Captain Samuel Brown patented his chain link suspension system and used it to build the first large-scale suspension bridge near Berwick-on-Tweed. The Union Bridge had a span of 437 feet, took less than a year to build and cost only £7,700. It attracted great attention and admiration. In 1822 the first design for a suspension bridge at Clifton by Hugh O'Neill was an exact copy.

Numerous suspension bridges were built at this time in Britain and France; many collapsed. But in 1826 Thomas Telford completed the magnificent Menai Bridge in North Wales using Brown's chains to support a record-breaking span of 570 feet. The new link permitted greatly increased levels of traffic to the deepwater port of Holyhead, the principal embarkation port for Dublin and the shortest route from London.

Bristol: a competition is announced

Meanwhile, trade in Bristol continued to decline, reflecting widespread economic depression and political unrest brought about by the Corn Laws and delays to the Reform Bill. A bridge was seen as 'one of those bold and animating schemes of improvement which would have given a spur and activity to the present parilized and torpid energies of the Citizens of Bristol'.[3] In 1829 a committee led by the mayor and involving other prominent citizens was formed to assess the feasibility of fulfilling Vick's bequest. An estimate of £90,000 proved that a stone bridge, as Vick had specified, was not viable. An iron suspension bridge 'to enhance the beauty and magnificence of the situation', though requiring substantial funds, was more likely to attract investors. In October 1829 the committee advertised a competition inviting designs for an iron suspension bridge, with a prize of 100 guineas. Entries were to be submitted by 19 November, just six weeks later.

Union Suspension Bridge over the Tweed, 1820. Engraved by William Read, drawn by George Buchanan, published by J McGowan (above left) (Adrian Andrews' collection).

The first design for a suspension bridge over the river Avon by Hugh O'Neill, detail from a pencil and wash drawing, 1822 (above centre and right) (Bristol's Museums, Galleries and Archives).

Isambard Kingdom Brunel was 23 years old. His only previous experience of suspension bridges had been six years earlier when first apprenticed to his father, Marc, who designed two short-span bridges for the Ile de Bourbon (now Réunion) off the African coast. To study the latest example, Isambard spent two days in a detailed examination of the Menai bridge, which had suffered repeated damage from gales. He considered the deck to be too light.

There were 22 entries for the competition, including two by leading suspension bridge engineers, Captain Brown and Bristol-born William Tierney Clark, who had recently built the first suspension bridge over the Thames at Hammersmith. Only the plans for the entries of James Meadows Rendel, 30 years old and a former pupil of Telford, and Isambard Kingdom Brunel seem to have survived. Brunel submitted four beautifully drawn sepia plans and elevations. His lengthy supporting statement revealed a profound desire to respond to the dramatic site, reflected in his choice of locations, requiring spans from 720 to over 1,000 feet. The most detailed design involved a 300-foot tunnel from the 'Giant's Cave' in the cliff on the Clifton side, from where the traveller would 'burst upon the splendid scene', cross a dramatic span of 980 feet, enter another short tunnel and then proceed along a natural ledge of a wooded valley.

Brunel later wrote: 'I thought that the effect... would have formed a work perfectly unique... the grandeur of which would have been consistent with the situation.'[4] He candidly acknowledged that the design had 'no striking outlines to attract the eye' and lacked the symmetry associated with suspension bridges. However, the absence of

Menai Bridge, North Wales, 1826, albumen photograph (Adrian Andrews' collection).

Elevation and plan of Marc Brunel's chain bridge over the River Sainte-Suzanne, Ile de Bourbon, 1823. Published in The Penny Magazine, *1833 (Adrian Andrews' collection).*

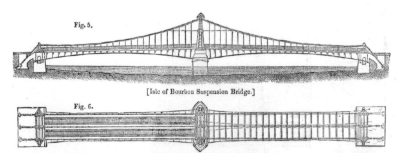

Fig. 5.

[Isle of Bourbon Suspension Bridge.]

Fig. 6.

[Plan of Isle of Bourbon Suspension Bridge.]

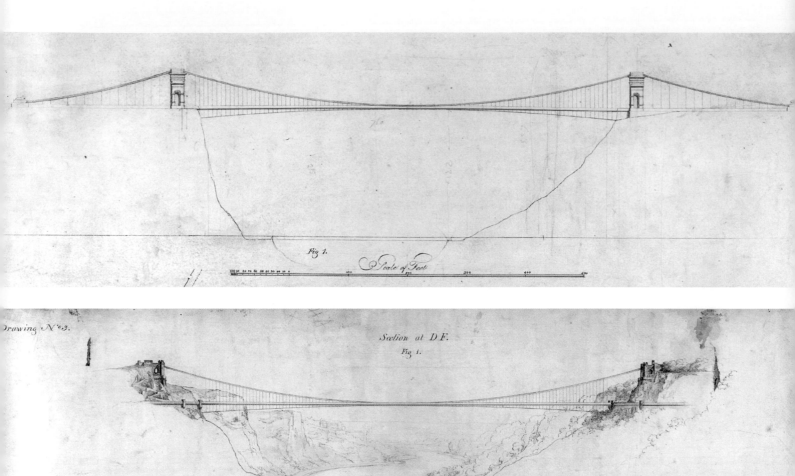

Fig 1.

Scale of Feet

Section at D.F.

Fig 1.

Fig 2.

Scale of Feet

Fig 2.

Section at D.G.

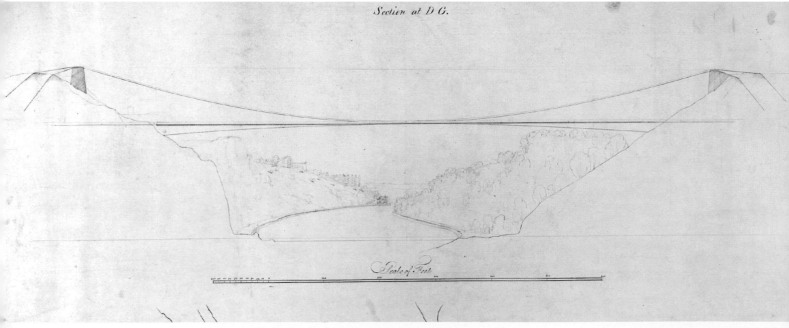

Scale of Feet

abutments and piers would have meant significant savings. All four designs show Marc Brunel's influence in using staying catenary cables below the deck, similar to those successfully used on the Ile de Bourbon bridges to provide stiffness to the deck structure.[5]

A contentious decision

The committee dismissed most of the entries on either non-compliance, cost or aesthetic grounds. Lacking technical experience, they approached Telford, then aged 70, acknowledged 'father of British civil engineering', first president of the Institution of Civil Engineers and internationally acclaimed for his Menai suspension bridge, to make the final decision. Telford was self-taught, from humble origins and renowned for encouraging his apprentice engineers in every way. In contrast, Isambard was the only son of a famous engineer and had received the best available education in France, especially in mathematics. Some have accused Telford of envy of the young Brunel's ideas. However, as Anthony Burton points out: 'Recent problems at Menai where lack of lateral wind resistance had caused serious damage, had made him cautious … he really thought that the bridge that Brunel designed would not stand up, it would collapse.'[6] Telford commented that 'Brunel's bridge, although very pretty and ingenious, would most certainly tumble down in a high wind'.[7] His confident proposal of 15 years earlier for a suspension bridge with a 1,000-foot span at Runcorn had been tempered by his experience at Menai and a number of recent bridge failures with loss of life.

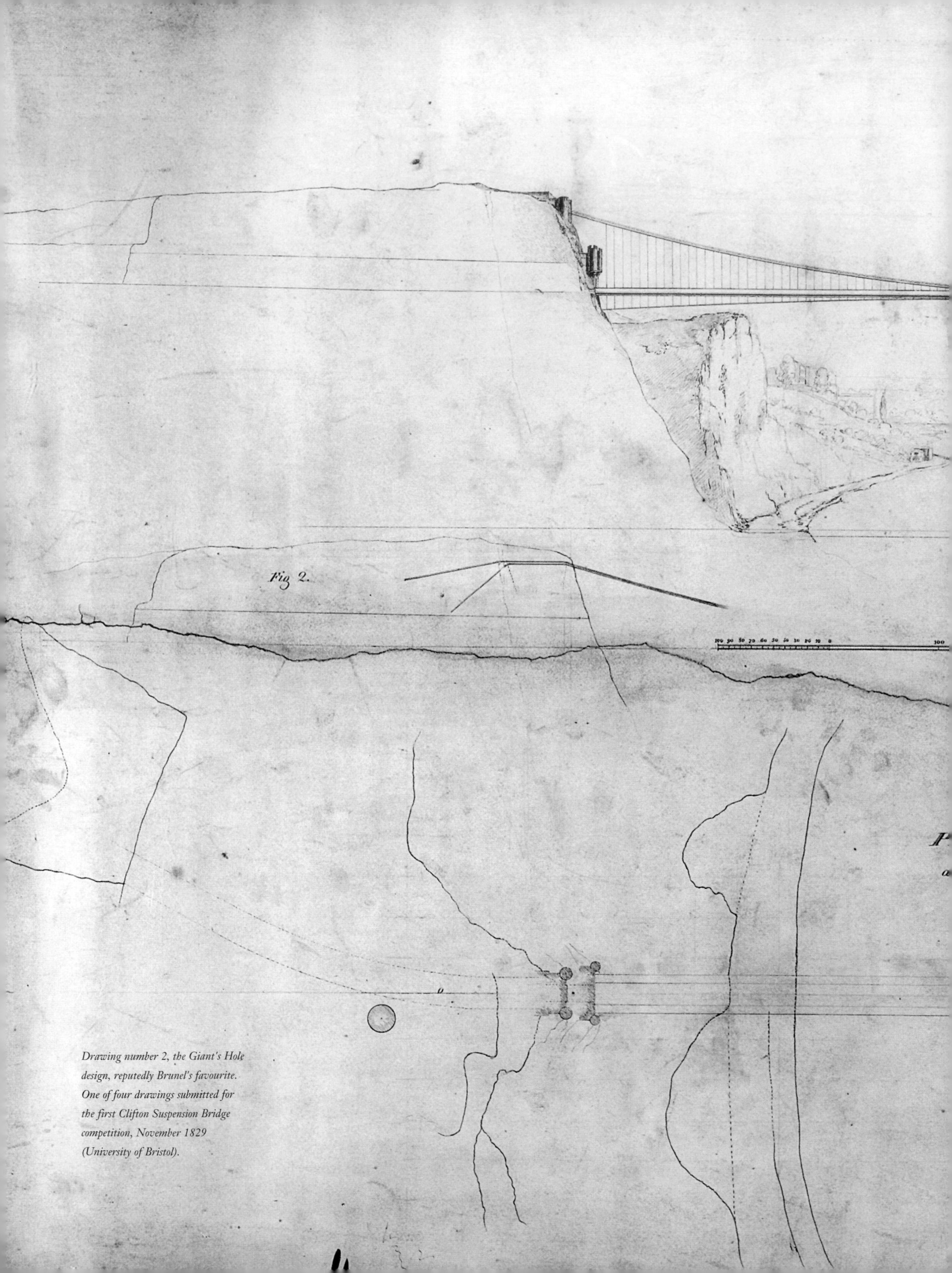

Fig 2.

Drawing number 2, the Giant's Hole
design, reputedly Brunel's favourite.
One of four drawings submitted for
the first Clifton Suspension Bridge
competition, November 1829
(University of Bristol).

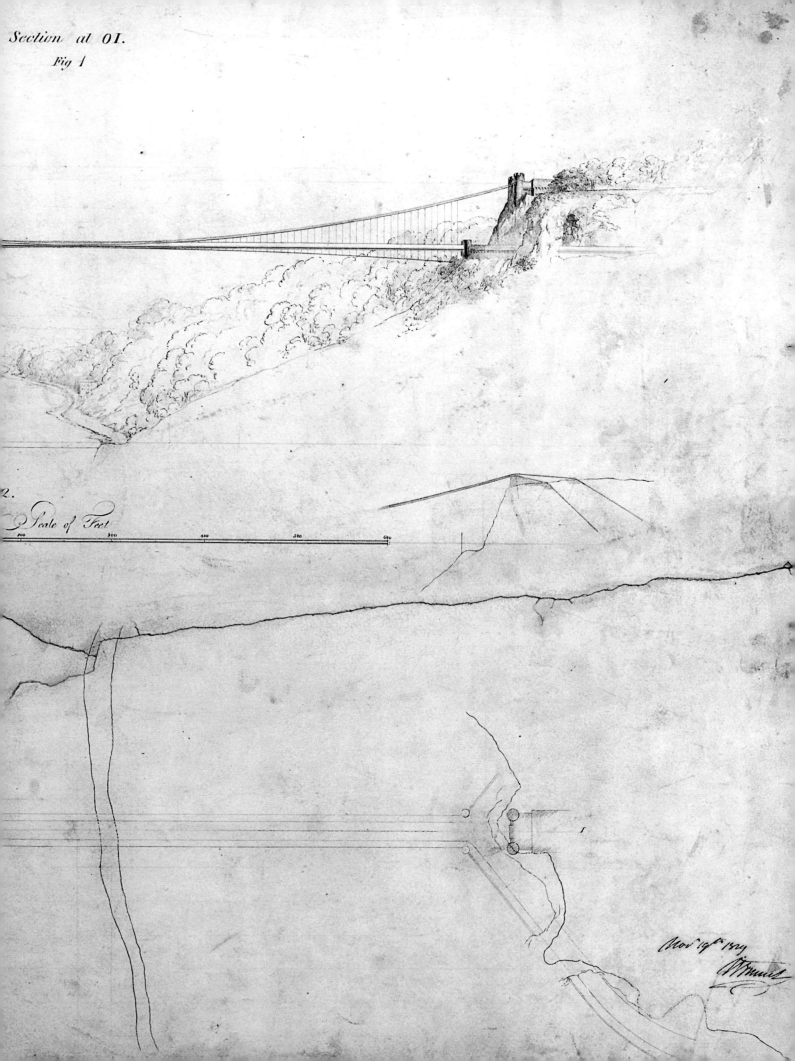

Section at OI.
Fig 1

Fig 2.

Scale of Feet

Nov 19th 1829

Thus, despite Brunel's meticulous supporting calculations demonstrating the feasibility of his designs, Telford rejected every entry, convinced that 600 feet was the maximum practical span. He criticised the long span designs as being particularly susceptible to wind. Telford was then invited by the committee to submit his own proposals and returned within a month with a design for a three-span suspension bridge, supported by two colossal hollow masonry Gothic towers that rose almost 260 feet from the banks of the Avon. The committee clearly thought the best result had been achieved, a magnificent bridge, almost on budget at £52,000, by the greatest engineer of the day – 'a monument of architectural taste and splendour, without parallel, not only in this country, but in any part of the globe'.[8] Brunel, 'disappointed at the turn matters had taken… smoked away his anger'.[9] He obtained permission to withdraw his plans from the competition and left Bristol on the mail coach for the North of England.

Portrait of Thomas Telford by Samuel Lane, 1822 (Institution of Civil Engineers).

On 23 January 1830 a prospectus was issued with a lithograph of Telford's design. Within a week, as the Bill for Parliamentary approval was about to be submitted, considerable public and professional criticism was voiced. Brunel wrote sarcastically to the committee:

> As the distance between the opposite rocks was considerably less than what had always been considered as within the limits to which suspension bridges might be carried, the idea of going to the bottom of such a valley for the purpose of raising at great expense two intermediate supporters hardly occurred to me.[10]

The committee rapidly back-pedalled and announced 'a fair and liberal discussion of the merits and demerits of the Design is all that they wish', claiming that they were not necessarily committed to Telford's proposal. Such confusion rapidly undermined investors' confidence. The committee decided to seek a second opinion. On 13

William Armstrong's design, May 1830, lithograph drawn by William West who lived in the Clifton Observatory (Bristol's Museums, Galleries and Archives).

February Marc Brunel wrote that he received a 'deputation of gentlemen from Bristol' who wished to consult him about the project of erecting a bridge across the Avon. This served as 'a cordial to my bruised spirit, I explained to them how the lateral agitation might be prevented and how the effects of the wind might be counteracted'.[11]

Artist William West, an enthusiastic supporter of the bridge project and tenant of the Clifton Observatory which overlooked the site, published a new proposal in May 1830. This design by local surveyor and architect William Armstrong was for a simple, under-chained girder bridge with a modest span of 450 feet, linked by masonry viaducts. That summer, local sign writer William Burge published his proposal for a truly monumental bridge with Tuscan Doric masonry columns supporting a 600-foot iron-ribbed roadway, clad by a masonry entablature – a Classical response to Telford's Gothic.

A second competition is held

By October 1830 the committee, still well short of funds, announced a second competition. This time Telford was a competitor 'on the same footing as the others' rather than referee, as an expert judge who was not an engineer by profession was appointed in advance – a fact that did not deter amateurs or minor architects. Davies Gilbert MP was a former president of the Royal Society, an eminent mathematician and theorist and a Parliamentary Commissioner who had advised on the Menai bridge and was described as 'a person of science and impartiality… qualified to judge of this principle as well as the specifications'.[12] His third choice of co-judge was John Seaward, whom Gilbert considered 'an excellent mathematician' and expert in iron-founding.

Four entrants ignored the terms of the competition by proposing arched structures. One daringly combined a 625-foot suspension span with a cantilevered arch. Only six

William Burge's design, August 1830, lithograph (Bristol's Museums, Galleries and Archives).

entrants adhered to the rules and submitted designs for suspension bridges. Charles Capper of Birmingham rendered Telford's approach as a picturesque Romantic ruin. Directly challenging Telford's caution, Rendel re-submitted his proposal for a bridge spanning the full width of the gorge, 780 feet, stating that 'as the site is unrivalled, the bridge should not be surpassed in magnitude or scientific contribution'.[13] The more experienced Brown re-submitted his scheme with a span of 780 feet too. But Rendel's was estimated at £92,885 in contrast to Brown's at £30,000. Smith & Hawkes proposed a span of only 630 feet.

Brunel again submitted four designs, commissioning the best Bristol artist, Samuel Jackson, to illustrate the landscape. The first was a re-working of his favourite 'Giant's Cave' design, with its audacious span of 900 feet, twice that of any existing suspension bridge. Designs two and three proposed moderate spans of 720 feet, made possible by a massive masonry abutment on the Leigh Woods side, but with elaborate Gothic masonry towers. The fourth reflected Telford's solution with a short span supported on Egyptian-style columns and piers, the first use of this style among any of the entries: functional, fashionable, economic, simple compared to Gothic, yet distinctive. Telford's design was included. Bristol-born William Tierney Clark was invited to re-submit his scheme but declined the invitation.

The 12 entrants to the second competition were reduced to a short list of five. On 16 March 1831 the judges tactfully set aside Telford's scheme due to:

> inadequacy of funds for meeting such high and massive towers… they abstained
> from any remarks on architectural effect or ornamental appearance, confining their

Lithograph of design for a suspension bridge across the river Avon by C H Capper, 1830 (University of Bristol).

Brunel's design No 1 for the Clifton Suspension Bridge, 18 December 1830, an ink and wash drawing by Samuel Jackson and Isambard Kingdom Brunel (GWR Museum, Swindon/National Railway Museum). The deck design differs from the 1829 submission, lacking Marc Brunel's staying catenary chains.

Detail of Brunel's design No 2 for the Clifton Suspension Bridge, 18 December 1830, artist as above (GWR Museum, Swindon/National Railway Museum). This view shows the large Leigh abutment. The Gothic towers were based on the entrance to Lancaster Castle.

Brunel's design No 4, 1830 (National Archives). This was his first Egyptian-style design and matched Telford's minimum 360-foot central span, but with more economic masonry columns.

Brown's 1830 competition entry, with a proposed span of 780 feet (Adrian Andrews' collection).

attention exclusively to those particulars which merely concerned the strength and durability of the proposed structure.[14]

The remaining entries were placed in order of merit: first, Messrs Smith & Hawkes, second, Brunel, third, Brown and fourth, Rendel. Significantly, these were the only competitors with experience of suspension bridge engineering. However, the judges criticised each design. They considered that with a maximum load the weight per inch of chain should not exceed five-and-a-half tons per square inch. Only Brunel's design at four-and-one-fifth tons per square inch fulfilled this. The judges had reservations concerning his chain design – where each bar was joined directly to the next without Brown's patented short intermediary coupling link – as well as the anchorages and suspension rods.[15] Smith & Hawkes were declared the winners.

Brunel promptly arranged a meeting with Gilbert and convinced the judge that his objections were unjustified. He promised to make detailed alterations to meet Gilbert's requirements, though this is not mentioned in his diary. Within two days the committee re-examined the plans and not only announced Brunel the winner, but also appointed him project engineer at a fee of 2,000 guineas. What Smith & Hawkes made of such dealings is not recorded.

The architectural style was chosen on 27 March. Brunel triumphantly wrote:

> Of all the wonderful feats I have performed… yesterday I performed the most wonderful. I produced unanimity amongst fifteen men who were actually quarrelling about the most ticklish subject – taste… The Egyptian thing I brought down was… unanimously adopted.

His proud father wrote in his diary:

> Isambard appointed Engineer to the Clifton Bridge. The more gratifying that Mr Telford, Captain Brown and Mr T Clark were his competitors and that, for my part, I have not influenced any of the Bristol people on his behalf either by letter or by interview with any of them.

A painting was commissioned for public display to attract investors. The piers were to be topped by pairs of sphinxes originally facing each other, but soon reversed to face the approaching traveller. Yet historical authenticity was compromised by the intention to clad the piers with cast-iron panels, illustrating every phase of the bridge's history, including quarrying of iron ore and every stage of the making of wrought iron – 'the very images of the industrial revolution'.[16]

Marc's role in the project has never been fully appreciated: though he played no part in trying to influence the judges, he did influence his son. Practically every entry in Marc's diary during January, February, March and April 1831 records him designing details and making drawings. Thus on 12 January: 'Devised this day, for Isambard's bridge, a new mode of carrying the heads of the chains'. On the 20th he wrote: 'Engaged this day on the mode of passing the chains of the bridge over the heads, with all the combinations necessary for repairing [a condition of Gilbert's] and likewise for the compensation against dilation.' Throughout March: 'Working on Isambard's bridge.' Such entries continue intermittently into 1832. This proud father was determined to ensure his only son's first commission would be a world-class piece of engineering.

View of the Avon Gorge with the Approved Design for the Clifton Suspension Bridge *by Samuel Jackson, 1831 (Bristol's Museums, Galleries and Archives).*

A false start

On 21 June 1831, even though the project was at least £20,000 short of the necessary funds, a small ceremony marking the beginning of construction was held. Sir Abraham Elton, a local dignitary, declared that the time would come when Brunel would be recognised in the streets of every city and 'the cry would be raised, "There goes the man who reared that stupendous work, the ornament of Bristol and the wonder of the age"'.[17] But such optimism was premature for by October, frustration at delays to the Reform Bill saw Bristol erupt in the bloodiest riots to take place in nineteenth-century England. Brunel enrolled as a special constable, helping the mayor to escape and later testifying against some of the rioters. There followed a severe loss of business confidence and work on the bridge ceased. Fresh appeals for subscribers brought few results due to competing investment opportunities, especially the Great Western Railway, to which Brunel was appointed chief engineer two years later.

The delay prompted new suggestions for cheaper schemes. William West had recently toured France and Switzerland inspecting suspension bridges. He was especially impressed by the newly completed Fribourg bridge in Switzerland, with a record-breaking span of 870 feet. West wrote to the committee, saying: '… the Fribourg bridge supports (cables) are formed wholly of wire; and from the facts I trust I shall be able to prove that the application of wire is equally strong and far more economical than chains.'[18] In an illustrated report, West claimed Fribourg had cost only £23,000. The committee promptly requested a meeting with West and Brunel, but in a forthright letter the latter stated that he had long studied the use of wire and was 'intimate' with the principal French manufacturers, Messrs Seguin. He concluded:

> It does not follow because wire is light and strong that it is cheap. Wire being upwards of four times as dear as bar-iron, and not being anything near four times as strong, you may have the same strength in a chain of iron-bar for much less than in wire.[19]

He also observed that wire was less reliable, and difficult to protect against corrosion. Brunel was clearly convinced of the superior quality of British iron manufacturing.

The committee accepted their engineer's opinion and instructed him to estimate the cost of a bridge 'of equal strength and durability, and capable of extension … when sufficient funds for that purpose shall be obtained'. West was tactfully thanked and paid 30 guineas for his trouble. Brunel produced a revised scheme estimated at £35,000, making savings by reducing the size of the towers, narrowing the roadway, abandoning the footways, specifying a bare timber deck without gravel and removing the ornamentation. This reduced the weight by half, thus requiring only single chains, although the design allowed for second chains and footways to be added later if funds permitted. Perhaps embarrassed by the bare economy of this solution, the committee quickly abandoned it and reverted to Brunel's original design.

While Brunel was attempting to reduce costs at Clifton, he was engaged to design a new suspension bridge over the Thames. Hungerford was the only pedestrian bridge in London and was completed in 1845 at a cost of £110,000 – twice the Clifton budget.

Chaley's record-beating Fribourg Bridge, over the Sarine River, Switzerland, 1834, from two stereoviews and a postcard (Adrian Andrews' collection).

Late in 1835 Brunel was instructed to prepare working plans for the Leigh Woods abutment. On Boxing Day he wrote: 'Clifton Bridge – my first child, my darling is actually going on – recommenced week last Monday – Glorious!!' He then listed projects totalling over £5,000,000 for which he was now responsible and which he believed had resulted solely from the prestige of winning the bridge competition.

A grand foundation ceremony

In 1836 advertisements invited tenders for excavating and transporting stone from a quarry owned by a trustee of the bridge, William Miles, on his estate downriver. A hauling-way up the side of the gorge was constructed with man-powered capstans set at the top and base. At 7am on 27 August a trumpet fanfare echoed through the gorge. Vast crowds of 'not less than 60,000' occupied every vantage point, and specially chartered vessels crowded the river. A long procession of eminent scientists, engineers and civic dignitaries made its way to the site, including Marc Brunel on his first recorded visit. Isambard 'held aloof … asserted his pre-eminence, as designer of the bridge, by walking alone behind the Marquess's carriage and ahead of the rest of the procession'.[20] A great cheer went up as the Marquess of Northampton, the president of the British Association, laid the foundation stone of the Leigh Woods abutment.

Progress remained slow, however. By spring 1837 Brunel reported delays due to financial disputes between the contractors. Shortly after, the workforce was laid off. In June the contractors were declared bankrupt. The committee took control and the labourers were re-engaged. Brunel was recorded as 'working like the Devil' and expected his assorted gang of Cornish miners and local stonemasons to do the same.[21] The massive abutment was clad in red sandstone ashlar enclosing a double-storey vaulted superstructure of cheaper local pennant stone. The cavernous chambers were not rediscovered until 2001. This approach reflected a time when materials were more expensive than labour, especially forged ironwork.

Photograph taken during the exploration and survey of Chamber No 4 of the Leigh abutment, 2002/2003 (photograph by Robert Lisney).

Detail from one of Brunel's 1838 sketch books showing the building of the Leigh abutment (University of Bristol).

The Ceremony of Laying the Foundation Stone of the Suspension Bridge *(detail) by Samuel Colman (Bristol Museum, Galleries and Archives).*

Lithograph showing view of the ceremony, 1836 (private collection).

Alternative cost-saving schemes for completing the project continued to be presented. In 1837 Thomas Motley proposed his new 'inverted bracket suspension system' but after close questioning 'Mr. Brunel was convinced that the design was objectionable'.[22] The following year his opinion was sought on 'Mr. Dredges's patent tapering chains', a design foreshadowing modern cable-stayed bridges. Dredge had powerful supporters, so the committee requested Brunel to meet them 'at a time most convenient to himself'.[23] By now Brunel was fully engaged with the Great Western Railway and failed to find a convenient time.

Clifton Suspension Bridge, 1840, Ironwork Contract and Specification – details. Elevation of the deck, plan and elevation of a saddle and a chain bolt cap. By kind permission of Network Rail (Adrian Andrews' collection).

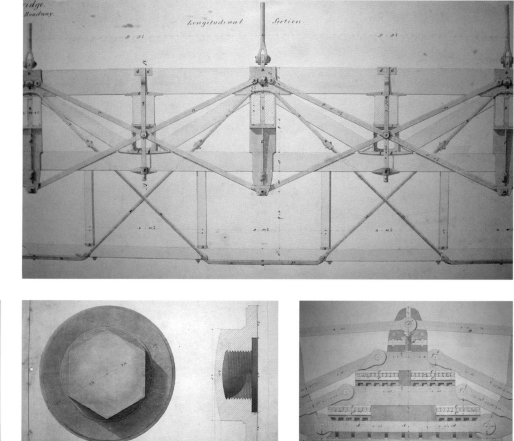

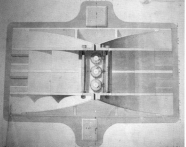

The Leigh Woods abutment was completed in 1840 but at a cost of £13,971, nearly one third of the available funds. The compromise, forced by Telford's caution, to reduce the span by using an abutment proved fatally costly, especially as difficulty was being experienced in securing promised funds. The 1840 specification and drawings for ironwork, discovered in 2004, illustrate the inventiveness and sheer quality of Brunel's detailed design. The order was placed with Messrs Sandys, Vivian & Co, at Copperhouse Foundry, Hayle in Cornwall, with very detailed specifications as to how the components were to be manufactured to the highest standard.[24]

Funding fails

By February 1843 the towers stood ready to support the chains, the anchorage and approach roads were finished, half the ironwork was on site but £45,000 had been spent. Hopelessly over-budget, the committee announced that a further £30,000 was needed and Brunel was instructed to stop work. Both he and the committee tried to raise funds by offering future toll income to the contractors, but to no avail. Creditors began to press for payment.

Avon Gorge with the bridge towers, 1850s photograph (Bristol's Museums, Galleries and Archives).

Finally, in October 1851, Brunel was directed to sell off the iron and plant on the best terms he could. The bar chains, machinery and building equipment were sold to the South Devon Railway Company for use on Brunel's Royal Albert Bridge at Saltash, linking Devon and Cornwall.

Hearing of the demise of the project, the consulting engineer to the US Government, Lieutenant Colonel Edward W Serrell, approached the trustees proposing to complete the bridge using wrought-iron wire cables in place of bar chains. He had just built the world's longest single-span suspension bridge (1,040 feet) in under 11 months, joining Canada and the US at Lewiston. It was similar to Brunel's design No 4 of 1829. Serrell claimed to be capable of making:

> wire cables impervious to the atmosphere and water, and of giving a uniform strain to each strand … that all the former disadvantages of the method are done away with, and a structure produced at a trifling cost.[25]

His initial estimate was £7,400, later revised to £17,000. In 1857 Serrell's design was adopted – although 'the whole construction proposed was of the slightest kind' and

*Clifton Suspension Bridge,
under construction, spring 1864,
photograph (Clifton Suspension
Bridge Trust).*

the deck was to be only 19½ feet wide[26] – and a new company formed. For the last time, Brunel was consulted and expressed strong disapproval of it, but said that he would not oppose the directors if this design were adopted. With consistent loyalty, the directors deferred to Brunel and paid Serrell £90 to defray expenses. In 1862 his Lewiston bridge collapsed in a storm.

At the height of his powers but worn out by unremitting hard work, on 9 September 1859 Brunel suffered a stroke and collapsed on board the ss *Great Eastern*. He died six days later, aged 53. In Bristol there was a campaign to remove the towers, known locally as 'Vicksville'. But ironically, Brunel's untimely death led to the completion of the bridge. The Institution of Civil Engineers resolved 'to finish it as a fitting monument to their late Friend and Colleague at the same time removing a slur from the engineering talent of the country'.[27] By coincidence, Hungerford Bridge was about to be demolished to make way for John Hawkshaw's Charing Cross railway bridge.

A fitting memorial

Early in 1860 Hawkshaw and leading fellow engineer, W H Barlow, wrote a feasibility study showing how the Hungerford ironwork might complete the Clifton bridge. Their report impressed the directors and in May a new company was formed. Completion of the project was estimated at £45,000, but it was stressed that it would not be a profitable investment opportunity. Nevertheless, appeals were successful and within the year £30,000 had been raised. A new Act of Parliament was sought and received royal assent on 28 June 1861.

The chains and ironwork of Hungerford Bridge were bought for £5,000 and Cochrane & Co of Dudley appointed as the new contractors under the supervision of Hawkshaw and Barlow. Project management was entrusted to the energetic and able Thomas Airey, who designed much of the machinery used to construct the bridge. The bridge as designed by Hawkshaw and Barlow differed in several ways from

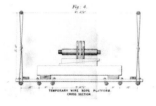

*Diagram of the Falsework,
Description of the Clifton
Suspension Bridge by W H Barlow,
Minutes and Proceedings, 1867
(Institution of Civil Engineers).*

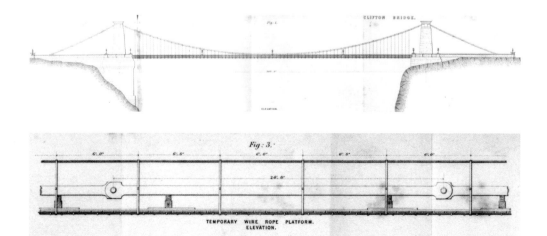

Clifton Suspension Bridge under construction,
1863-4, stereoviews (by kind permission of
Robert M Vogel, Washington DC, USA.
Clifton Suspension Bridge Trust archive)

Brunel's. Based on thorough research and in order to stiffen the deck, Barlow added two longitudinal girders which, like the transverse girders, were made of wrought iron, replacing Brunel's combination of timber and iron.[28] This provided stability but increased weight. At the insistence of a local landowner and investor, the deck was widened from 24 to 30 feet. To support this heavier deck and to spread the stresses evenly, each suspension rod was bolted through the joint of each chain, unlike at Hungerford, necessitating an additional third chain and alterations to the Hungerford saddles.[29] These changes required an extra 500 tons of ironwork. The design of the anchorages was also altered and the land chains on the landward side shortened, thus eliminating the need for suspension rods. Finally, Brunel's cast-iron decorative panels and sphinxes were omitted. By November 1862 the chains from Hungerford arrived, fittingly transported by the GWR.

Construction began early in 1863 with the erection of massive timber scaffolding around both towers. The Hungerford saddles (weighing four tons each) were installed on top of the piers. By June the new anchorages had been excavated to a depth of 70 feet and were ready for the chains to be secured. The base of each tunnel was cut to fan out in all directions so that the infill of blue Staffordshire brick formed an immovable plug.

Next, a wrought-iron wire cable was fixed between both towers forming the first link. Eight further cables were then added, each 1,100 feet long and weighing two tons. Six of the cables covered by stout planks bound with iron hoops formed a walkway. Two cables served as handrails and tracks for the grooved wheels of a 'cradle' to support the bar chains during assembly. The ninth cable was set at head height and from this a 'traveller' or light frame carried each link of the chains. This 'falsework' formed a temporary bridge of considerable strength. However, its movement was said 'to try the nerves of the most persons in passing over it, oscillating as it did with every breeze'.[30] By August this work had been completed 'reportedly without accident', though in a violent storm later that year the restraining cables anchoring the temporary bridge to the gorge broke loose and it was flung 70 feet upwards. Remarkably, it survived intact and the cables were re-secured.

Work to assemble the chains now began. The traveller running on the upper cable carried each half-ton bar into position, where it was bolted to its neighbours. As each new link was added, wooden blocks were placed underneath for stability, the falsework supporting the chain as it lengthened. Throughout the winter months the construction team worked simultaneously from both sides, assembling up to 40 links a day. Once the first chain was complete, the second was built above it, followed by the third. When the chains on the south side were finished, a new falsework was rigged for the chains on the north side, 4,200 links in all.

By May 1864 the chains were complete and ready for the road deck to be suspended from them. One hundred and sixty-two vertical wrought-iron suspension rods set eight feet apart were fastened to the chains by bolts. To equalise the stress, and to

View of the grand opening ceremony, 8 December 1864 (Bristol's Museums, Galleries and Archives).

Illustration from supplement to the Western Daily Press, 9 December 1864 (University of Bristol).

facilitate replacement, each rod was fastened to only one of the three chains: top, middle or bottom. Two wrought-iron girders running the full span of the bridge provided the main support for the road deck and also divided the roadway from the footpaths. To construct these a narrow rail track was laid to the edge of each abutment, on which travelled a long jib crane, the jib overhanging the 245-foot drop. The cranes then hoisted each of the 40 one-ton girder sections into position. These were attached to the rods by bolted plates. One of the 81 cross-girders was bolted beneath the girders at right angles making the structure rigid, then temporarily covered by timber planks so that the cranes could move forward. On 2 July the last cross-girder was fixed and the centre of the bridge joined.

The road deck comprised interlocking sleepers of five-inch, well-seasoned Memel Baltic pine, a high quality 'engineering' timber with a life expectancy of at least 30 years. Over this, at right angles, planks two inches thick were laid, avoiding the need to disturb the main flooring during repairs. By the end of August work on the approach roads and the tollhouses and the painting of the bridge were nearly complete. In November the Board of Trade inspected safety. Five hundred tons of stone were placed on the deck and footpaths, which lowered the deck by seven inches. When the stone was removed, the deck immediately returned to its former position. The inspection was declared as being 'in every respect perfectly satisfactory'.[31]

By December the bridge was ready for its official opening. It had been hoped that the Prince of Wales or Lord Palmerston, the Prime Minister, might perform the ceremony but in the event neither attended. Nor did any member of the Brunel family, 'because they felt that Brunel's name did not figure with sufficient prominence'.[32] His younger son, Henry, wrote 'the whole thing feels quite independent of any question of honouring memory'.[33]

So on 8 December 1864 a rather provincial opening took place. The day was fine and at 10am a mile-long procession accompanied by at least 16 bands set off from the city centre through streets lined with cheering crowds. Flags flew and church bells rang. The invited guests joined the procession for the final approach to the bridge at midday. The stands and every viewpoint were crammed with spectators. After the ceremonial first crossing by the procession, field guns fired a salute and the parade returned to the Clifton side. Various dignitaries gave speeches, the Bishop of Bristol said prayers and the two Lords-Lieutenant of Somerset and Gloucestershire formally opened the bridge. Three cheers were given for the engineers and the National

The newly-completed Clifton Suspension Bridge, spring 1865, stereoview (Adrian Andrews' collection).

Menai Suspension Bridge, 1826, albumen print c 1850s (Adrian Andrews' collection).

Clifton Suspension Bridge, spring 1865, albumen print (Adrian Andrews' collection). The structural simplicity of the Clifton bridge contrasts with the complexity of the earlier Menai bridge.

Anthem was played. Afterwards, invited guests attended a banquet and the workmen were given beer and sandwiches. In the evening the bridge was illuminated, but partly due to stormy weather 'failed to afford the amount of gratification to the public which had been anticipated...'.[34]

A world-class icon

Brunel's inventiveness and tenacity were reflected in the number of changes he made to the design of the bridge structure. However, the complex combination of timber and iron components for the deck would have required high maintenance and certain replacement. No doubt, given adequate funding and a longer life, Brunel's technical genius would have harnessed the new metal technology of steel and galvanised steel wire.

Hawkshaw and Barlow completed the bridge in 18 months, on time and to budget. Their design alterations produced what was claimed by the first guide book as 'the strongest chain suspension bridge ever constructed'. They were respectful of Brunel's concept, but their modifications meant that Clifton Suspension Bridge has survived virtually unaltered. The sheer quality of materials, manufacture and construction make it a fitting memorial to Brunel, a world-class icon of the Victorian age and symbol of Bristol.

All other major suspension bridges of that era have either been removed or replaced, or have undergone considerable reconstruction. Telford's Menai Bridge suffered frequent and considerable storm damage and by 1940 required the replacement of the chains and deck structure with steel components. Chaley's Fribourg bridge was replaced in 1923 by a concrete viaduct; Roebling's Niagara railway bridge lasted 50 years but was then replaced; Ellet's Wheeling bridge, Ohio, of 1847, the first 1,000-foot span, was damaged and repaired many times before restoration in the 1980s; the other losses are too numerous to mention.[35] Clifton's bridge is a unique survivor of the first generation of great suspension bridges and has survived into the twenty-first century carrying four million vehicles a year, thanks to meticulous maintenance by the Clifton Suspension Bridge Trust and dedicated staff.[36] Although much credit lies with Hawkshaw and Barlow, who brought both structural refinements and highly professional financial and project management practices to the enterprise and whose contribution has been all too frequently underestimated, Brunel's engineering inventiveness and remarkable accuracy in his assessment of chain-link design contributed to its success.

On the cap of the Leigh Woods pier are inscribed the words: 'Suspensa vix via fit'. Adapted from Horace, and with a play on the name Vick, they loosely translate as 'A suspended way built with difficulty'. Rarely can a more apt inscription have been placed on a structure.

BRISTOL RIOTS 1831

The worst political riots of the nineteenth century were centred on Queen Square, a key site in the city for assembly and protest. They were partly sparked by the national chaos that surrounded the defeat of the Reform Bill (the attempt to widen enfranchisement and remove some of the country's rotten boroughs).[1] However, of greater significance was local resentment at the perceived arrogance and corruption of the Bristol Corporation and concerns regarding the economic stagnation of Bristol as a port. Military incompetence also contributed to the scale of the destruction.

Events began with the arrival in Bristol of Sir Charles Wetherell, recorder of the city, who came to open the assizes at the Guildhall on 29 October 1831. His coach was stoned on his way into the city. Wetherell, a London resident and MP for Boroughbridge in Yorkshire, would have lost his seat if the Reform Bill had become law. Although he had been fêted in Bristol for his opposition to Catholic emancipation two years earlier, his selfish stance against reform had made him unpopular.

Having opened the assizes Wetherell dined with the mayor, Charles Pinney, at the Mansion House in Queen Square. A crowd of 3,000 pro-reformers had gathered outside and refused to leave when the Riot Act was read. They began to pull up railings as weapons and were on the point of setting fire to the Mansion House when troops arrived. The reprieve was short-lived and the following day the crowd made a successful attack on the Mansion House, ransacking the wine cellar and drinking heavily before turning their anger on surrounding buildings. The rioting spread beyond the square towards the harbour.

Lieutenant Colonel Brereton, leader of the troops and sympathetic to reform, refused to give the order to fire on the crowd. With fewer than 100 regular constables to call upon, special constables were recruited to help restore order along with hired thugs, referred to by the *Bristol Mercury* as 'hireling Tory-bludgeon men'.[2]

Brunel arrived in Bristol the day after the riots started to supervise work on the Clifton Suspension Bridge. He was sworn in as a special constable and arrested a man, though he was tricked into handing him over to a fellow rioter disguised as another constable. He later gave evidence at the trial of Pinney, who was charged with neglect of duty. Brereton committed suicide during his own trial.

During the three days of rioting twelve people are known to have died. Troops eventually quelled the riot. Afterwards, five of the rioters were hanged and 77 were either imprisoned or deported. Private housing on the whole of the north side and most of the west of the square was destroyed by fire. Other buildings lost included the Mansion House, the Custom House, the Excise Office, the Bridewell, two prisons, six warehouses and the Bishop's Palace. Total damage exceeded £300,000. More importantly, much of the confidence in Bristol was destroyed, damaging the local economy and the development of the city. One casualty was work on the suspension bridge.

The Burning of the Bishop's Palace *by W J Müller, c 1831 (Bristol's Museums, Galleries and Archives).*

Bishop Butler's window in Bristol Cathedral and part of the Bishop's Palace by Hugh O'Neill, 1822 (Bristol's Museums, Galleries and Archives).

Ruins of the Chapel in the Bishop's Palace, after the Riots, *1832 by W J Müller (Bristol's Museums, Galleries and Archives).*

HUNGERFORD BRIDGE

In his diary entry of 26 December 1835, in which he listed the capital projects he was then engaged upon, Brunel wrote that he had 'condescended' to design a footbridge across the Thames and that 'I shan't give myself much trouble about it'. The bridge provided a convenient connection between the densely populated boroughs of Lambeth and Wandsworth and the fresh vegetable and fruit market at Hungerford.

Brunel had been suggested for the project on the recommendation of his brother-in-law, Sir Benjamin Hawes. The contractor was William Chadwick who also worked on the Maidenhead bridge. Work began in 1841 and the bridge was opened at noon on 1 May 1845. Twenty-five thousand people crossed on the first day, paying a halfpenny toll.[1]

The bridge was 14 feet wide and had three spans: a central span of 676 feet with two side spans of 343 feet each. The two redbrick and stone piers were hollow in structure to make them light and were set on large footings that also served as landing piers for steamers. On top of each pier stood an Italianate tower, reminiscent of a campanile, that was constructed from four solid pillars of brick joined with lighter brickwork.

Brunel experimented with the strains that could be taken by the four rows of chain holding the footway, finding that he could use thinner links nearer the centre, thereby cutting costs. The weight of the chains was taken by the two towers. To ensure that the pressure they exerted was always vertical upon the piers, the saddles rested on oiled rollers (a method originally designed by his father for Clifton). The chains could also accommodate unequal loads on the spans, although Brunel later remarked that he observed the bridge had a pronounced swing in windy conditions or when it was crowded with people. At the land abutments the chains passed over a fixed saddle. These abutments were spread over a large area, resting on numerous piles that had been driven obliquely into the ground, the spaces between them filled with concrete.

It was said of the bridge:

> … with the exception of the wire bridge at Fribourg, in Switzerland, which is 870 feet, it will be by far the largest in existence; and will, with the Thames Tunnel, the block-machinery at Portsmouth, the Great Western Railway, &c. &c., assist to hand down to posterity the enterprising spirit and genius of a Brunel.[2]

Brunel's proposal to widen the bridge so it might take carriage traffic was not taken forward. It was thought that the 'Great Stink' rising from the heavily polluted river might put people off crossing, but the leasing of the piers to steamboat companies was profitable and after the completion of Waterloo station in 1848 the footbridge carried a considerable number of pedestrians, becoming a vital link for the capital.

Hungerford Market closed in 1859, the land was purchased by the South Eastern Railway and the bridge was demolished to make way for the company's Charing Cross railway bridge. The original river piers were incorporated in the new design and the chains were used at Clifton. A £50-million Millennium Project, begun in 2000, has seen the building of a new footbridge on each side of the railway bridge and the restoration of Brunel's Surrey pier.

Buckingham Street, Strand, 1854 (© Museum of London), a view towards the Thames showing a tower of the Hungerford Bridge in the distance. (facing page)

Hungerford Bridge, pencil drawing, c. 1850 (Elton Collection: Ironbridge Gorge Museum Trust).

Brunel's Hungerford Suspension Bridge, London by William Henry Fox Talbot, c 1845 (National Museum of Photography, Film and Television/Science and Society Picture Library).

EGYPTIAN REVIVAL

The revival of ancient Egyptian ornamentation in furniture, silverware, pottery and interior and architectural design reached its peak in the late eighteenth and early nineteenth centuries. The term 'Egyptomania' was coined to describe this phenomenon.

Although there had been sporadic interest in reviving the Egyptian style since as far back as the Roman Emperor Hadrian's decoration of his villa at Tivoli around AD 130, it was the publication of Giovanni Battista Piranesi's *Diverse Manniere d'Adornare i Cammini* in 1769 that provided a systematic approach to its use in contemporary design. This volume had included illustrations of Egyptian decorative motifs derived from artefacts in the Vatican's collection and designs for Egyptian revival fireplaces. In the accompanying essay, Piranesi discussed the variety of

Egyptian designs and 'the process whereby their art had been abstracted from nature, keenly observed'.[1] The supple natural forms incorporated in man-made objects and structures included animals, lotus buds, papyrus flowers and palms.

In Britain in the 1770s, Josiah Wedgwood began producing Egyptian ware in black basalt, and his son later developed a group of designs inspired by the hieroglyphics of the recently discovered Rosetta Stone. The first Egyptian revival interior in Britain was the billiard room at Cairness House in Grampian, which was devised by James Playfair in 1793. In 1801 designer Thomas Hope completed his Egyptian room at his London home where he displayed his collection of Egyptian antiquities. Illustrations of the room with its Egyptian-style furnishings, frieze and colour scheme featured in his book *Household Furniture and Interior Decoration Executed from Designs by Thomas Hope*, published in 1807.

It was Napoleon's Egyptian campaign of 1798 that gave the revival wider popularity and topicality. The French army had been accompanied by around 150 artists, scholars, scientists and surveyors. Their depictions of the ancient Egyptian monuments could be seen in Baron Vivan Denon's *Voyage dans la Basse et la Haute Egypte Pendant les Campagnes du Général Bonaparte* (1802) and, even more significantly, in the landmark 20-volume

Description de l'Egypte published between 1809 and 1828. The campaign itself had been a disaster militarily.

Egyptian revival architecture could be seen in Britain in such buildings as William Bullock's Egyptian Hall (1811) at Piccadilly, John Foulston's Egyptian Hall (1823) in Devonport Library and Joseph Bonomi's Temple Mill (1842) in Leeds. Masonic lodges such as P L B Henderson's Chapter Room (1900) in Edinburgh continued to use Egyptian symbolism into the early twentieth century. Egyptian-style tombs, obelisks and funerary monuments at Kensal Green and Highgate cemeteries in London were also part of the revival, drawing criticism from those who thought their pagan origins inappropriate in a Christian setting.

The Egyptian influence is evident in Brunel's winning design for the second Clifton Suspension Bridge, elaborated upon in Samuel Jackson's beautiful paintings showing the proposed hieroglyphics and sphinxes. Following the unanimous decision by the committee to award Brunel the project in March 1831, the engineer wrote to his brother-in-law, Sir Benjamin Hawes, saying: 'The Egyptian thing I brought down was quite extravagantly admired by all and unanimously adopted.'[2] The chequered history of the bridge's construction meant this design was not carried through.

Bristol artist W J Müller travelled to Egypt in November 1838, arriving in Alexandria by the recently introduced steamer service. He was 'among the first established European artists to set foot in the country'.[3] Müller would have been familiar with Egyptian imagery from the publications inspired by Napoleon's campaign and the mummy cases and other artefacts being sent back by European archaeologists, but was keen to have direct experience of the monuments and architecture himself. He journeyed 400 miles up the Nile from Cairo, arriving in Thebes in early January 1839 where he saw the temples at Luxor, the Ramesseum, the Valley of the Kings and other ancient sites. The drawings and on-the-spot watercolours of Egyptian figures and scenes he brought back to Bristol were admired by his friends,

and his subsequent oil paintings proved popular with buyers and critics when shown at the Royal Academy between 1840 and 1843. Scottish artist David Roberts had also travelled in Egypt in 1838, just prior to Müller's arrival, as part of a Near-Eastern tour. He was an acclaimed painter of European topography and architectural views but felt in need of new subject matter. The sketches and notes he brought back from Egypt provided him with enough material to last the rest of his career.

Detail from Samuel Jackson's View of the Avon Gorge with the Approved Design for the Clifton Suspension Bridge, *1831, showing sphinx facing inward (facing page, bottom) (Bristol's Museums, Galleries and Archives).*

Gateway of Clifton Suspension Bridge drawn according to Brunel's design by Samuel Jackson, 1831, showing sphinxes facing outward (facing page, top) (University of Bristol).

Plates from Owen Jones' Grammar of Ornament, *1856 (top) (University of Bristol).*

The Ramesseum at Thebes, Sunset *by W J Müller, 1840 (Bristol's Museums, Galleries and Archives).*

The Temple of Dendera: upper Egypt *by David Roberts, 1841 (Bristol's Museums, Galleries and Archives).*

BRUNEL IN BRISTOL DOCKS

Angus Buchanan

Cleaning the Floating Harbour *by Thomas L Rowbotham, 1828, a reminder of what the harbour looked like when it was tidal, before the construction of the Floating Harbour in 1809 (Bristol's Museums, Galleries and Archives).*

From the Middle Ages to the eighteenth century Bristol had possessed, after London, the second most important port in Britain. Despite its inland character, some ten miles up the River Avon from the Severn estuary, the harbour around the confluence of the River Frome with the Avon in the heart of the city provided secure wharves for the sailing ships of the period. The exceptionally high tidal range in the river courses had the advantage of helping ships coming up or going down, provided that they moved with the tide, but it also had the disadvantage of leaving vessels stranded on tidal mud banks twice in every 24 hours.

As ships became larger the inconveniences of the harbour were accentuated, so that other ports on the western approaches, and especially Liverpool and Glasgow, were able to challenge Bristol for the lucrative Atlantic trade. With the onset of rapid industrialisation in the eighteenth century, Bristol could no longer sustain its regional dominance as 'the Metropolis of the West', and it began to lose ground to its rivals.[1]

The response of Bristol to this mounting crisis was slow and indecisive. For several decades from the 1740s various schemes were considered for improving the port facilities, but all of them were abandoned before the end of the century, and the situation continued to deteriorate.[2] Eventually, in 1803, the city council and Merchant Venturers – a society that had evolved from the Guild of Merchants in the fifteenth century and which controlled the city docks until 1848 – agreed on a plan presented

Western Wapping Dock *by Thomas L Rowbotham, 1826. Today, this is the site of the Bristol Industrial Museum (Bristol's Museums, Galleries and Archives).*

by William Jessop, one of the leading canal engineers of his period, and secured an Act of Parliament. This created the Bristol Dock Company (BDC), with powers to undertake major hydraulic works including the construction of dams at Netham and Rownham, and the excavation of the 'New Cut' to keep the tidal flow separate from the enclosed docks. The harbour of the old river courses in the centre of the city was thus kept at permanent high-water level as the 'Floating Harbour', with the help of a supply of river water through the 'Feeder Canal' from above the weir at Netham. The system was provided with entrance docks at Cumberland Basin and Bathurst Basin, which could be isolated from the harbour and the tidal river by lock gates.[3]

This ingenious arrangement had been completed in 1809, and the new harbour facilities had given a boost to the trade and the morale of Bristol. Unfortunately, the euphoria was short-lived, both because the BDC felt obliged to charge high harbour dues in order to recoup some of the construction costs of around £600,000 – almost three times more than Jessop's original estimate of £212,470 – and because serious faults developed in the hydraulic system. In particular, the harbour began silting up alarmingly, and the sewage that had traditionally been deposited in the rivers was no longer able to flow away easily on the tide. The sewage problem was partially solved by the construction of a culvert from the head of the Frome arm of the harbour, under the Floating Harbour and discharging into the New Cut. But the mud and sand banks continued to grow in the harbour, and it was at this point that Brunel was consulted.

Still recuperating from the injuries received in the Thames Tunnel disaster of January 1828, the young Brunel had visited Bristol and become involved in the competition to build a bridge over the Avon Gorge at Clifton. In the course of winning this competition he came into contact with a network of Bristol merchants and industrialists, most of whom welcomed his charm and dynamism. One of them, Nicholas Roch, was a member of the board of the BDC, and he suggested that they should consult him about the engineering difficulties that they were encountering with the operation of the Port of Bristol. Brunel, fretting for work to keep him busy, was glad to receive this commission, and applied himself immediately to a thorough examination of the harbour and to an analysis of its problems. He presented his report in 17 manuscript pages on 31 August 1832.[4]

The Overfall Dam by Thomas L Rowbotham, 1827. View looking across Bristol's New Cut and showing excess water leaving the Floating Harbour (Bristol's Museums, Galleries and Archives).

Brunel tackled the problems of Bristol Docks with characteristic briskness and confidence. He saw immediately that the main source of trouble was the inadequate supply of fresh water through the harbour:

> A constant stream, though nearly imperceptible in its motion, will carry with it the lighter particles of mud, which form the principal parts of such deposits [i.e. the shoals that had developed in the harbour]. If the whole of the River Avon were at all times running through the Float, and which I have no doubt Mr. Jessop originally intended should be the case, such a stream might generally be obtained…

The remedies were thus apparent: raise the height of the dam at Netham, so that more river water would flow through the Feeder Canal into the Harbour, while at the same time providing for culverts at Prince Street and Rownham so that the Harbour could be regularly scoured. These were substantial measures designed to provide the best balance in the hydraulic regime of the Harbour, but Brunel also realised that the removal of the existing shoals called for more drastic action, and for this he recommended the ingenious device of scraper-dredgers:

> The simplest way of effecting this will be by means of flat bottomed barges with large hoes or scrapers, strongly attached to them and suspended in such a manner as to be capable of being raised or lowered according to the depth of water or the consistency of the mud.[5]

This is the genesis of the idea of the 'drag-boat', which Brunel envisaged as consisting of two vessels dragging themselves together and so scraping a hoe-full of mud into the middle of the basin, from which it could be swept away by more conventional scouring. In practice, the first drag-boat operated on its own, warping itself across the dock basin by hawsers attached to bollards on the opposite side, and this arrangement worked so well for ten years from 1834 that it was then replaced by a slightly larger boat. This replacement, which became known as the *BD6* – 'Bristol Dredger No.6' – remained in service until 1961, and when it was then scrapped parts were preserved in Bristol City Museum. It was powered by a steam engine with a single inclined cylinder of 16 inches bore and 36 inches stroke.[6] Brunel appears to have been asked to provide a similar scraper-dredger for Bridgwater Dock in Somerset, and this vessel survived the closure of its dock, being transferred first to Exeter Maritime Museum and having a chequered history thereafter.

The BDC directors were constantly short of cash and anxious to make economies, so they had little enthusiasm for any remedies likely to be expensive. They approved the drag-boat idea, and they also gave the go-ahead on the culverts at Prince Street Bridge and Rownham, but they procrastinated about changes to Netham weir. The culvert at Prince Street was finished first, but depended on the lower part of the Floating Harbour being sealed off and then drained so that mud from the upper harbour could be scoured through the culvert. This would have resulted in loss of trade – something the directors were anxious to avoid. It is hard to believe the culvert was ever very effective. The culvert at Rownham was a different matter, as this was through the dam forming the Floating Harbour and it could be operated every time the tide fell in the New Cut, without disrupting trade. Mud shovelled from the drag-boat or from conventional dredging was deposited in an open trunk in the bottom of the dock and then, as the water level fell in the New Cut with each tide, the culvert was opened and the mud sluiced through into the tidal stream. It proved to be effective, and has continued in use to the present day.

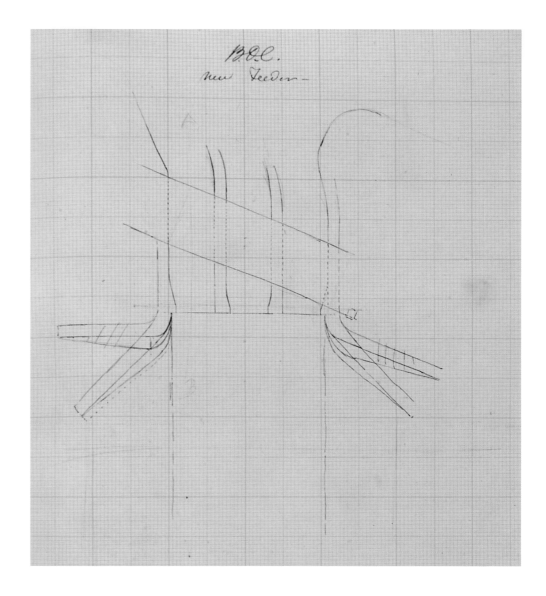

Brunel sketch of feeder sluice for Bristol, 1835 (University of Bristol).

These works were completed in 1833-34, and the success of the drag-boat and the Rownham culvert brought significant relief to the hard-pressed directors. The conversion of the Rownham weir into an underfall dam had a useful by-product, moreover, as it allowed the BDC to begin the process of filling in part of the old river

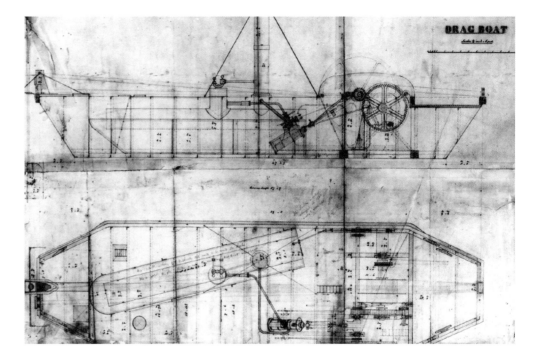

*The BD6 drag boat photographed in 1952
(opposite top) and 1961 (opposite bottom)
with original contractor's drawing (above)
(PBA Collection at Bristol's Museums,
Galleries and Archives).*

course behind the dam, which became the 'Underfall Yard'. This was after Brunel's time, but it enabled the dock engineers to establish a workshop adjacent to the culvert controls, and close to the new hydraulic engine house that was installed in the 1880s. It proved to be a compact and convenient arrangement for as long as the Floating Harbour remained in operation.[7]

From Brunel's point of view, the response of the directors to his recommendations was less than satisfactory, because they had opted for the short-term palliatives rather than the long-term remedy of increasing the flow of river water through the dock system. He emphasised this deficiency, despite the considerable improvements in the overall depth of the Floating Harbour, in his next report in April 1835, and the directors reluctantly agreed to some modification to the height of the Netham Dam in order to direct more water into the Harbour. However, he was still having to stress the point in subsequent reports, and it seems only to have been tackled after 1842.

Meanwhile, a different anxiety had emerged to preoccupy the attention of the directors in the shape of the two Entrance Locks to Cumberland Basin. Jessop had constructed these side-by-side at the western end of the Basin, separated by a narrow central pier and opening into the river at an awkward right-angle. The northern lock was slightly larger than the southern one, but both were too small to accommodate 'the increasing Demands of the Steam Packets', so the directors asked Brunel to advise them on what should be done.[8] He responded with a report in January 1836, setting out a series of options.[9] The first was 'simply repairing the present Locks', without increasing their dimensions, but he regarded this as being 'almost inadmissible'. Second was the possibility of repairing the North Lock while reconstructing the South Lock to 225 feet length by 54 feet breadth. And third, both locks could be widened, with the North Lock being extended in the ground to the north.

Brunel made it clear that he preferred the thorough-going third alternative, but the directors opted immediately for the 'inadmissible' first choice. In fact this option turned out to be less simple than Brunel or the directors had anticipated, due to problems with settlements in the narrow pier between the two locks. Eventually, in 1844, it was the second option that was adopted, and Brunel was commissioned to design a new South Lock.

By this time Brunel had acquired a pressing interest in the size of the entrance locks because his second steam ship, the ss *Great Britain*, had been launched on 19 July 1843 from the dry dock on the southern side of the Floating Harbour in which she had been built, only to be trapped in the Floating Harbour by her inability to pass through the entrance locks. It seems that Brunel had miscalculated his influence with the directors of the BDC, and had been over-confident that his measures recommended in 1836 would have been sufficiently adopted to make the passage of the new ship straightforward. This, however, did not happen, and all the tinkering with repairs to the locks failed to produce a lock adequate for the purpose. The ss *Great Western*, Brunel's first steam ship, launched in 1837 from Patterson's shipyard near Prince Street Bridge, had already got into trouble at the entrance locks, being obliged to dismantle her paddle wheels in order to get in and out of the Floating Harbour. Such a practice was clearly unacceptable in normal operation, so the ship had taken to berthing outside the harbour and eventually shifted its base entirely to Southampton.[10]

In view of this experience it is strange that Brunel failed to make adequate provision with the BDC to allow the *Great Britain* to leave the Harbour in 1844, but when approached by the directors in June of that year to re-build the South Lock he was not inclined to do a rush-job. He stated his conviction that 'the repair of the Lock is a most serious business and will probably involve a *very heavy* expense'. In a letter to his friend, Captain Claxton, written about the same time, Brunel said: 'I think of recommending a thoroughly good lock'[11], and within a couple of weeks he submitted his report to the directors, advising the complete reconstruction of the lock to dimensions of 240 feet by 52 feet, for which work he estimated the cost as £22,000. Although alarmed at the cost of the project, the directors could see no alternative and agreed on a new lock of 245 feet by 54 feet.[12]

The pressure on Brunel to act quickly was relieved by 11 December, when the *Great Britain* was squeezed through the North Entrance Lock on a particularly high tide, having been stripped of as much weight as possible, and with masonry removed in order to let her pass. She never returned to Bristol Docks in her long operating career, but she was brought home as a rusting hulk from the Falkland Islands in the summer of 1970 and installed in the dock in which she had been built. Meanwhile, most of the work on the reconstruction of the South Lock had been put out to tender. Contracts had been signed by early in 1845, but thereafter progress was distressingly slow. The BDC was under mounting attack from mercantile interests objecting to its high charges and poor services, and it surrendered in 1848 to a combination of parties which ran the docks as a municipal enterprise under a Docks Committee. This body protested to Brunel about the slow rate of progress on the new lock, but it was not until 1849 that it came into operation and even longer before it had overcome its teething problems.

After all the frustrations and delays that had accompanied its construction, the new lock served its purpose well. With its two single-leaf caisson gates closing against a sill in the masonry of the carefully constructed half-oval cross-section basin, the lock gave access to substantially larger vessels. The single-leaf arrangement had the additional advantage of relieving the central pier between the two locks. The lock gates, of bulbous cross-section, contained air chambers that made them partially buoyant when the tide came in, so that they could be moved relatively easily despite their size. They were controlled by chains running through recesses in the lock masonry to capstans

Two views of Brunel's Lock, c 1860s (PBA Collection at Bristol's Museums, Galleries and Archives).

Brunel's Lock (facing page, top) showing the masonry of the South Entrance as reconstructed by Brunel in 1844-49 (photograph by author).

One of Brunel's wrought-iron lock gates being removed for scrap, c 1890s (Author's collection).

on either side. Within a few years this arrangement had become obsolete as a result of the development of hydraulic power, but when they were installed the operation of such equipment still depended on manual power.[13]

The last feature of the South Entrance Lock to be put in place was the bridge across it to the central pier. This was a swing-bridge, and it was – and it remains – significant as the first indication of Brunel's solution to one of the great engineering problems of the railway era: that of constructing wide-span bridges relatively cheaply and safely. The problem had become acute with the collapse of the railway bridge over the River Dee in Chester in 1847. This had been one of many composite cast-iron and wrought-iron bridges designed by Robert Stephenson for his railways, and it made engineers extremely careful about the use of cast-iron in tension thereafter. Brunel had already tried to avoid using the material as much as possible, although he did build a number of short-span cast-iron bridges. The significance of the South Entrance Lock bridge, however, lies in its complete dependence on wrought-iron plates rivetted together to form two tubes or trusses from which the bridge platform was suspended. From this small beginning, Brunel went on to evolve the huge trusses around which his railway bridges at Chepstow and Saltash were constructed. The Chepstow Bridge was replaced in the 1960s, but the Royal Albert Bridge over the Tamar at Saltash is still in full working order.

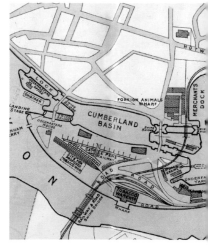

Aerial view of Cumberland Basin, c 1930 with plan of the docks taken from the 1925 Port of Bristol Authority handbook (PBA Collection at Bristol's Museums, Galleries and Archives).

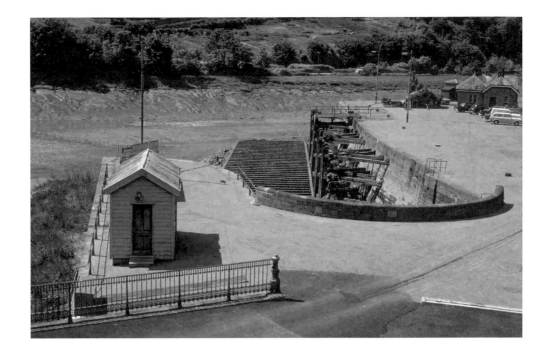

The North Entrance Lock: the original lock was sealed when the new lock was installed in 1873. Brunel's lock is immediately to the left (photograph by author).

The Bristol prototype also survives, although not quite as it appears. When the North Entrance Lock was eventually re-built by Brunel's successor as port engineer, Thomas Howard, he appropriated Brunel's bridge for his own lock because it was his intention to close down the South Lock once the North Lock was in operation. However, local mercantile interests insisted on keeping open the Brunel Lock as an alternative to the new lock, so Howard was obliged to provide it with a swing bridge. He seems to have done this by getting out Brunel's drawings and replicating the original bridge, and it is this replica, shortened at the land end by the removal of the swinging equipment, that is now fixed in place across the disused Brunel Lock. The original bridge is also no longer in use, having been made redundant by the modern flyover of the Cumberland Basin road system, but it still rests in its open position on the south side of the North Entrance Lock. It has been reasonably well maintained, in recognition of its distinction in the history of technology.

The completion of the South Entrance Lock was effectively Brunel's last contribution to the Bristol Docks, for even though he lived another ten years it was a tumultuous period in his life and he became heavily preoccupied with the construction of the ss *Great Eastern*. There was never any question of this mighty vessel, the largest ship until the end of the nineteenth century, being built in Bristol. Instead, it was built on the banks of the Thames at Millwall in London, and even there it had to be launched sideways into the river. Brunel remained deeply attached to Bristol, however, telling one of the Clifton Bridge trustees that 'all hands must pull together for our revered parent the City of Bristol – and I am always ready to help'.[14] But his altercations with Bristol clients, and especially the Docks Committee, made it increasingly difficult for him to maintain harmonious relations with them, and as most of his closest friends in the city either died or moved away he lost the powerful network of contacts that had been so influential in finding commissions for him in the city. To some extent, also, Brunel lost heart in the possibilities of Bristol, admitting in negotiations for the South Entrance Lock that:

Brunel's swing bridge was moved to this position in 1873 and remained here until the construction of the Cumberland Basin Flyover in the 1960s (Author's collection).

I have recommended these dimensions because I believe they would be sufficient to accommodate all ordinary Steam boats built for the Irish trade – and this I now think is sufficient for the Port of Bristol.[15]

His difficulties with the BDC and the Docks Committee, the failure of his scheme for a pier at Portishead, and the full stop to which the Clifton Bridge had come, combined to convince him that Bristol was no longer a place where great projects were welcome.

If Brunel appeared to lose patience with Bristol, it is understandable considering that he had little time left in which to complete all the ambitious schemes that filled his mind and his sketch books. But Bristol reacted favourably, even if slowly, to his inspiration and within a couple of decades of his death in 1859 Clifton Bridge had been completed as a memorial to Brunel, the North Entrance Lock had been enlarged and re-aligned to improve access to the Harbour for larger vessels, and dock works were in hand at the mouth of the River Avon, at both Portishead, where Brunel had projected a pier, and at Avonmouth. The heightening of the weir at Netham ensured the supply of river water through the Floating Harbour, and Brunel's drag-boat and culverts gave excellent service in keeping it free from shoals. Elaborate plans for 'dockizing' the whole course of the River Avon between the Floating Harbour and the

Workmen engaged in construction work in the Bristol docks, c 1900 (PBA Collection at Bristol's Museums, Galleries and Archives).

sea were canvassed but came to nothing, as it became apparent that the success of the new docks at Avonmouth made such schemes redundant. This did mean, however, that access to the City Docks remained tidal, and as ships continued to grow in size not even the improved access at Cumberland Basin could persuade companies to use these facilities rather than those at the mouth of the river. The Floating Harbour consequently went into a slow decline, being saved from extinction only by its recognition as an outstanding heritage feature for a maritime city, and by its re-invention as a major amenity for the City of Bristol. It is likely that Brunel would have approved of this transformation of the Port of Bristol, even though it has taken rather longer than he envisaged to achieve.

PORTBURY, AVONMOUTH AND PORTISHEAD

In 1851 the wooden-hulled paddle steamship *Demerara* was launched in Bristol by the firm of William Patterson. She was the largest ship to have been built in the city since the *Great Britain*. During her first journey down the Avon, en route to Glasgow for the installation of her engines, she ran aground off Round Point, just outside the entrance locks at Cumberland Basin. She was written off by the insurers because of the scale of the damage. Patterson salvaged her and she was rebuilt as a sailing ship and renamed *British Empire*.

This episode served to demonstrate what many already knew: that Bristol's docks were unsuitable for the new, large vessels now in operation. After protracted discussions, new private docks were built at the mouth of the river at Avonmouth (1877) and Portishead (1879), in competition with the city centre. These were purchased by the Corporation in 1884 and a period of prosperity for the Port of Bristol followed.

In 1908 the Royal Edward Dock was completed close to Avonmouth and in August 1978 the deepwater Royal Portbury Dock, capable of accommodating ships of 70,000 tons, opened near Portishead, adding to the

The Demerara, The Illustrated London News, *1851.*

extensive complex of berthing facilities on the Avon. In 1991 Terence Mordaunt and David Ord acquired the port from the Council, thus ending public ownership that had lasted nearly 150 years.

The long-awaited facilities outside of the city centre had been recommended by Brunel in a report to the Bristol City Council in 1839 in which he had suggested constructing a large, floating pier at Portishead. He wrote:

> I propose two or three vessels of 300 or 200 feet of length, built of iron, as the material cheapest and best adapted to the purpose, of 16-feet and 20-feet draft of water, and about 30 feet beam, moored close stem and stern, so as to form one continuous floating body. Any steamboat or other vessel alongside will of course be on the same level as the pier; the passengers, on disembarking, will at once be on a level platform or deck, under shelter, where the luggage or goods can also be placed; and the communication with the shore will be effected

Artist's impression of Brunel's proposed floating pier, landing place and steam packet harbour at Portbury (private collection).

without steps… Such a pier would afford stowage for almost any quantity of coals, fresh water, and general goods, which could be stored here for embarkation.[1]

This proposal would have been a boon for the transatlantic passenger trade. Brunel assumed that cargo vessels would still continue down to Bristol. In anticipation of proceeding with this scheme, the Portbury Pier and Railway Company was established in 1846 and the following year an Act was passed permitting a line to be built from Bristol to Portishead (the proximity of Portbury and Portishead led the names to be used interchangeably). Brunel presented designs for a pier as described in his report, but the project was abandoned.

Photograph of the Gypsy, which ran aground in the Avon in May 1878 as the tide receded, breaking her back and blocking access to and from the docks (PBA Collection, Bristol's Museums, Galleries and Archives). The grounding of this ship and the difficulties encountered in moving her further damaged Bristol's reputation.

Dawn near Reading, *watercolour, 1870, showing GWR broad gauge express on mixed gauge track, artist unknown (Elton Collection: Ironbridge Gorge Museum Trust).*

THE GREAT WESTERN RAILWAY

Steven Brindle

The names of Isambard Kingdom Brunel and the Great
Western Railway (GWR) are indissolubly linked. They have
a special place in railway history, and indeed in world history,
as is acknowledged by the GWR's main line from London to
Bristol being a candidate for the status of World Heritage
Site. Brunel's role as the GWR's chief engineer represented
the central, continuing theme of his career, from his
appointment on 7 March 1833 right through to his untimely
death in 1859. His work for the GWR and its empire
covered a huge range of projects, of civil, structural and
mechanical engineering and architecture. It remains the basis
for a large part of Britain's railway network today.

A Train of the First C

LIVERPOOL   TRAVELLER   MANCHESTER   LIVERPOOL   TIMES   MANCHESTER

Drawn by I. Shaw, Liverpool.

A Train of the Second

TRAVELLING ON THE LIVER

London, Published by

Travelling on the Liverpool &
Manchester Railway, *aquatint,
1833 (Elton Collection: Ironbridge
Gorge Museum Trust).*

The GWR originated as the Bristol Railway, and at the outset it was very much a
Bristol venture. In this it was distinctive, for railways began in the north of England
and the first generation was dominated to a remarkable degree by northern engineers
and by Lancastrian – in particular, Liverpudlian – money. Though Bristol had lost its
pre-eminent role in the Atlantic trade by the 1830s, the city still had a rich and
vigorous merchant elite, who realised that one of the keys to recovery was to improve
their connections with their markets at home. The opening of the Liverpool &
Manchester Railway in 1830, and that fact that their Liverpudlian rivals were already
planning railways to Birmingham, London and other places, galvanised the Bristolians
into action. There had already been several failed attempts to found a London to
Bristol railway: the huge prospective cost was intimidating. However, in the autumn of
1832 a company was successfully established. In January 1833 it held its first public
meeting, and by March it was ready to appoint an engineer to carry out a survey.[1]

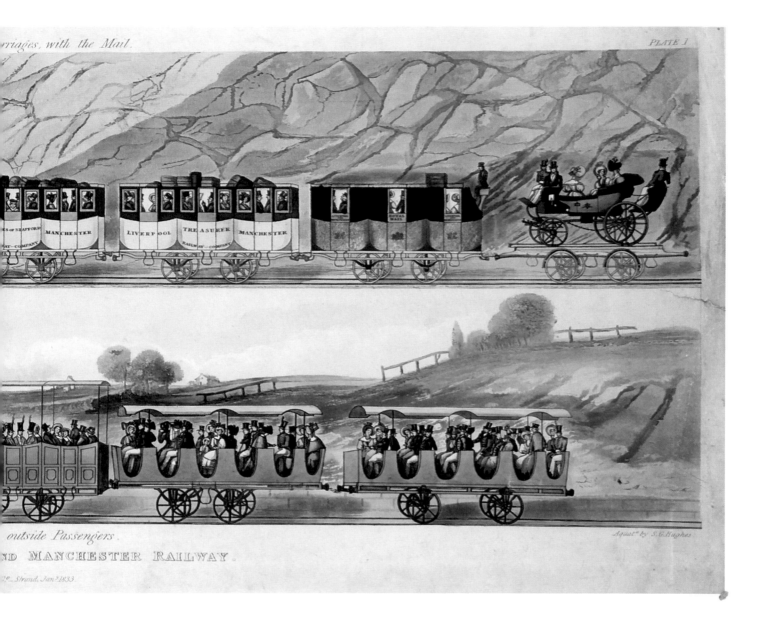

...rriages, with the Mail.   PLATE I

outside Passengers.   *Aquat.ᵈ by S.G.Hughes*

...ND MANCHESTER RAILWAY.

...e. Strand, Jan.ʸ 1833

Brunel was already well known in Bristol through his Clifton Bridge competition design and his work in the city's docks. He was only 27 years old and had no experience of designing railways, but at this date few people did. It is disconcerting, though, to reflect that his whole experience of railways had been limited to one look at the Stockton & Darlington Railway and a return journey on the Liverpool & Manchester, on a trip north in 1831. On 6 March 1833 he travelled on the Night Mail from London to Bristol (on top of the coach, to save money), arriving exhausted to attend the crucial meeting of the new company. At 2pm the next day he was told that he had been awarded the job – by a majority of one. On 9 March Brunel set out on horseback with a local surveyor, William Townsend, to commence his survey of the Bristol end of the line.[2]

Railways were born out of several different ideas and technologies that developed in late Georgian Britain. On the civil engineering side – making the line – there were precedents in the superb turnpike roads and canals which had spread all over the country. Building these had combined great accuracy of surveying with great feats of earth moving and of sheer human and animal muscle-power. In terms of structural engineering, they had required the construction of enormous numbers of bridges, and bridge design had leapt forward as a result. The concept of having vehicles run on rails for a smooth and controlled ride had already been established for many years in the use of horse-drawn tramways at quarries and mines, mostly in the north of England and in Wales. In terms of passenger services, there were precedents in the mail-coach services, which ran over the new turnpikes.

The great new idea was in mechanical engineering – the use of steam locomotives. Stationary steam engines, mostly used to keep mines dry, had been developing for over a century, while the steam locomotive was invented in a rapid series of developments at a number of ironworks and collieries between about 1810 and 1815. These ideas were put together by a tiny number of businessmen and engineers, and the first results were the Stockton & Darlington Railway (1824), the Canterbury & Whitstable Railway (1825) and, above all, the Liverpool & Manchester Railway (1830). A social and industrial revolution had been launched, which arguably had wider long-term consequences for humanity than anything that had come out of revolutionary France.

The uncrowned king of the northern engineers, the man who dominated this first phase, was the Northumbrian George Stephenson. He had risen from being an apprentice colliery engineer at the age of 14 to designing the Liverpool & Manchester line, probably the most expensive single engineering project yet undertaken in Britain. It says much for Brunel's self-confidence, as well as his perceptiveness, that his reaction on taking his first ride on it in 1831 was not to be over-awed; it was that he could do better. Sitting in the carriage, he drew circles to illustrate its swaying motion:

> I record this specimen of the shaking on the Manchester Railway. The time is not far off when we shall be able to take our coffee and write whilst going noiselessly and smoothly at 45 mph – let me try.[3]

It was an early indication of Brunel's independence of mind, and determination not simply to follow established practices.

These qualities were demonstrated in his route for the new line. Brunel perceived that there were two basic alternatives, of which one was clearly preferable. This ran up the Avon to Bath, then up out of the steep-sided valley and across the relatively level countryside of north Wiltshire and the Vale of the White Horse, descending very gradually into the Thames Valley and so to London. A remarkably straight and level route could be devised, but it would have to cover one very difficult stretch between Bath and Chippenham. The other route, as proposed by two Bristolian surveyors, Brunton and Price, ran from Bath up the Avon valley to Bradford on Avon, and then via Marlborough, Hungerford and Newbury to the Thames at Reading. This would deliver the business of the major towns between London and Bristol, but at the cost of a winding route through hilly country. It speaks volumes for Brunel's perceptiveness that he realised from the first that a national railway network should be based on fast trunk routes linking the major cities. The branch lines could come later, and it would not matter so much if they were winding and the trains slower as a result. This sounds obvious now: in 1833, it represented a profound insight, one not shared by the many critics who would have preferred the 'southern' route.[4]

Prior Park, Bath, engraving by Anthony Walker, 1752, thought to be the earliest depiction of a railway in Britain (Elton Collection: Ironbridge Gorge Museum Trust).

Brunel completed his survey, working from horseback, sometimes for 20 hours at day, in just nine weeks. On 30 July he presented his proposals: the line would be 116 miles long and cost £2½ million, including stations and locomotives. (In fact, it turned out to cost two-and-a-half times that.) The proposals were accepted, and on 27 August the newly formed London Committee of the company confirmed Brunel's appointment as the company's engineer. The same meeting resolved to change the company's name, and that evening Brunel wrote the magic initials 'GWR' in his diary for the first time.[5]

He was to receive a salary of £2,000 a year, which sounds generous until one realises that, from this, he was expected to find all his office expenses, including the salaries of his clerk and draughtsmen. Brunel hired a house and office at 54 Parliament Street, in an area of Westminster already favoured by the engineering profession. By the end of the year he had moved to the nearby 18 Duke Street, which was to be both his home and office until the end of his life. He took on a chief clerk, Joseph Bennett, who was to run the office until Brunel's death in 1859, and a number of draughtsmen, and began to build up his team of assistant engineers.[6] Brunel had a large coach built of a type known as a 'britschka', big enough to house a bed and a desk, so that he could live on the job. It is hard to imagine how, but by the end of the year the overall design for the whole line was completed. It was only nine months since Brunel had first been appointed.

Brunel's office in Duke Street, c 1859 London (Elton Engineering Books).

In the spring of 1834 the GWR submitted Brunel's designs with its parliamentary bill. Building a railway, like a turnpike road or a canal, required Parliament's authorisation. While railway companies were expected to negotiate for the purchase of land, and to pay a market rate for it, it was clear that they would need powers of entry into property, and in the last resort of compulsory purchase, to be sure of success. For late Georgian Britain, a society dedicated to the sanctity of private property, this was a major issue which only Parliament could sanction. Parliament's authority was also needed to raise the large sums of capital required. (The GWR and the London & Birmingham, the two largest companies, were each authorised to raise £3.3 million in shares and loans.)[7] Finally, there was the crucial point that a railway would inevitably enjoy a privileged, if not a monopoly status in its area. Parliamentary approval was not something which any company could take for granted: the GWR's first bill ran into heavy opposition and was defeated in July 1834. A second attempt succeeded, with Brunel performing as the company's main witness: the GWR's Act of Parliament received royal assent on 31 August 1835.[8]

While Brunel was developing his designs, he was also involved in negotiating with landowners, helping to sell shares in the company, and getting the bill through Parliament. In addition, he was wrestling with the pressing problem of where the London terminus should be. Ideas came and went, from the Bishop of London's estate at Paddington, to Vauxhall on the South Bank, to a joint station with the London & Birmingham Railway at Euston. This latter idea was what the GWR actually received parliamentary approval for: it seemed to make sense as the companies would not be in direct competition and both needed to control their soaring costs. But the negotiations broke down, partly over the issue of Brunel having chosen the seven-foot 'broad gauge' for his railway, and by 1836 the GWR had turned back to Paddington, which meant they needed another Act of Parliament.[9] The politics and finance of railway building were every bit as complex and demanding as the engineering, and while Brunel had total responsibility for the third, he was heavily drawn into the first two as well.

From the outset Brunel intended the GWR to be 'the best of all possible railways', an integrated system designed to provide fast, smooth-running services. His 'northern route' had been chosen to keep the line level, and it was (and is) remarkably so: none of it is steeper than 1 in 660, except for two stretches of 1 in 100 at Wootton Bassett and the Box Tunnel. It was also remarkably straight: most of the curves have a radius of over half a mile. George Stephenson praised Brunel's planning before the parliamentary committee: '… if a better route exists, I do not know of it.'

Brunel's design for the permanent way – the roadbed of the finished railway – was, if anything, even more original. His experience of travelling on the Liverpool & Manchester had led him to question the Stephensons' approach to track-design. They laid stone blocks carrying cast-iron shoes, which carried the rails, but the blocks tended to subside slightly unevenly under the weight of the trains, giving an up-and-down motion to the ride. The GWR would have a quite different system, the rails laid directly on continuous timber bearers, which would sit on sleepers at 15-foot intervals. The sleepers would be fixed to timber piles, driven 15 feet into the ground, to hold the assemblage down, on a bed of ballast and rammed sand.[10]

Some time between the GWR's first and second parliamentary bills, Brunel had made a yet more fundamental decision, to space his rails seven feet apart. This was a marked change from the four-foot-8½-inch gauge used on all the other railways then planned or in existence. In the spring of 1835 he persuaded the parliamentary committee to omit any reference to the gauge from the GWR's bill, leaving him free to do this, but it was not until June that he formally presented the idea to his directors. It was a

remarkable exercise in thinking from first principles. So that the GWR's services would be smooth and fast, he explained, he had designed a line with broad curves and gentle gradients. However, this depended on the design of the rolling stock as well: to run trains safely at high speed, it was desirable to keep the centre of gravity as low as possible. Furthermore, to design a stable, smooth-running vehicle, it was desirable to have the wheels as big as possible, to reduce friction between wheel and track. The problem with the four-foot-8½-inch gauge was that, with the wheels so closely spaced, the body of the carriage or truck had to sit up above them. This raised the centre of gravity of the vehicle, and limited the possible wheel-size. It also limited the size and weight of the locomotives. A broader gauge would allow all these problems to be solved.[11]

Brunel considered the objections to his plan: broader tracks and bigger vehicles would require a wider and more expensive line (the GWR's tunnels were 30 feet wide, as compared to 24 feet on the London & Birmingham Railway, representing a huge difference in the volume of rock to be extracted). There would also be 'some inconvenience' in making a junction with a standard-gauge railway. Nevertheless, Brunel carried the board with him, and the stage was set for one of the great industrial controversies of the modern age. The broad gauge was a brilliant conception, which might well have given Britain a superior railway network. The essential problems were that the four-foot-8½-inch gauge was already the industry standard, there was far more mileage of it, and the inconvenience caused where the two systems met was something dreadful.[12] Brunel spent the rest of his life defending his system, and trying to argue his way round the blunt and unpalatable fact that the Stephensons had got there first. Perhaps the most remarkable thing about the broad gauge was that he was allowed to build it in the first place: the GWR's directors and shareholders were hard-headed businessmen. This is one of many testimonies to Brunel's force of personality, as well as his originality as a designer.

The GWR was going to be built starting from either end, with the London Committee and the Bristol Committee supervising their respective halves of the line. Brunel divided the route into sections to be let as contracts, coded by area: L for London, R for Reading, S for Swindon, C for Chippenham, B for Bristol.[13] From the autumn of 1836 on, Brunel and his staff were kept busy developing the Act of Parliament designs into the huge volume of precisely detailed contract drawings needed, as the railway grew west from London and east from Bristol. The staff were led by six resident engineers, three each for the eastern and western ends of the line, with a number of assistant engineers.[14] They cannot be regarded as a 'design team' in any real sense. Brunel was not a good delegator, jealously guarding his absolute responsibility for the design process: the residents and assistants were there to interpret and execute his will. Designs were worked up from Brunel's first sketches, developed by his assistants, corrected, then sent to his office to be drawn up, and corrected again, before being engraved (or rendered as beautiful watercolour designs). Then Brunel produced his own estimate for every contract, and ran the tender process, usually inviting three or four contractors to bid for each job.

In February 1836 the first contract, for the viaduct over the Brent valley at Hanwell in West London, was let to the contractors Grissell & Peto, and the great work was under way.[15] The brick viaduct, with its beautiful elliptical arches carried on tapering pylons of vaguely Egyptian shape, was a first indication of Brunel's superb talents as a designer, though there is evidence that his father had provided a good deal of assistance with this, as he had with the Clifton Bridge competition entry.[16] The viaduct, which still stands, was subsequently named after Lord Wharncliffe, chairman of the House of Lords committee that had passed the GWR's Act of Parliament.

Much of the most difficult work was in bridge design: time and again this was complicated by the need to keep the railway gradients as gentle as possible, while providing adequate headroom over roads, rivers and canals. Then, there were the problems created when the line crossed a road or a canal at a skewed angle, increasing the width of the opening that had to be provided. Sometimes, bridges presented both problems at once. For this reason, in the first stages of work, Brunel found himself

Views of Box Tunnel and Wootton Bassett Incline, illustrated by J C Bourne, published 1846.

Wharncliffe viaduct, illustrated by J C Bourne, published 1846.

Contractor's drawing of Maidenhead Bridge, c 1835 (Adrian Vaughan Collection).

Maidenhead Bridge, lithograph by Edwin Dolby, c 1840 (Elton Collection: Ironbridge Gorge Museum Trust).

having to design three bridges using cast-iron beams, as the only way of spanning the gap with a sufficiently slender (or shallow) structure. The material was new to him, and the largest of these bridges, crossing the Uxbridge Road at Hanwell, was to give serious trouble, with one of the beams breaking before it was complete, then breaking again a year after the line opened. It was a rare instance of a design problem getting the better of him.[17]

The bridge over the Thames at Maidenhead, however, drew forth Brunel's talents at their most magisterial. There was 100 yards of river to span, and a tight clearance above the navigation way, but there was also a helpfully placed islet. Brunel planted a central pier on it and leaped the river with the two widest brick arches, of 128 feet each, that had ever been built. The level issues required the arches to be very shallow, rising by just 28 feet, and this posed major questions about the weight-distribution and transmission of the arching forces within the structure: a meticulously worked-out drawing in one of his notebooks shows how he reached his solution.[18]

The new railway was being built through an entirely rural landscape: like the canals, the railways were to an overwhelming extent created by the manual labour of hundreds of navvies, armed only with spades, picks, shovels and wheelbarrows, assisted by teams of horses. Brunel's role as the designer should not be allowed to obscure the contribution of his contractors and their hundreds of men, who endured great hardships. Yet Brunel could behave in a notoriously high-handed, even arbitrary way to his contractors: he would add extra work to their contract, withhold payments, and make judgments always in the GWR's favour. At its worst this led to bitter and protracted litigation, between the GWR and the firms of William Ranger, Hugh and David McIntosh, and William Marchant.

In August 1837, with several of the contracts at the London end complete, the directors announced that the line would open to Maidenhead in November. This proved completely unrealistic, and the opening was deferred until the following May.

This was still ambitious, since only a mile-and-a-half of track had been laid by Christmas 1837. But the board was anxious to produce results since there was no doubt that Brunel's initial estimate of £2½ million would prove to be far too low.[19] Economies had to be made, and one of them was the grand classical terminus that he had envisaged at Paddington: all that could be afforded was a series of timber platforms and simple shed roofs, fitted in and around the brick arches of the Bishop's Road Bridge.[20] The spring of 1838 saw frantic activity to finish the London end of the line, and on 31 May the first train ran from Paddington to Maidenhead. At 11am the directors and 300 invited guests, including Brunel, his wife and parents, gathered at Paddington. They travelled on a special train to Maidenhead, inspected work on the bridge there, ate a cold lunch in a tent and travelled back to Paddington at a brisk 33 miles per hour.[21]

May 31 must have been a triumphant day, but as soon as the GWR started to run public services, from 4 June, problems started to mount quickly. Brunel's expensive track design was anything but a success, making for a rough, jolting ride: in some places the ballasting was not deep enough, in others the ballast was too fine. There were problems with the piles which he had designed to hold the track down: in places, this seemed to result in the line being 'over-fixed'. There were doubts about his carriage-designs, and there were rumours flying around about the supposed instability of the Maidenhead bridge.[22]

Worse, and less easily addressed, were the problems with the GWR's first locomotives. Brunel had insisted on controlling this, like all other matters, and had written specifications for engines unlike anyone else's, with large driving wheels but relatively

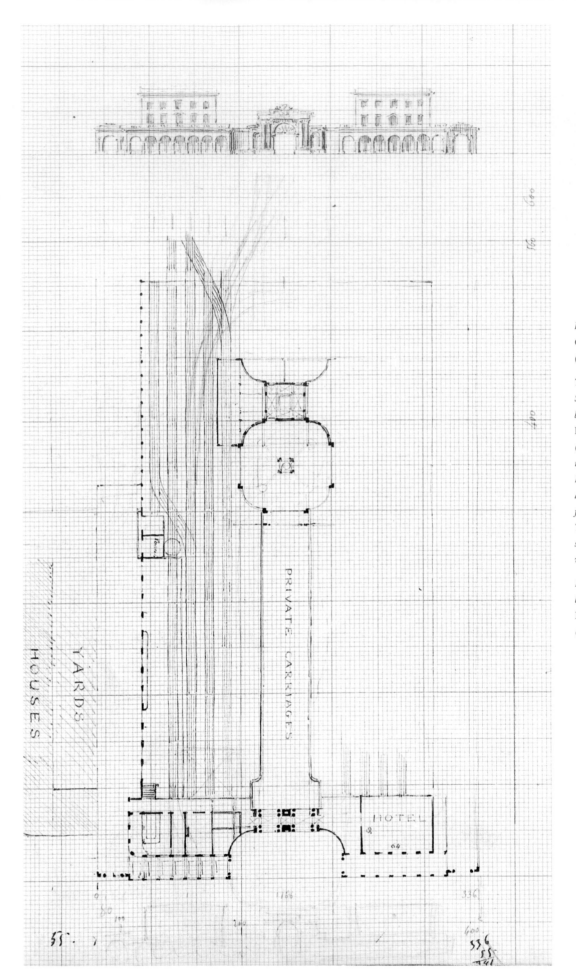

PRIVATE CARRIAGES

HOUSES

YARDS

HOTEL

Paddington plan and elevation, GWR Sketchbook 1836 (University of Bristol).

Two views of the first Paddington Station, (facing page, top and bottom left) The Illustrated London News, *22 July 1843 (University of Bristol). An uninviting exterior beneath the Bishop's Road bridge led through to a wooden shed that provided shelter for first and second class passengers. Those travelling third class waited separately alongside the goods that were also being transported.*

Paddington Station, (facing page, bottom right) The Illustrated London News, *8 July 1854 (University of Bristol).*

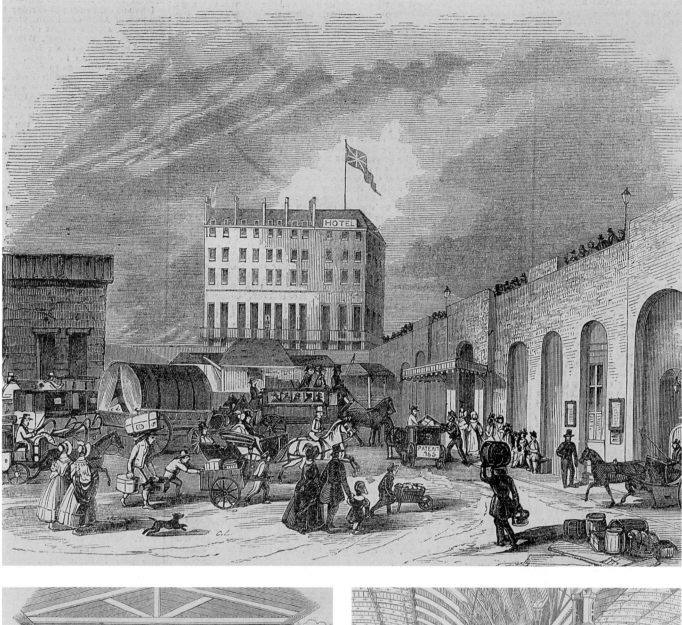

small boilers. When the first engines were delivered in the spring of 1838 they all turned out to be underpowered and unreliable. Fortunately, the company had two crucial assets. One was the *North Star*, an engine built by Robert Stephenson & Company for an American company, the New Orleans Railway, which had gone out of business before it opened. Bought by the GWR and converted to the seven-foot gauge, it was their only really reliable locomotive in the summer of 1838. The second asset was a young engineer, Daniel Gooch, who had been appointed as locomotive superintendent at the age of 21 the previous year. Gooch's superiority to Brunel as a mechanical engineer swiftly became clear, causing a good deal of tension between them, yet Gooch, the company's secretary, Charles Saunders, and the director George Henry Gibbs were to be among Brunel's most important allies in the trials that still lay ahead.

As the costs of building the GWR escalated, more capital had to be raised. By 1838 the GWR, like most early railway companies, had a strong group of shareholders from Liverpool. This 'Liverpool party' tended to support the Stephensons' way of doing things, and for most of 1838 they lobbied against Brunel and his different approach.[23] The problems shook even Brunel's confidence and in July 1838 he briefly contemplated resignation.[24] He fought back, defending the broad gauge and proposing that outside experts should be brought in to assess his work.[25] The Liverpool party kept up the pressure, and the outside consultants turned out to be hostile to the broad gauge as well.[26] Speed trials with the *North Star* were organised in December 1838.

View of Bath station from Beechen Cliff, c 1850s (with detail) (National Trust).

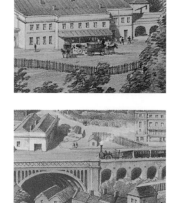

*Skew Bridge at Bath, tinted lithograph
by J C Bourne, 1846 (Elton Collection:
Ironbridge Gorge Museum Trust).*

*Contractors' drawing of Twerton
Tunnel, west end, c 1835
(Adrian Vaughan Collection).*

When the results came out badly, Brunel's career hung by a thread. He and Gooch, working on Christmas Eve, found that by increasing the size of the blast pipe opening they could produce a marked improvement in the engine's performance. Fresh speed trials turned the debate in their favour, and at the crucial vote on 9 January 1839 the 'northern' opposition was defeated by 7,792 votes to 6,145. Brunel was saved and so, for the time being, was the broad gauge.[27]

Meanwhile, the line had been growing rather more slowly up the Avon valley from Bristol towards Bath. The Bristol Committee seems to have been more generous in its approach to architecture than its London counterpart and at the western end of the line there was an abundance of good building stone. While at Paddington Brunel had to make do with temporary platforms and buildings, at Bristol Temple Meads the company paid for a splendid terminus in Tudor Gothic style, elevated over a viaduct.

From Temple Meads to Bath and beyond, Brunel created a series of magnificent set-pieces which united architecture and engineering. Just outside Temple Meads the line crossed the Floating Harbour on a two-arch bridge, then crossed the Feeder Canal on great timber trusses, then crossed the Avon on a mighty stone arch 100 feet wide (the latter still exists, though concealed by later widenings to either side). Going east up the Avon valley the line had to run through a dramatic series of deep cuttings and short tunnels at Saltford and Twerton: Brunel gave them a series of grand, castellated portals.[28] At Bath, despite the city's Georgian architecture, he designed the station in Elizabethan style, and gave the long viaduct that skirts the city centre gothic arches, battlements and turrets. This viaduct had to cross the Avon at an awkward skew: Brunel's suspicion, or inexperience, of cast iron seems to have been reflected in another masterly design for a timber bridge, this time with twin arches of 89 feet each, again with a gothic treatment.[29] The line opened from Bristol to Bath in August 1840. By this time, it was also open from London to Steventon in Wiltshire.

The most difficult section of the line remained to be finished. For the 13 miles from Bath to Chippenham, scarcely one mile of the line is within ten feet of the natural ground level: to keep the track at the right level it had to be raised on embankments, sunk in cuttings, or tunnelled. The hardest part of all was the one-and-three-quarter-mile tunnel beneath Box Hill, the longest that had ever been attempted. Begun early in 1836, this remarkable, terrible feat of engineering was only completed in April 1841, at the cost of over 100 lives. By 31 May the track was laid throughout and inspected by Sir Frederick Smith for the Board of Trade. Another month was needed to iron out the final problems, but the first through-train ran from London to Bristol, with a surprising lack of ceremony, on 30 June.[30]

In fact, the GWR's trains could already travel a lot farther over the allied Bristol & Exeter line, to Bridgwater. The GWR's empire was expanding, and Brunel was in charge of all of it. To understand this, we must briefly go back to 1835. In that year the Bristol & Exeter Railway (B&ER) was set up as the natural continuation of the GWR: it too appointed Brunel as its engineer, and adopted the broad gauge. The GWR, in its first flush of enthusiasm, was already promoting branch lines too. In December 1835 Brunel exultantly listed his new responsibilities in his diary. These started with the GWR itself, and went on via the Clifton Bridge and Bristol Docks to the Merthyr and Cardiff Railway ('I owe this to the GWR. I care not about it.'); the Cheltenham Railway ('do not feel much interested in this'); and the Bristol & Exeter ('This survey was done in the grand style – it's a good line too - and I feel an interest as connected to Bristol to which I really owe much.'). He concluded with the 'Newbury Branch' ('a little go, almost beneath my notice'); and, 'I forgot also the Bristol and Gloster Railway.' The entry says more for Brunel's self-confidence and industry than for his humility.[31] For each of these lines a resident engineer was appointed, but Brunel retained a high degree of overall control. On the Bristol & Exeter, this led to conflict with William Gravatt over his bridge-designs, which Brunel disapproved of, and several of which gave trouble. A painful wrangle ensued, in the course of which Brunel demanded Gravatt's resignation; the board ordered an investigation, and Brunel himself resigned.[32]

Despite these problems, the first main phase of the history of the Great Western's empire ended with the opening of the B&ER to Exeter in 1844. They had taken over the Bristol & Gloucester Railway and the Cheltenham & Great Western Union (C&GWU).[33] They were completing the C&GWU's line from Swindon to Gloucester and they were building a ten-mile branch line from Didcot up to Oxford, but these were minor works compared with what had been completed. The GWR seems to have slowed down somewhat, but in fact it, like its standard-gauge rivals, was winding itself up for one of the most frantic outbursts of economic energy in British history, the Railway Mania of 1845-6.

The Railway Mania was driven by three main factors. First was the profitability of the lines which had just opened.[34] Second, all the companies had grasped the importance of covering territory and getting their claims in first. Third, the competition for territory was bringing the gauge issue to a head. Brunel had defended the seven-foot gauge on the grounds that it would serve a distinct, separate area of England. This had always seemed questionable and in July 1844, when the two different systems met for the first time at Gloucester, the consequences for passengers and goods services alike were chaotic.[35]

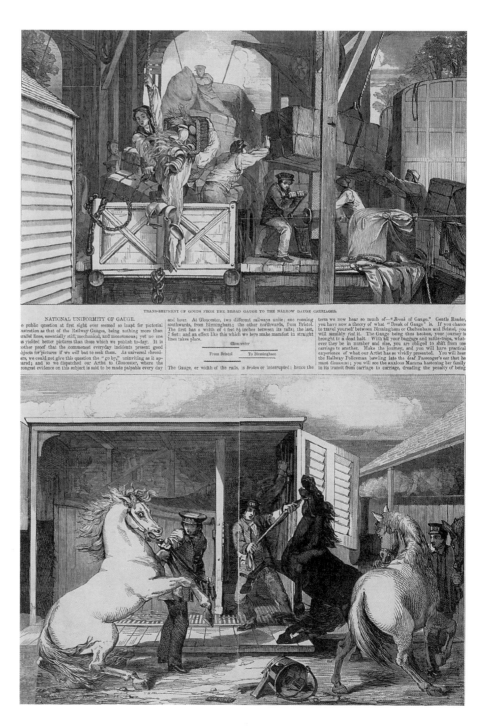

Opening page of The Illustrated London News *article calling for national uniformity of the gauge, 6 June 1846 (University of Bristol).*

In 1843-5 the GWR sponsored the establishment of several more broad-gauge railway lines to extend its network. The story becomes very complicated here, as the GWR had different relationships with these new companies and, in a couple of cases, great difficulty retaining control. A number of companies were set up to build branches off

Viaducts on the South Devon Railway, 1848 (National Trust).

the GWR's main line, most importantly the Wiltshire, Somerset & Weymouth Railway (148 miles) and the Berkshire & Hampshire Railway (a 39-mile branch from Reading to Basingstoke). Two companies were set up to extend their network over the Severn and into Wales: the Gloucester & Dean Forest Railway (15 miles) and the South Wales Railway (211 miles from Gloucester to Fishguard). Three companies were set up to extend the main line to the South West: the South Devon Railway (52 miles from Exeter to Plymouth), the Cornwall Railway (Plymouth to Truro, 81 miles with its branches) and the West Cornwall Railway (the last 31 miles to Penzance). Further companies were set up to carry the broad gauge into the Midlands, including the Oxford & Rugby Railway (50 miles), the Oxford, Worcester & Wolverhampton Railway (97 miles) and the Monmouth & Hereford Railway (45 miles).[36] Brunel was appointed chief engineer to all of them, which illustrates the huge extent of his responsibilities between 1844 and 1854. He supervised and checked the surveys, gave evidence before Parliament, supervised the letting of the contracts and retained overall control through his resident and assistant engineers.

In 1844-5 the 'battle of the gauges' broke out, centred on the issue of whether the GWR would be allowed to run broad-gauge lines into the Midlands. Round one went to Brunel's opponents: the GWR had taken a stake in the standard-gauge Birmingham to Gloucester line (opened in 1844), but in January 1845 the new Midland Railway bought up the outstanding shares and snatched it from under the GWR's nose. The company seemed to be about to lose round two as well. Anticipating a rush of railway business in Parliament, the Board of Trade appointed five commissioners – the Five Kings – to advise on the whole question. Their report, also in January 1845, favoured the standard gauge, and said that the broad gauge should not be allowed to spread north. Despite this, the GWR managed to get its bills for the Oxford & Rugby and the Oxford, Worcester & Wolverhampton lines passed, and given royal assent, by August. Round two to the broad gauge.

The opposition were re-grouping, and already in July 1845 a Royal Commission had been set up to investigate – and decide. Dozens of expert witnesses were called; the opponents, led by Robert Stephenson, were vocal; and on 25 October Brunel himself answered 200 questions. Challenged whether, if he were starting again, he would still design his railways to a broad gauge, he replied: 'I should rather be above than under seven feet now.' Eventually he persuaded the commissioners to agree to a speed trial: the *Ixion*, one of Daniel Gooch's *Firefly*-class locomotives, was tested against a Stephenson standard-gauge engine. *Ixion* easily outperformed its rival on average and top speeds.

In 1846 the commissioners reported. At first, there was good news for Brunel: they acknowledged the broad gauge's superiority as to speed and comfort for passengers.

Last broad-gauge train leaving Paddington for Penzance, 10.15am, 20 May 1892 (private collection).

Then came the bad news. These qualities were seen as primarily benefiting the well-heeled passengers of express trains, a relatively small number of people. For the 'general commercial traffic' of the country, the standard gauge would do perfectly well. Furthermore, 1,901 miles of standard gauge had been built, as against 274 miles of broad gauge. If it were imperative to produce uniformity, the commissioners said, they would have to decide in favour of four feet 8½ inches. When the 'Act for Regulating the Gauge of Railways' was passed in July 1846, it stated that henceforth four feet 8½ inches would be regarded as the standard. However, the GWR got a let-out clause: railway Acts could still have a 'special enactment defining the gauge', if a case could be made. Round three in the battle of the gauges was, at best, a draw for the broad gauge.[37] Even so, from this point the writing was on the wall: for the GWR to grow, it would have to take over standard gauge lines, and would be unlikely to be allowed to convert them. They started to lay 'mixed' track that could take both kinds of traffic, and in 1861, just two years after Brunel's death, the first four-foot-8½-inch-gauge train rolled into Paddington Station itself.[38]

As railway design and construction were becoming better established and understood, and more junior engineers were trained, Brunel was able to leave more of the routine work to his assistants, and though he continued to try and control matters from a distance, he also found time to do highly innovative and wide-ranging work during the railway mania and its aftermath (c 1844-55). The beginning of this period found him working on one of his most ambitious experiments: the South Devon Atmospheric Railway was certainly innovative, but it has long been regarded as Brunel's most notorious failure.

There is no room for doubt, though, about Brunel's work as a bridge designer in this period. The rapid growth of the GWR's sprawling empire required great numbers of new bridges to be built and a remarkable number of them were designed by Brunel himself. To get cash-strapped railway companies up and running, he had to keep costs down and one way was to build bridges in timber, in which he attained an absolute mastery. First, there were two bridges over the Thames on the Oxford Railway.[39] Then came a series of bridges on the Swindon to Gloucester line, with spans of up to 71 feet: Brunel devised king-post trusses tied with wrought-iron rods to span the gaps. For the South Devon Railway, five large viaducts were needed: they could not afford stone, so again they were in timber, the largest at Ivybridge with five spans of 61 feet each. The South Wales Railway, with deep valleys and estuaries to span, presented even bigger challenges. The biggest of all was the Landore valley outside Swansea: Brunel's viaduct, 580 yards and 37 spans long, with a main span 100 feet wide, must have been among the biggest timber bridges ever built.[40]

His most famous achievements in this material, though, were the Cornish viaducts. Brunel had been appointed engineer to the Cornwall Railway in 1845, displacing

Captain W S Moorsom, whose proposed route for the line had just been thrown out by Parliament. Brunel produced a vastly superior design, reducing most of the 1:60 and 1:50 gradients that took up over half of Moorsom's route. He had to contend with Cornwall's picturesquely complex topography: there were eight estuaries and 34 deep valleys to cross. Brunel produced a number of standard-design timber viaducts made of 'kyanised' pine beams tied with iron rods, carried on slender stone piers. The highest, at St Pinnock, had piers up to 118 feet high, the timber structure rising another 35 feet. The structures looked impossibly slender and terrified some of the passengers, but Brunel knew what he was doing. He warned the company that they would be expensive to maintain, and indeed they cost £10,000 a year to keep in repair, but they got the railway running and making money. In due course, all of Brunel's timber bridges were replaced, in iron or (as with most of the Cornish ones) in stone. It is sad that none of this remarkable body of work survives, but actually that was the whole point: they were only intended to be cheap and temporary. They are one of the most effective ripostes to the frequently laid charge that Brunel paid little heed to cost and the interests of the shareholders.[41]

Brunel built comparatively few bridges in cast iron. It was used by almost all railway engineers in the 1830s and 1840s, but Brunel distrusted it because of its weakness in

Gover Viaduct on the Cornwall Railway, watercolour by H Geach. This timber structure was replaced by a masonry one in 1898. (Elton Collection: Ironbridge Gorge Museum Trust).

tension and his difficulties with the Uxbridge Road bridge in 1837-8. In response, and with his characteristic unwillingness to accept other people's thoughts at face value, he began to devise his own designs in the material. The first result seems to have been the canal bridge at Paddington of 1838-9, recently rediscovered and saved from destruction in 2003-4.[42] He carried out several further rounds of experimental and design work during the rest of the 1840s, though this area of Brunel's activity remains little known and requires further research. Brunel retained reservations about the use of cast iron, which may have been borne out after the collapse of Robert Stephenson's Dee Bridge at Chester in May 1847. In the ensuing parliamentary enquiry, Brunel set out his views in some detail.[43]

By this time, major improvements in techniques for rolling and fabricating wrought iron were under way. Wrought iron, with its much higher tensile strength, was the bridging material of the future, and the years 1847-59 saw a remarkable series of advances. Robert Stephenson and his team were in the lead, having developed and built tubular girders of wrought-iron plate to span 400 feet at Conway (1847-8), and two spans of 460 feet in the Britannia Bridge over the Menai Straits (1848-9). Brunel attended the floating and lifting of the giant tubes for the latter, to lend moral support to his friend in this most nerve-wracking of operations.[44]

Chepstow Bridge, hand-coloured and tinted lithograph by W Richardson, published August 1851 (Elton Collection: Ironbridge Gorge Museum Trust).

Characteristically, though, Brunel's own ideas were developing along quite different lines. More research is needed to understand his first steps in mastering the use of wrought iron, but by 1850 he had built large 'bowstring' trusses spanning 100 feet (over the Usk at Newport) and 200 feet (over the Thames at Windsor). Much wider spans were needed to take his railways across the Wye at Chepstow and the estuary of the Tamar at Saltash. The Chepstow bridge had to cross 600 feet of river, allowing a clear span of 300 feet over the navigation channel, on a river with a huge tidal range.

Brunel's concept seems to have been totally unprecedented. He would clear the 300-foot gap with a pair of giant, parallel trusses, each carrying one railway track. The lower chord of each truss would be formed by a wrought-iron trough, within which the railway track sat. The upper chord would be formed by a great, wrought-iron tube or cylinder. Upper and lower chords would be anchored in iron towers at either end, and braced apart by two big uprights in between. The lower chord would be suspended from wrought-iron chains, anchored to the towers and the upper chord. It might be described as a form of 'closed' suspension bridge, in which the tension forces are all balanced, and all contained within the structure, rather than being transmitted to anchoring points in the earth. The two great trusses were floated into position, raised and fixed in meticulously arranged operations planned with the help of Brunel's friend, Captain Claxton, in 1852-3.[45] Sadly, this engineering *tour de force* was demolished and replaced in 1962. Chepstow was, however, the preparation for Brunel's greatest achievement of all as a bridge-builder: the Royal Albert Bridge at Saltash linking the Cornwall Railway to the rest of the GWR's empire, in which he brought his concept of the 'closed' suspension bridge to perfection, and which happily survives.

In the last week of 1850, Brunel worked up sketch designs for a great new station at Paddington: the astounding growth in the GWR's empire, which he had done so much to plan and bring about, had meant that his temporary station there could not cope with the demand any longer. He was sitting on the building committee for the Great Exhibition at the time, and was so impressed by Paxton's design for the 'Crystal Palace' and by the contractors who built it, Fox, Henderson & Company, that he insisted that they should turn his sketches into reality, in what would now be called a 'design and build' contract. The overall conception, though, remains Brunel's, and his great three-span roof at Paddington also survives as one of his greatest remaining legacies.[46] It is deservedly one of the best-loved of railway stations, and is perhaps the most potent symbol of the Great Western Railway, the company which provided Brunel with his greatest opportunities, and represented the central theme of his astonishing career.

THE BRISTOL TERMINUS

Brunel's GWR station in Bristol, now home to the British Empire and Commonwealth Museum, opened in 1841 and is thought to be the first true railway terminus, with trains and people all inhabiting the same integrated space beneath a single roof. A viaduct was built first and the station constructed around it. The booking hall was at ground level and passengers reached the platforms by climbing an internal staircase to the first floor, the track coming in 15 feet above ground-level on brick vaults.

The building, with its impressive three-storey entrance in Bath stone, is in a hybrid English revival style, showing Tudor influences. This distinguishes it from the more classical detailing found at other GWR projects, such as the Greek revival Royal Western Hotel – now Brunel House – in Bristol. The architectural treatment was carried through from the booking offices to the body of the station, providing a unity not always seen in such structures. At William Henry Barlow's St Pancras (completed 1868), for example, there is a stylistic separation between the industrial train shed to the rear and the Gothic hotel in front. The beautiful hammer-beam roof that spans the 72 feet of the Bristol passenger shed is an elaborate 'set-dressing to add architectural presence' as most of the roof's weight is actually supported by the iron columns of the aisles.[1]

Brunel's terminus has not been used as a railway station since 1965. Work on what has become Bristol Temple Meads station commenced after Brunel's death and was completed in 1878. Its architect was Brunel's colleague, Sir Matthew Digby Wyatt, who had co-designed Paddington station. It was originally built to serve the Bristol & Exeter, Midland and Great Western Railway companies, and has a neo-Gothic design.

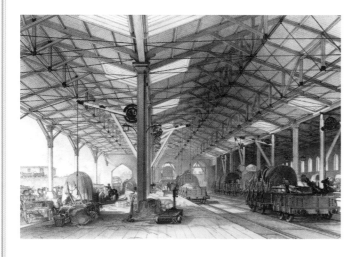

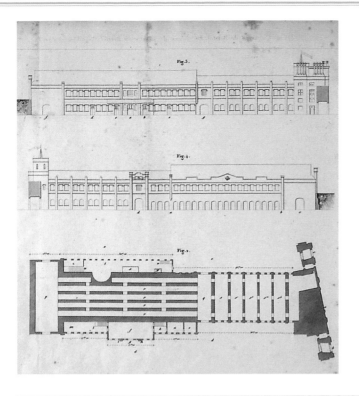

Goods shed at Bristol (facing page, bottom) illustrated by J C Bourne, published 1846 (Bristol's Museums, Galleries and Archives).

Brunel's terminus (facing page, top, and this page below), now housing the British Empire and Commonwealth Museum (photograph by Ian Blantern).

Details of plans and elevations of terminus from book of engineering drawings (left) (University of Bristol).

Passenger shed at Bristol, c 1840s (National Trust).

THE ATMOSPHERIC RAILWAY

Brunel originally intended parts of his route through South Devon to be covered by an atmospheric railway. Brunel had thought this was the most effective and cheapest means of negotiating the stiff gradients through the difficult coastal terrain.

An atmospheric system had been patented in 1839 by Jacob and Joseph Samuda, who had been intrigued by a short, experimental line laid by the Birmingham, Bristol & Thames Junction Company called 'Mr Clegg's Pneumatic Railway'. The first full-scale trial version of their system, developed in partnership with Samuel Clegg, opened on the Dalkey branch of the Dublin & Kingstown Railway in August 1843. It was later used on the London & Croydon Railway.

A continuous line of nine-inch diameter pipe with a slotted top was laid between the rails of the permanent way. The lead carriage of a train carried a piston arm that ran inside the pipe. A leather flap, backed by iron, opened just ahead of the piston arm and closed behind it, forming an airtight valve. A large, stationary trackside steam engine pumped the air out of the pipe ahead of the train, exerting atmospheric pressure on the back of the piston, and effectively sucking the train along up a gradient. The train could descend by gravity if necessary.

Brunel had visited the Samuda brothers' test track and thought it had potential for working on a much bigger scale. Experimental trials took place on the Exeter-Teignmouth section of the South Devon Railway in August 1847 and a regular service was started in September. In January of the following year this was extended to Newton Abbot. Eight Italianate atmospheric pumping stations were to be constructed at four-mile intervals along the route. Of these, two remain today, albeit adapted for other uses: the station at The Strand, Starcross near Dawlish and the station at Newton Road in Torquay.

The atmospheric experiment proved short-lived. The leather valve soon began to rot in the damp Devon climate and it was also eaten by rats. As a result, more energy had to be produced by the pumping stations to compensate for the leaks, escalating fuel costs. At a meeting of shareholders at the Royal Hotel in Plymouth on 29 August 1848, Brunel announced that the system was to be abandoned. The developments in the capabilities of steam locomotives meant more conventional power could now be harnessed to tackle the gradients, thereby removing one of the system's key advantages.

The Atmospheric Railway at Dawlish, South Devon Railway, *watercolour by Nicholas Condy, 1848 (Elton Collection: Ironbridge Gorge Museum Trust).*

View of the railway at Dawlish c 1848, with detail (above and left) (private collection).

Illustrations from William Dawson's South Devon Railway album published 1848 (Newton Abbot Town Council).

THE ROYAL ALBERT BRIDGE

The Royal Albert Bridge, carrying the Cornwall Railway across the Tamar at Saltash, was Brunel's last and arguably his greatest railway work, mitigating some of the disappointments associated with the ill-fated ss *Great Eastern*. It is thought to be Brunel's only major project to have been completed on time, to budget and without loss of life during construction. It brought together Brunel's mastery of tubular construction, suspension principles, riveted wrought iron and maritime engineering.

The Royal Albert Bridge, c 1860 (private collection).

Brunel had begun surveying the deep tidal estuary on the boundary of Devon and Cornwall in 1847. A bridge constructed from cast-iron girders or with wrought-iron suspension chains would have been unable to bear the weight of a railway train across such a wide span. Brunel began experimenting with wrought-iron girders, looking for a simple, efficient and economical tubular structure. He also learned from the experiences of shipbuilders engaged in the construction of large, ocean-going hulls and in Robert Stephenson's work at Conway and the Menai Straits. Brunel's Chepstow bridge over the river Wye provided a useful means for trying out his ideas, but the much wider span at Saltash, the need to allow clearance of at least 100 feet above high tide (which was 70 feet deep) and the limited number of piers that could be set in deep water called for new innovations.

Brunel completed his designs in October 1852. Of the four possibilities, he chose a 19-span structure, 2,220 feet long and 30 feet wide, which rose 260 feet above its foundations, carrying a single rail track. There were two main spans of 455 feet each, requiring just one central pier with one at each bank. The 17 side spans varied in length between 70 and 90 feet.

Following further experimentation and a survey of the riverbed, the Great Cylinder was constructed measuring approximately 37 feet in diameter and 90 feet in length and comprising an inner and outer tube. It was towed out into position in the middle of the estuary in June 1854, upended and sunk. The space between the tubes consisted of compartments that could be filled with compressed air. Workers entered the compartments to dig away the 20 feet of mud that lay beneath the fast-flowing water so the cylinder would sink down to a solid foundation. It reached bedrock in February 1855 and the laborious work then began of levelling the rock and filling the inner tube with masonry to form the stone pier. By the end of 1856, the pier was above the high water mark.

In the meantime, the first of the two trusses that formed the main spans was being built on the banks of the Tamar. The truss was formed from a wrought-iron oval

the track was suspended below the tubes, its supports connected to the sets of chains that hung down vertically from them. The chains had originally been made for the Clifton Suspension Bridge.

When the bridge was opened by the Prince Consort in May 1859, Brunel was unable to attend because of his ill-health. He had only been able to see the completed work by making a slow crossing of the bridge while lying upon a couch resting on a flatbed railway wagon. When he died in September, the directors of the Cornwall Railway voted to place a memorial to him on the bridge. It reads: 'I K Brunel, Engineer, 1859.'

Opening of the bridge viewed from Saltash station, The Illustrated London News, *14 May 1859 (University of Bristol).*

'The Diving Bell used at the Thames Tunnel after the Irruption of Water', engraving by George Cooke after an original drawing by Clarkson Stansfield, 1827 (Science Museum/Science and Society Picture Library). Brunel himself had been first down in the diving bell to inspect the damage at the tunnel. He later used the bell as the basis for his design of the Great Cylinder at Saltash.

tube curved into an arch. It was floated out on pontoons on 1 September 1857 and manoeuvred into position upon the temporary tops of the piers, forming the beginning of the crossing from the Cornwall side. The band of the Royal Marines played 'See the Conquering Hero Comes', the crowds cheered, the flags flew and a general holiday was declared to celebrate the achievement, another example of Brunel's superb combination of engineering skill and showmanship.

In the coming months hydraulic jacks slowly raised the truss as the piers were built up underneath it. Each pier was formed from four cast-iron octagonal columns. It reached its final position in May 1858. In June the Devon truss was floated out, reaching its final position in February the following year. The plate girder to carry

View of the bridge today alongside the Tamar road bridge (photograph by Simon Lewis).

Saltash Bridge, c 1860s, stereoview (Adrian Andrews' collection).

The Great Britain *dropping her pilot off
Anglesey, August 1852, by Joseph Walter
(ss* Great Britain Trust*).*

SS GREAT BRITAIN

Andrew Lambert

Introduction

Nineteenth-century industrial technology created almost limitless opportunities to revolutionise long-distance communications and transport. While most engineers and innovators crept forward slowly, making small incremental adjustments that deployed a single novel concept to improve their chosen system, one man had the intellect, education and daring to assemble all the latest technical advances, reduce them to order, and integrate them into a new design, far in advance of anything that had come before. Along the way every detail, from the titanic to trifling, was subjected to the searching insight of a unique genius.

Nowhere was this process more obvious than in the development of the intercontinental steamship. Having conceived and begun to execute the world's first inter-city railway, Brunel's restless mind sought new opportunities and his remarks about extending the railway from Bristol to New York were quickly turned into reality. While his first ship, the ss *Great Western*, was the finest example of the wooden paddle-wheel steamer and the model for all future vessels of her species, Brunel could see both problems and opportunities.

Brunel created the modern ship and that first prototype survived the vicissitudes and vagaries of oceanic life, by accident and design, to return to her birthplace 127 years later and undergo a long-term restoration that reminded the world of how mass transport was born. The ss *Great Britain* is probably the most important ship ever built: she opened up the modern world, her descendants carry the vast bulk of all international trade, and the approach to transport economics that she enshrines has been re-used countless times, most recently with the Airbus A380. She is among the crowning achievements of human genius.

*Photograph of William Patterson
(PBA Collection at Bristol's Museums,
Galleries and Archives).*

Brunel and the steamship

It has long been accepted that Brunel made a major contribution to the development of the steamship. What is less well known is why. Ships and the sea were in his blood, and they occupied a unique place among his interests. Born in Portsea, while his father, Marc, was working on his block-making mills at Portsmouth, the younger Brunel would have known all about steamships by the time he was ten from Marc's extensive interest in the early development of these revolutionary craft. Throughout his life Brunel invested his own money in steamship projects and took no fee as consulting engineer,[1] while his constant use of paternal imagery during the protracted rescue of the *Great Britain* from Dundrum Bay reveals a strong emotional attachment. For Brunel the steamship was both a legacy from his father and the ultimate test of his ability to translate new materials and technologies into cutting-edge engineering. He would continue to work on ships until his death.

Marc Brunel provided his son with a unique, bi-lingual scientific, technical and engineering education, which culminated in a vital period in the Lambeth workshops of Henry Maudslay, the leading constructor of marine steam engines. Marc Brunel also introduced his son to a group of scientists, patrons and politicians who would play a critical part in his life, notably Charles Babbage, the pioneer of the mechanical analogue computer.[2] Babbage combined advanced liberal politics with a unique understanding of mathematics and science and engineering. He would play an important part in Brunel's life, from his earliest attempts to find work to the beginnings of operational research on railway gauges. Brunel regularly sought his advice. Unlike the majority of his contemporaries, Brunel had the intellectual equipment to work with Babbage. Brunel's peculiar genius lay in straddling the then largely separate worlds of science and engineering.

The Great Western Steamship Company as a business venture

In late 1835 the directors of the Great Western Railway met in London to discuss the line, and Brunel is reputed to have suggested that it be extended to New York. The ambition and drive of the company turned this witticism into a practical project, and by early 1836 a Brunellian concept had emerged. A very big ship with powerful engines would be the most economical method of crossing the Atlantic under steam. The new ship was built of wood, and launched as the *Great Western* on 19 July 1837.

This was Brunel's first ship, and it benefited greatly from contact with the Admiralty and with Maudslay, Son & Field, where he had completed his education. Her structure and form were provided by the Surveyor of the Navy. In return the Admiralty received regular reports on her voyages. The engines, the largest yet built, were entrusted to the London firm, which reduced the possibility of technical failure and ensured reliability in service. They were expensive, but Brunel was never much concerned with cost, only quality.

As he had no experience of wooden ship construction he relied heavily on the experienced steamship constructor, William Patterson, to transform his ideas into a ship. The *Great Western* was Brunel's university education in ship design and construction. She crossed the Atlantic in April 1838, the first commercially viable steam passage, in the process inaugurating timetabled inter-continental transport, the basis of the modern world. The steamship was a legacy from his father and the ultimate challenge for his ever-active mind. Consequently he had no interest in repeating himself, only in extending his repertoire and meeting new challenges.

Aquatint of the Great Western *after an original painting by Joseph Walter, c 1838 (private collection).*

Eastern Wapping Dock *by Thomas L Rowbotham, 1826. This was the site of Patterson's Yard (Bristol's Museums, Galleries and Archives).*

Having created the definitive wooden paddle-wheel ocean-steamer with his first ship, Brunel demonstrated the underlying obsession that propelled his work by moving rapidly and persuasively with the latest technology to create an entirely different ship to serve alongside the *Great Western*. The two ships would be different in every area save one: their ability to cross the Atlantic under steam in speed and safety.

The new ship and shipyard

The success of the *Great Western*, coming on top of the enormous critical success of the railway project that spawned her, provided Brunel with the status he craved. He was now *the* Engineer. The steamship project provided him with another opportunity to stand on the national stage at the head of his profession.

If the Great Western Steamship Company was going to provide a regular transatlantic service and turn a profit, it needed another ship. In 1838 the company began to collect materials for two more wooden paddle-wheel vessels. However, by March 1839 the directors had decided to build the ship of iron. This was a remarkable decision and one that could only have been taken on Brunel's advice. In October 1838 the iron-hulled paddle-wheel vessel *Rainbow* had called at Bristol, enabling the key figures in the company to go to sea, examine the structure, the impact of sea conditions on iron hulls and, critically, the impact they had on the compass. Captain Christopher Claxton RN, Brunel's close associate, was satisfied that the new system of compass correction devised by the Astronomer Royal was reliable. No one had constructed a ship of the *Great Western*'s size in iron or used one on the open ocean. On the safety of the compass system, Claxton's opinion counted: he was a naval officer and experienced navigator. The iron ship was begun in July 1839, only ten months after the *Rainbow*'s visit.

Any sensible business leader would have followed the *Great Western* with a sister ship, restricting any changes to detail improvements and possibly slightly enlarged dimensions to profit from the success of the original vessel. This was the initial plan, and much money was invested in timber and engineering resources to execute the task. However, Brunel was not content to rely on proven systems, as his uniquely scientific approach to engineering had brought to light some fundamental problems, and he soon found that innovative technologies were being developed that might enable him to transform the very concept of the ocean-going steamship. This was a challenge he could not resist.

The company was already looking at building a dry dock for the *Great Western* and an engine shop for repairs or new-build. It was therefore but a small step, for Brunel at least, to add in the capacity to manufacture engines and build iron ships, as no such facilities existed in Bristol. The dock would be built to hold the new ship and it is there that she now lies. The decision on engines was less satisfactory. The obvious choice would have been to order a new set from Maudslay, whose directors were

Great Britain, *returned to the Bristol dockyard, 1971 (Bristol United Press).*

Brunel's personal friends. However, dramatic developments in engine design and the high cost of Maudslay's tender led the company, largely against Brunel's wishes, to elect to build for itself. This proved to be both out of step with other steamship companies and a serious mistake. Within two years the whole facility had been put up for sale. The decision to opt for screw propulsion forced the company to change engine designs, sacrificing much of the initial investment in tools and patents. In their place, Brunel conceived a four-cylinder inverted V engine, based on a patent taken out by his father in 1822, with a massive chain to take power from the crankshaft to the propeller shaft. The engine was designed to fit the ship, long after such decisions should have been taken. The fact that it worked well, and was for some years the largest steam engine in the world, was yet another testament to Brunel's flexibility and speed of thought.

The decision to change the propulsion system was even more remarkable than the shift to an iron hull. It should be stressed that Brunel did not invent the screw propeller or the iron ship. Instead he took the latest ideas and opportunities generated by others, on a far smaller scale and with little of his astonishing ambition, and used them to transform the rules of engineering in each new sector of the profession that fell under his inspection. By 1840 the pioneering work of Francis Pettit Smith had established that the screw was an effective propeller, but Smith lacked the engineering knowledge to make it an efficient one. His wealthy backers funded the construction of a trials ship, the 200-ton *Archimedes*, to demonstrate the new technology to the Royal Navy and to the shipbuilding industry around Britain. She arrived in Bristol in May 1840. Brunel was then looking for improvements in the paddle-wheel system for his new ship. After going to sea in the novel vessel, Brunel advised the directors of the company in June to consider the screw for their new ship. The company's shipbuilding committee, which Brunel dominated, sent an engineer on the *Archimedes* to Liverpool, a voyage on which she encountered rough weather. His report persuaded the committee to suspend work on the stern of the new ship and conduct further trials. Brunel overcame internal opposition to conduct a series of remarkable trials in which he determined, with scientific precision, the propulsive efficiency of the system, the best method of driving the propeller shaft and the vital sea-going performance of the screw. Brunel's mastery of engineering and dynamic theory was fully displayed in his magnificent paper of 18 October 1840, which called for the new system to be adopted forthwith.[3] In the process, Brunel provided the first scientific assessment of propeller design. With data collected from the *Archimedes* and the *Great Western*, Brunel was able to demonstrate, not merely assert, that the new system was incontestably superior in key areas. The hydrodynamic measurement of ship resistance had been pioneered by his father a quarter of a century before; now it was being used for the first time to make engineering and commercial judgements. The key to his argument was the recognition that the *Archimedes* was an inefficient compromise and that he could make a far better screw-propeller ship.

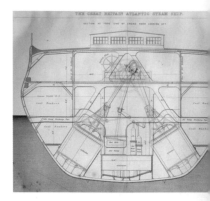

Section of Great Britain *engine room viewed from fore-end, designed by Thomas Guppy, drawn by John Weale, 1843 (Institution of Civil Engineers).*

Screw propulsion reduced weight, he argued, used a simpler hull form and:

> being unaffected by the trim or the rolling of the vessel… allowing of the free use of sails, with the capability of entirely disconnecting the screw or of varying the multiplying motion so as to adapt the power of the engine to the circumstances either of strong adverse winds or scudding.

The integration of steam and sail power was vital for transatlantic travel.

After adding a list of other advantages, including improved cargo carrying, improved steering and reduced beam (breadth), Brunel concluded:

> My opinion is strong and decided in favour of the advantage of employing the screw in the new ship… I am fully aware of the responsibility I take upon myself by giving this advice. But my conviction of the wisdom, I may almost say the necessity of our adopting the improvement I now recommend is too strong, and I feel it is too well founded for me to hesitate or to shrink from the responsibility. [4]

He knew that 'taking into consideration all the advantages of the screw it was better than any paddle and would soon supersede the paddle'.[5] In December 1840 the Board accepted his argument: the *Great Britain* would be a screw-propeller ship.

The combination of the iron hull with screw-steam propulsion created the modern ship. While everyone else saw the *Archimedes* as the ideal compromise for the 1840s, using a low-powered screw engine to assist the sailing ship, Brunel created the world's first full-powered screw steamship. That it was also the largest ship in the world and

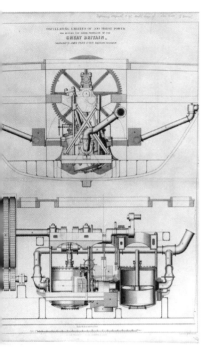

Drawing of oscillating engine for driving the screw propeller used in the 1850s' refit (PBA Collection at Bristol's Museums, Galleries and Archives).

Launch of the Great Britain, *1843, by Joseph Walter (ss* Great Britain *Trust).*

built of iron demonstrated the need to integrate innovations, not treat them as distinct items. The name *Great Britain* was finally adopted in 1842, in place of earlier favourites *City of New York* and *Mammoth*.

To assist him in developing the screw to meet his needs, Brunel shared information with the Navy, which was equally concerned to exploit the system, although in an auxiliary role. By acting as engineering consultant on the first screw warship, HMS *Rattler*, he secured the use of warships and naval facilities to develop his ideas. The *Rattler* was built before the *Great Britain* was finished and her trials, which Brunel controlled, provided all the data he needed to design the propeller of the *Great Britain*. Acting as a high-profile adviser to government on a matter of profound national importance was flattering. It demonstrated Brunel's pre-eminence among engineers. That he used that connection to secure vital information for the company revealed that his intellect was just as sharp in other ways. The *Great Britain*'s propeller was only designed after the *Rattler* trials had been concluded. Little wonder his six-bladed form was 'extraordinarily efficient', far ahead of any contemporary design.[6] It was also beyond the material and manufacturing abilities of the age, and was replaced by a more durable four-bladed bronze casting after the second voyage.

The ss *Great Britain* Leaving Bristol
for her Trials *by Joseph Walter, 1844*
(*ss* Great Britain *Trust*).

Like any master designer, Brunel was concerned to address the finer points of the
ship. His innovative six-masted rig, which featured deck-stepped masts, fore and aft
sails and labour-saving devices, was a direct response to the needs of an Atlantic liner
for sails that worked alongside the steam plant and could best handle the apparent
wind as she steamed ahead at speed. Where conventional sailing ships spread their
canvas on three square-rigged masts, requiring large amounts of skilled labour, Brunel
used only one square-rigged mast, and five masts each of which carried fore-and-aft
sails. The skilled men on his ship would be stokers and engineers, with a relatively
small deck crew for sail-handling. Similarly, Brunel developed the balanced rudder to
lighten and improve steering, recognising that with a propeller the rudder would be
critical for safe manoeuvring. The *Great Britain* was floated out of the building dock
on 19 July 1843, with Prince Albert performing the honours. It was a pioneering
Royal visit, the Prince travelling down from Windsor to Bristol on the footplate of a
Great Western locomotive and returning that evening. The ship was a national and
international celebrity icon long before putting to sea, being the largest ship in the
world, the first iron-screw liner and a symbol of British industrial pre-eminence at sea.

Brunel's fellow scientist, William Fox Talbot, with whom he had corresponded
extensively about the railway passing through Fox Talbot's west country estate, was on
hand to photograph the ship moored opposite her building dock in 1844. Getting the
ship out of Bristol proved to be rather fraught and involved many of the arguments
about damage to local facilities that would arise again in the twentieth century. Bristol
failed to appreciate the opportunity of transatlantic steam communications.
Navigational difficulties, harbour costs and parochial attitudes allowed the lead to pass
to Liverpool. The ship finally squeaked through the Cumberland Basin and out of
Bristol with a few courses of masonry removed late one December night in 1844.

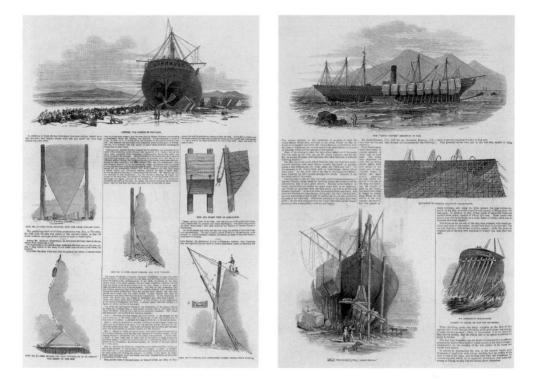

Great Britain *at Dundrum Bay,*
The Illustrated London News,
31 October 1846 and 21 August 1847
(University of Bristol).

When the *Great Britain* had her trials in January 1845, she performed almost exactly as Brunel had intended – a remarkable event for an entirely novel vessel far larger, more powerful and ambitious than anything hitherto attempted. The value of scientific research, data collection and advanced engineering knowledge was never more apparent. Then she steamed to London, where for several months she was the object of universal admiration, raising awareness and funds for the company and receiving many distinguished visitors, notably the Queen and the Prince. Finally, she arrived at Liverpool in July to make two transatlantic crossings before a winter lay-up and the modification of her rudder, rig and boilers. Brunel continued to view the ship with zealous parental concern. He would allow no one outside the company to work on her and, despite the many and various demands on his time, was always at hand to advise on his ship.

After a promising start to the 1846 season the *Great Britain*, having missed her turning point off the Isle of Man in fog, ran aground in Dundrum Bay on the Coast of Ireland on 23 September 1846. The only plausible explanation is the failure of the master, James Hosken, to make the necessary alteration of course at the proper time, through inattention or error.

Brunel was unable to visit the site until 8 December, when he took charge and devised a breakwater system of flexible materials that absorbed the worst winter gales. He was determined to save his ship, while the company directors seemed to lose heart and prepared for the worst. Brunel's reaction was personal. This was his baby and he was

going to save her. He complained that the directors were prepared to see the best ship in the world 'left… lying like a useless saucepan kicking about on the most exposed shore that you can imagine, with no more effort than the said saucepan would have received on the beach at Brighton'.[7] He won over the company and by August 1847 the ship was afloat. The salvage added another laurel to his crowded brow, but it was also the end of his connection with the ship. The company was underinsured and had sold the Great Western in 1847 to pay for the salvage. It now lacked the funds to repair its flagship. In 1850 the Great Britain was sold to Gibbs, Bright & Co of Liverpool for £18,000. The Great Western Steamship Company was wound up in 1852.

The tortured history of the company demonstrated that men of genius cannot command success. On its behalf, Brunel produced the two best liners in the world, but the failure to secure a mail contract, a subsidy paid for carrying mail overseas, and the costly experiment with engine construction doomed it to failure. With better business sense, Samuel Cunard had won the contract and built the necessary ships. The failure to secure the mail contract had been a massive setback: Cunard's unremarkable ships turned a profit purely from that contract, and his company has survived. That said, the failure of the Bristol civic leaders to invest in suitable facilities doomed the city as an Atlantic terminal and ruined the integrated vision of a Great Western service from Paddington to New York long before the incident in Dundrum Bay destroyed the steamship company.

Australia

Dundrum Bay could not destroy the *Great Britain*. The pounding she received in the heavy gales, which would have quickly destroyed most wooden ships, demonstrated just how robust and tough iron ships could be and gave great encouragement to the massive British investment in iron shipyards, particularly in the north of England and on the Clyde. The new owners converted the *Great Britain* into a sail and steam liner for the England to Australia service, installing new, more efficient machinery and modifying the sailing rig. This was a different service from that for which Brunel had conceived her and it called for a different ship. It was a testament to his enlarged vision and quality engineering that she was relatively easily altered for the task. She returned to service in 1852, but the modifications had not gone far enough. After the first voyage, it was clear that her masts would have to be altered again, so that she was now converted from a steam ship reliant on occasional use of sail, to a ship in which steam power was used only as an auxiliary to sail power.

Returning to England in February 1855 the ship was immediately hired as a troopship for service in the Crimean War (1854-56). For the next 14 months she carried British, French, Sardinian and Turkish troops to the war zone and back. The financial rewards

Great Britain *at Port Stanley, late 1960s (Bristol United Press).*

had been excellent and her owners seized the opportunity to rebuild the ship. She emerged with an additional upper deck, three massive square-rigged masts and a screw that could be hoisted out of the way inside the ship to reduce drag when the winds were good. In this form she proved ideal for the Australian run, carrying 600 passengers and enough sail to make a speedy 15 knots through the Roaring Forties between Cape Town and Tasmania, combined with an efficient steam plant to master the windless doldrums. In this service she would carry heavy loads, and travel vast distances. This period was probably her heyday as a successful passenger liner, although it is often mistakenly undervalued in standard histories. The *Great Britain* was the crack ship on the route to Australia and her success marked the start of the quantum leap in global communications that has seen the world seem to shrink rapidly.

After only one return trip the *Great Britain* was recalled for further war service, carrying troops to Bombay during the Indian Mutiny. Then from 1858 to 1875 she made 26 round-the-world trips from Liverpool to Melbourne, passing both the Cape of Good Hope and Cape Horn – an awe-inspiring record of service for a ship of her age and novelty, made all the more striking by the failure of newer ships to match her performance or her durability. Brunel's masterpiece had pioneered the Atlantic crossing and now opened the furthest ends of the globe to steam power. On this demanding circular route she carried thousands of emigrants, the first touring England cricket team and much else besides. Her homeward cargoes included gold, wool, cotton and wildlife. Then, in 1876, the old ship was refused the necessary classification to continue carrying passengers by Lloyds Register of Shipping, but her

Selection of images during return from the Falklands to Bristol, 1970 (Bristol United Press).

stout hull and massive rig saw a third return to service six years later, this time as a pure sailing ship on the Liverpool to San Francisco route. There was a certain irony to her career, starting as a radical, full-powered steamer, becoming a steam auxiliary and then a pure sailing ship. Her latter days as a sailing cargo ship – a 'windjammer' – have often been dismissed as a sad ending of her working days. In fact, as a bulk cargo sailing ship she was one of the earliest of a new generation of large sailing ships that came to dominate certain trades and routes in the early twentieth century. These great iron and steel windjammers are now treasured and preserved as representatives of the last golden age of sail.

This time her luck ran out. The long, dangerous route round Cape Horn finally caught up with her in 1886. She lost her fore topgallant masts in a gale off the Cape and put back to Port Stanley in the Falkland Islands, the nearest port of refuge. In such an isolated spot the damage proved to be beyond economic repair. Once again her stout construction, allied to the sheer isolation of the Falklands, saved her from the destruction that befell Brunel's other two great ships, both broken up in Britain. She was soon turned into a wool store afloat in Port Stanley harbour. There she remained until 1937 when, after a first attempt to salvage and return her to the UK failed to secure all the funds required, she was towed into a remote cove and scuttled. The ocean and the elements had done their worst, but this remarkable ship seemed to resist the passage of time, to be re-discovered in the 1960s, part of the growing appreciation of Brunel. A small group of experts and enthusiasts led by Dr Corlett and Eric Gadd realised that the greatest ship in modern history was still in existence, if no longer afloat. She had been saved by the fact that she was 8,000 miles from home.

Recovery and restoration

A massive effort to recover the ship, bring her home on a barge and return her down the Avon River to re-enter her original building dock culminated on 19 July 1970, the 127th anniversary of her floating out. This most dramatic of voyages caught the imagination of the world, and massive television and press coverage gave the old ship something of the celebrity she had enjoyed in her early days. Many thousands

Selection of images showing early restoration work on the ship, early 1970s to early 1980s (Bristol United Press).

attended her launch in Bristol in the summer of 1843; a similarly vast number are thought to have welcomed her back up the river 127 years later.

The ss *Great Britain* Project was formed to pursue the restoration of the ship and place her on display, but not necessarily in Bristol. The Corporation of Bristol City (later Bristol City Council) was not at all convinced that historic ships and visitor attractions were the way forward for the commercial docks, even though it was clear that their economic viability as docks was drawing to a close. It was to take 20 years for the city to understand the huge asset it had in the Floating Harbour and the docks for leisure and tourism, with the *Great Britain* at its heart. Meanwhile, a short lease of the building dock gave the project only six months' notice on which to remove the ship to another port. The lease allowed the city council to plan a new motorway bridge on the dockyard site. Intra-city motorways may have been the spirit of the age, but the city and its residents are fortunate that these dramatic plans did not come to fruition. The ship remained, and traditional restoration sought to return her as far as possible to her external visual appearance when she was first launched by Prince Albert.

Good progress was initially made as far as funds and gifts in kind allowed, but by the early 1990s a funding shortage brought the original 'restoration' programme to an end. Fortunately, the creation of the Heritage Lottery Fund provided new and strong

support for delivering the goals of the charity that owns the ship and is responsible for her preservation and her presentation to the public. The ship trustees took the bold decision to become a Registered Museum, with the ship as the main object on display, and to seek to conserve her to the highest standards. A new programme of conservation, interpretation and education for visitors in an exciting and fun environment was required.

At about the same time, good relations with Bristol City Council allowed the whole of the original Great Western Dockyard around the ship to be brought back under one lease held by the charitable trust. This was a significant step. The surviving elements of the shipyard created by the Great Western Steamship Company, including the company offices and drawing office, the dry dock itself and the remains of the large marine steam-engine works, are now the subject of discussions with English Heritage about possible World Heritage Site status, in association with the GWR line from Paddington to Bristol.

On the back of a new conservation plan, a review of future work on the ship was carried out. Its conclusions led the trustees to take the ship in an innovative direction for her preservation and the enjoyment of her many visitors well into the twenty-first century. The plan revealed the wealth of material within the ship that related not only to Brunel but also to the later phases of her long working life. In particular, the material from the Australian voyages was identified as a real treasure. All this strongly suggested the need to move away from the old assumptions about 'restoration' and 'taking her back to the original'. To regard the form of the ship at one fixed point early in her history as the original and the only one of value is now recognised as simplistic. It was clear that the whole of the ship was original, and that the multi-layered additions and changes that she underwent through her working lifetime were all potentially significant for telling the story of the past. In that sense, the ship is like an English cathedral where there is great danger in pursuing restoration back to a single time too fixedly and throwing out much of real value in the process. Fortunately, the original plan to restore the ship to a fixed date had not gone so far as to frustrate a more multi-layered approach. Today, material from all periods of her working life is treasured, right up to her return to Bristol in 1970, and its preservation is vital to the preservation of the ship as a whole.

The creation of the Heritage Lottery Fund, with a clear remit to support projects for the preservation of the national heritage, chimed well with the new approach, and a major programme of preservation and development has been instigated with the Fund's assistance. Following the conservation plan, a detailed condition report revealed the state of the iron structure and concluded that severe corrosion was continuing at an unsustainable rate. This corrosion stems from the action of the sea water salts within which the hull has effectively been pickled for much of its life. The trustees did not want simply to replace the defective plating in the commonplace manner of a shipyard-style restoration, since this would lead merely to the eventual creation of a replica ship by default, and the loss of the original ship. Nevertheless, the effective removal of chloride contamination from wrought iron is extremely difficult. The only viable solution for long-term preservation of the ship without damaging the surviving ironwork proved to be the provision of a controlled, low-humidity environment around the hull. Research has demonstrated that this environment is most likely to arrest the chloride corrosion activity almost indefinitely.

To create such an environment requires a sealed envelope around the ship. Fortunately, the vast majority of the chloride problems occur in the lower hull. Therefore, the trust has built a flat glass sheet across the dry dock with water running over it so that the ship appears to be afloat, but underneath the glass, and through the interior of the ship, a dry environment will stop the corrosion while still maintaining full access for visitors. Coupled with significant enhancements to the visitor interpretation of the ship and dockyard and a new museum, the whole makes an exciting development that should secure the long-term future of the ship for everyone to enjoy.

Conclusion

To have built the *Great Britain* would have been claim enough for most men, but Brunel was not most men. He had entered the fast-moving, chaotic world of steamship design just as the technologies his father had pioneered came to fruition. His superior education and all-round engineering vision were critical to the success of the steamship. It was no accident that he built the first true intercontinental steamship and the first ocean-going, iron-hulled screw propeller. Both were the largest ships afloat when completed. That he moved on from the *Great Britain* to build something twice the size of anything yet conceived reflected his unique vision of modern transport. In the process he applied the fundamentals of engineering to a specific problem, transforming a collection of ideas, novelties and concepts into highly effective, scientifically exact ships. In 1840 Brunel conceived the iron-hulled, oceanic screw steamship as an integrated machine: the ship, her structure, form and machinery were combined to create a harmonious and efficient vessel. Brunel could see the value of each part of his design in perfecting the whole. The *Great Britain* was widely imitated, but no-one would ever make such a bold leap in the development of ship design again.

Every iron or steel ocean-going commercial vessel owes its genesis to this one ship. She is the beginning of the modern world and she was born in Bristol. The survival of this epoch-making ship, in her original building dock, alongside the remains of the factory where the engines were made, is unique. The ship and the surrounding facilities give the modern visitor an opportunity to see one man's genius in the original metal.

Bow of the Great Britain *from under the glass sea, 2005 (ss* Great Britain *Trust).*

ERIC GADD MBE AND DR EWAN CORLETT

A leading advocate for the return of the ss *Great Britain* to Bristol was Eric Gadd, a self-employed master printer and former commodore of the Cabot Cruising Club with sidelines in taxi driving, after-dinner speaking and hospital radio. Eric was also Captain Courage on Radio Bristol for many years, telling stories for children on a Saturday morning.

He used to take school children on tours of the docks and give talks about Bristol. One day, a child asked where the *Great Britain* was now: Eric did not know but promised he would find out. From this initial conversation grew Eric's involvement in the Brunel Society and the successful campaign to bring the ship home.

The locally-based society was formed in February 1968, its members being drawn from across the country. It aimed to ensure that the work of Brunel and his father was preserved, to promote research and to foster the men's imaginative spirit. Its immediate objective was to return the *Great Britain* from the Falklands, and a separate committee was established to take responsibility for this initiative. Sir Humphrey Brunel Noble, great-grandson of Isambard Kingdom Brunel, was the society's first president. Founder member Eric Gadd was its vice-chairman.

In September 1969 a hospital patient bet Eric £5 not to have his hair cut until the ship returned. Accompanied by a guard of honour of Sea Cadets, his shoulder-length hair was finally cut on 7 July 1970, the £5 donated to the Hospital Broadcasts Society and the locks sold at 5s

each to aid the ship fund. Among his other promotional activity, he wrote the lyrics of a song recorded by pupils at St Michael's School as part of the Tiddlers campaign to raise funds for an original *Great Britain* print to be presented to Jack Hayward, who had financed the cost of bringing the ship back. For several years he marked Brunel's birthday by presenting red carnations to the *Great Britain* staff and the toll collectors at the Clifton Suspension Bridge.

Eric Gadd was awarded an MBE in 1988 for his charity work.

Another key figure instrumental to the success of the campaign was the versatile naval architect, Dr Ewan Corlett. His interest in the *Great Britain* was triggered when he received the gift of a print of the ship on his leaving British Aluminium Company in 1953 to set up his own consulting company.

In 1967 Dr Corlett wrote to *The Times* drawing attention to the ship's plight in Sparrow Cove and highlighting her national and international importance. His letter sparked the interest of Richard Goold-Adams, an influential writer and journalist, Basil Greenhill, the director of the National Maritime Museum, and a number of MPs. Two years later, accompanied by Lord Chalfont, Dr Corlett surveyed the ship, establishing the feasibility of her salvage. He was on board when she returned up the Avon to her original dock in Bristol and continued to provide technical advice on her conservation and restoration.

Dr Corlett wrote many papers and articles on naval matters and his book *The Iron Ship*, published in 1975, provided a detailed insight into the *Great Britain*'s technical history. He was a trustee of the National Maritime Museum, was a member of a number of distinguished institutions and was awarded an OBE in 1985. He was particularly proud of the award he received in March 1996 from the World Ship Trust in recognition of his work in saving the *Great Britain*.

On his retirement in 1988, Dr Corlett took up a vocation in the church and was ordained as a priest in 1992.

Eric Gadd with his decorated van used to promote the campaign, c 1970 (Port of Bristol Authority).

SS GREAT WESTERN

Brunel's friend, Thomas Guppy, a Bristol manufacturing engineer and business man who had been instrumental in establishing the Great Western Railway Company, brought together the partners to form the Great Western Steamship Company. This followed Brunel's half-jocular remark at a GWR board meeting where they had discussed the length of the line between London and Bristol: 'Why not make it longer, and have a steamboat go from Bristol to New York…?'

The ss *Great Western* was Brunel's first ship-building project. She was an oak-hulled paddle steamer, the largest steamship built to date and the first to provide a regular transatlantic service, heralding a new era of ocean-going transport. She was built by the firm of William Patterson, a leading Bristol shipbuilder, who gave additional strength to her timber hull by using copper sheathing, iron bolts, and wood and iron trusses. Her two-cylinder engines were provided by Maudslay, Son & Field in London.

The ship was launched on 19 July 1837 from Wapping Wharf and sailed to London for fitting out in August. During her return trip to Bristol, fire broke out in the boiler room and Brunel was injured when he fell 18 feet through a hatchway from a burning ladder. His fall was broken and his life saved by Captain Christopher Claxton, managing director of the company, who had been hosing down the deck. News of the fire, exaggerated in the telling, resulted in 50 of the ship's passengers cancelling their booking for her maiden voyage.

The Great Western Steamship Company's main competitors in the race to develop a transatlantic steam line were the British and American Steam Navigation, a London-based group led by an expatriate American businessman called Junius Smith, and the Transatlantic Steamship Company of Liverpool. As a publicity stunt, Smith chartered the *Sirius*, a channel steamer, to cross the Atlantic in competition with the *Great Western*. The *Sirius* started her journey in Cork on 4 April 1838. The *Great Western* left Bristol on 8 April. She arrived in New York 15 days later, only 12 hours behind *Sirius* despite the rival ship's four-day head start and shorter route, and with over 200 tons of coal to spare. A sailing ship would have taken a month to make the journey.

The *Great Western* made 74 crossings of the Atlantic during her career. She was later purchased by the Royal Mail Steam Packet Company and for ten years operated out of Southampton on the West Indies run. She was broken up at Millbank in 1857.

Launch of the Great Western *from William Patterson's Yard, 19 July 1837, painting by Arthur W Parsons, 1919 (Bristol Museum, Galleries and Archive).*

NURSING AND THE CRIMEAN WAR

In March 1854 Britain and France declared war on Russia in support of Turkey, who had been fighting with its European neighbour since October 1853 following years of unresolved disputes. Sardinia was already fighting alongside the Turks. The conflict was sparked by concerns that Russia was looking to expand its empire by reclaiming Turkish territory in the Crimea on the Black Sea. For the first time in an overseas conflict, the British public had ready access to front-line news, as reports were despatched by telegraph. *The Times* correspondent, William H Russell, spent nearly two years with the troops and Roger Fenton, who went out in 1855, was the first official British war photographer.

In his reports, Russell drew attention to the inadequacies of the medical provision for the soldiers. Among the new measures taken in response to this coverage and other unfavourable reports, Brunel was asked by the War Department to design a pre-fabricated hospital that could be sent out in sections to the

Photographic van used by Roger Fenton during his documentation of the Crimean War, 1855 (National Museum of Photography, Film and Television/Science and Society Picture Library).

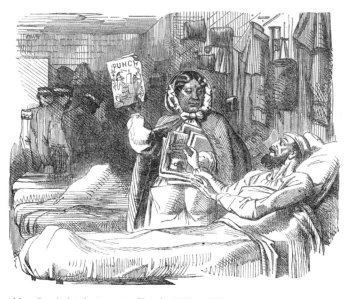

Mary Seacole distributing copies of Punch, *30 May 1857 (Punch Cartoon Library & Archive).*

Crimea. He replied that his 'time and best exertions would be, without any limitations, entirely at the service of Government'.[1] He took up the challenge on 16 February 1855 and in early March an experimental ward was set up near Paddington to test the feasibility of Brunel's plans. The extensive complex of buildings required for the hospital had to be adaptable to any plot of ground, be capable of being easily extended, contain every comfort possible under the circumstances, and be highly portable and economically constructed. In all, 23 ships were used to carry out the pre-fabricated sections along with the stores and other materials needed to erect and equip the hospital, the first arriving in May 1855. It was set up at Renkioi in the Dardanelles, under the supervision of John Brunton, and was ready to take in 300 wounded by 12 July 1855. Its capacity had reached 500 by August and 1,000 by December. When its work was discontinued in March 1856, it had the potential for admitting over 2,000 patients.

Two nursing heroines who had defied convention emerged from the war: Florence Nightingale and Mary Seacole. Nightingale had faced years of parental

Ambulance at Sebastapol and soldiers washing at Balaclava hospital, The Illustrated London News, *2 June 1855 (University of Bristol).*

Florence Nightingale at Balaclava, The Illustrated London News, *23 June 1855 (University of Bristol).*

opposition to her ambition to be a nurse before she was finally allowed to undertake basic training in 1851. In 1853 she was appointed superintendent (unpaid) of the Institute for the Care of Sick Gentlewomen in London. The following year she was invited to take a team of nurses to the Crimea by Sir Sidney Herbert, the War Minister. Her team was based at the military hospital in Scutari in Asian Turkey, where injured soldiers were brought from across the Black Sea for treatment. During one of her visits to inspect the hospital facilities nearer the front line, she contracted typhus, which left her physically weak for the rest of her life. On her return to Britain at the end of the war, she was involved in the Royal Commission set up on her recommendation to review health provision for the British army and in 1860 she established the Nightingale Training School for Nurses.

Mary Seacole (born Mary Jane Grant) was the daughter of a freed black slave and a Scottish soldier. She grew up in Jamaica, where her mother ran a boarding house for invalided British soldiers. She acquired her medical knowledge from her mother and set up her own boarding houses in Jamaica and Panama. In 1854 troopships began to leave Jamaica, carrying soldiers to the Crimea. Seacole came to London, offering her medical services, but was rebuffed, presumably because of her colour. Undeterred, she went to the Crimea at her own expense and set up a hotel outside Balaclava. Soldiers came to the hotel for hot meals, clothing and other supplies, while the injured were treated at a small hospital upstairs. Seacole also followed the battles, riding on a mule, to treat the injured on the frontline. When she returned to Britain she faced bankruptcy, but was saved by a public appeal that was supported by the national press.

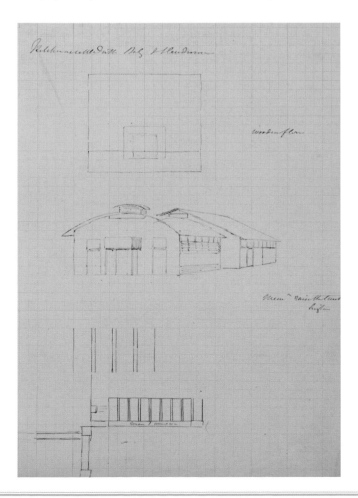

Sketch and plan for Brunel's pre-fabricated hospital, 1855 (University of Bristol).

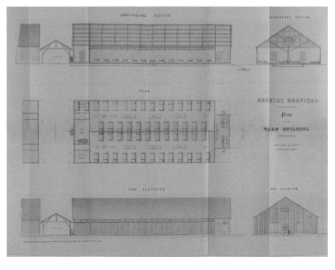

CRIMEAN TROOPSHIP

A selection of views of the ss Great Britain *made while she was serving as a troop-carrier in the Crimean War, 1854-56 (ss* Great Britain *Trust).*

SS GREAT EASTERN

The ss *Great Eastern* was conceived as the biggest steamship yet to be built, one that would be capable of carrying 4,000 passengers at a time on a non-stop trip to Australia. Brunel was partnered in the venture by John Scott Russell, who he had met during preparations for the Great Exhibition of 1851. Their proposal was offered to the Eastern Steam Navigation Company who were looking for new business interests. Brunel placed a number of his own nominees on the company board and contributed considerable personal sums to the capital needed. Russell's engineering firm was awarded the contract to construct the ship and work began at Napier Yard in Millwall in 1853.

The ship, originally called *Leviathan*, was nearly 700 feet long, six times the size of the largest ship built to date. Brunel designed her to be unsinkable, extending the watertight double plating to five feet above the ship's deepest load line. In addition, iron bulkheads divided the ship into ten watertight compartments. To move such a large ship required two engines, one attached to a screw propeller built by James Watt & Co, the other used to drive two enormous paddle wheels built by Russell. The ship was constructed parallel to the riverbank with the intention of easing her sideways down specially constructed slipways and floating her off on high tide: her size meant a conventional launch was impossible.

Robert Howlett photograph of the Great Eastern *showing the chains used to control the launch, 1857 (Institution of Civil Engineers).*

Illustration by Robert Dudley of Great Eastern *during the laying of the Atlantic cable, 1865 (Institution of Civil Engineers).*

On 4 February 1856 Russell was declared bankrupt, having kept the precariousness of his financial situation hidden from his partner. Work came to a halt and the accountants and bankers argued over the fate of the unfinished ship. With considerable effort, more money was raised and Brunel assumed overall control of the project. However, progress was slow as he struggled to win over Russell's men, and there were endless contractual arguments.

Under threat of another bankruptcy should there be more delays, there was an attempt to launch the hull on 3 November 1857 using untried and inadequate equipment. Brunel had originally planned to use hydraulic launching gear specially built for the project, but because of the financial difficulties this was judged by the company to be too expensive and he had to use steam winches and capstans worked by teams of men. To further complicate an already risky procedure, the company, seeking to make money, had sold tickets to the public to attend the launch event and the yard was overrun with onlookers, getting in the way of Brunel's system of signals. The launch was a disaster and the attempt abandoned.

After further unsuccessful attempts made later that month and into December, new hydraulic rams were

Illustration from Jules Verne's A Floating City *(1871), a novel based on his journey to New York on the* Great Eastern *(Nikky and René Paul).*

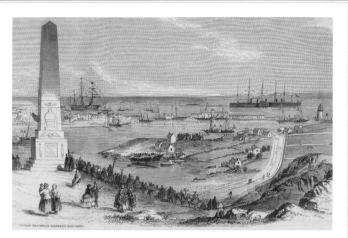

Arrival at Holyhead, The Illustrated London News, *22 October 1859 (University of Bristol).*

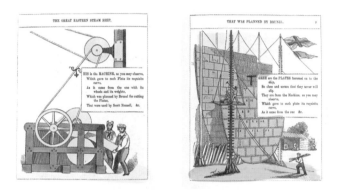

From The Wonderful Ship that was Planned by Isambard Kingdom Brunel, *children's verse book, c 1858 (private collection).*

ordered. The ship finally floated on 31 January 1858. She was moved to Deptford for fitting out and a new company, the Great Ship Company, was formed in November 1858 to undertake completion of the work.

Sea trials were planned for 7 September 1859 but were postponed for two days. In the rush to be ready, two temporary stop cocks fitted to the heaters on two of the ship's funnels were left in place, causing a devastating explosion that killed five men and injured several others. News of the disaster is thought to have hastened Brunel's death.

The *Great Eastern* proved uneconomic as a passenger ship but had a new lease of life when she was used to lay the first successful transatlantic cable, at the suggestion of Brunel's friend and colleague, Daniel Gooch. The ship first set out from Ireland in July 1865 on behalf of the Telegraph Construction and Maintenance Company. On 2 August, the cable end was lost overboard and the attempt abandoned. A new company was formed – the Anglo-American Telegraph Company – and a fresh attempt with an improved cable reached Newfoundland on 27 July 1866. John Gordon Steele, in his account of the cable laying, wrote:

> It is by no means the least of her ironies that the one thing the *Great Eastern* turned out to be good for – indeed perfect for – was perhaps the only use her creator had not foreseen from the beginning.[1]

Portrait of John Scott Russell (University of Bristol).

Great Eastern *commemorative medallion (Elton Collection: Ironbridge Gorge Museum Trust).*

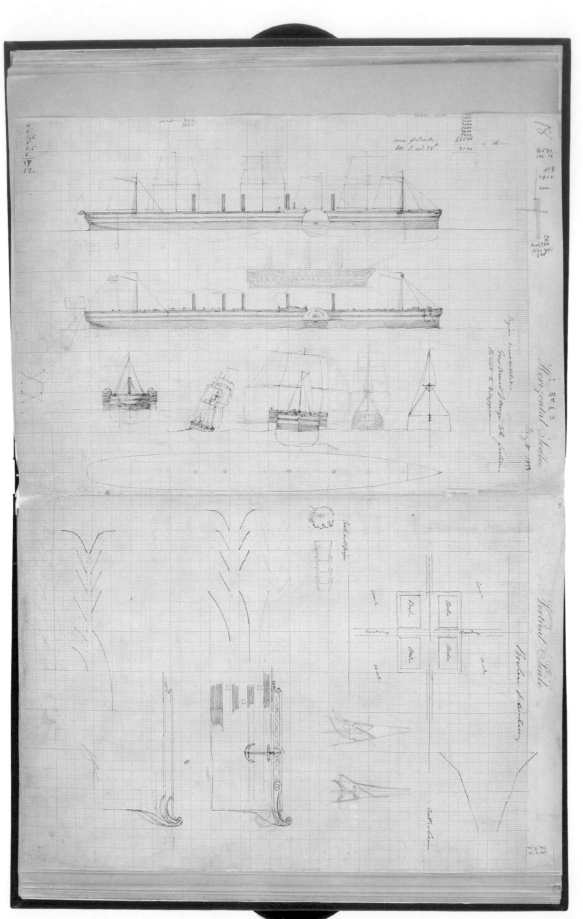

Large sketchbook 1852/4 showing early proposals for the Great Eastern *(University of Bristol).*

THE BRUNEL COLLECTION, BRISTOL UNIVERSITY LIBRARY

Nick Lee

In 1950 the University of Bristol received a gift of great significance, especially to engineers and historians of technology. The Library's Diary of Manuscript Accessions records it flatly enough '8th December, 1950: DM 162. Account books, sketch books, diaries and letters of I.K. & H.M. Brunel. 7 boxes received from the Dowager Lady Noble, 23 Royal Crescent, Bath'. Celia Brunel Noble was the grand-daughter of Isambard Kingdom Brunel and the author of a biography still used today containing information which can be found in no other source, *The Brunels, Father and Son*.[1]

The boxes were lightweight wooden document chests used in her grandfather's office and home, 17-18 Duke Street, Westminster. Three of them are still used to house plans and drawing instruments (some by Gardiner of Bristol) once owned by Brunel and *his* son, Henry Marc, and engraved with their names or initials. News of this astonishingly generous gift seems to have been received quietly enough by the city which had been the scene of so many of Brunel's activities.

Something of the excitement no doubt felt at the time, though absent from the record, can be detected in an article written 25 years later by the librarian, Norman Higham, 'Brunel: the Man and his Manuscripts'.[2] This conveys a little of the drama of Brunel's life, the extraordinary fertility of his invention, and the breadth of his interests:

> The sketch books show the creative mind at work; the fifteen volumes of letter books and the numerous original letters contain a record of the organisational effort which was a necessary accompaniment to his creative energy.

What most impressed Norman Higham was the *variety* of material present in the sketch books, and it is true that they provide an incomparable overview of his engineering work, from the Italianate Hungerford Bridge over the Thames to the brilliant prefabricated hospital at Renkioi. But further inspection reveals some surprising gaps: for instance, several sketchbooks are devoted to the *Great Eastern*,

*Brunel's drawing instruments
(University of Bristol).*

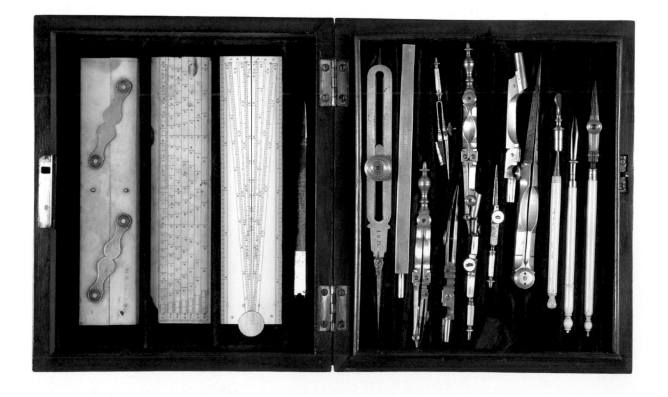

while there is next to nothing on the *Great Western* and surprisingly little on the *Great Britain*, though in the latter case there is at least the compensating presence of the ship itself only a mile away. There is also less than expected on the Clifton Suspension Bridge. Another absentee is the Royal Albert Bridge, Saltash, Brunel's most striking iron bridge, still working well and carrying loads undreamed of at the time it was

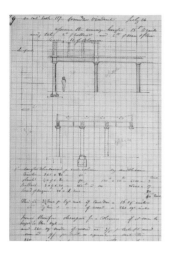

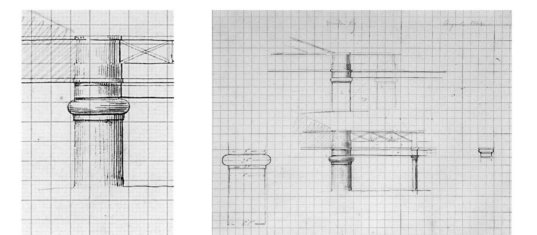

Calculations and sketch
(with detail) for Windsor Viaduct
(University of Bristol).

built. With regard to the influential bow and string girder bridge over the Thames at Windsor, there is plenty of material on the adjacent viaduct and nearby station, but only a small sketch of a two-span bridge of conventional form, quite unlike the one eventually put up. Fortunately, all three bridges are well described and illustrated in William Humber's *A Complete Treatise on Cast and Wrought Iron Bridge Construction*.[3] What, in fact, the sketchbooks document most fully and graphically is the Great Western Railway in all its aspects, the best conceived and executed railway of its time.

The Brunel papers were not allowed to gather dust. L T C Rolt was soon using them to provide information for his remarkable biography, *Isambard Kingdom Brunel*.[4] He used other sources too, including the papers still at Walwick Hall, others at the Institution of Civil Engineers and those (now in the National Archives) that Lady Celia's husband, Sir Saxton Noble, had already donated to the Great Western Railway, but much of his book was based on the collection at Bristol. There was no catalogue and little space in which to work on the documents, but Rolt was helped by George Maby, who would later become the library's first full-time archivist, producing the catalogue and précis which is still in use today.

Soon Bristol University Library began to attract gifts of material about Brunel and his achievements. The first such record I can find is of 8 April 1960 when Mr A P Bolland gave 49 letters from Brunel to Charles Alexander Saunders, Secretary to the Great Western Railway. On 15 July 1965 Mr J Hollingworth donated a copy of the fire insurance policy for the *Great Eastern*, and a week later the library bought an 1851

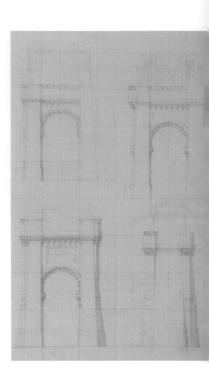

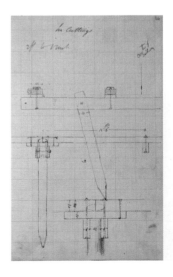

engraving of the Hungerford Suspension Bridge. Joan Eyles, who was later to bequeath to the University the best private geological library in the country, then added *Sketches and Memoranda of the Works for the Tunnel under the Thames*, and Mrs J Holme gave the tinted portrait photograph of Thomas Richard Guppy which now hangs in the Special Collections Reading Room beside an engraving of Brunel at his desk. Guppy was Brunel's friend and colleague on the *Great Britain* and one of the promoters of the Great Western Railway.

A start was made on building up a small library of books, pamphlets and prints which, though not strictly part of the collection, nevertheless shed welcome light on Brunel and his activities: William Humber's work on iron bridges, John C Bourne's *The History & Description of the Great Western Railway*, Isambard Brunel's *The Life of Isambard Kingdom Brunel, Civil Engineer*, and John Scott Russell's monumental *The Modern System of Naval Architecture*; smaller items, too, such as Christopher Claxton's *History and Description of the Steam-ship Great Britain*, James Pim's *The Atmospheric Railway* and the *Leviathan* number of *The Illustrated Times*.[5] These are sometimes more informative than the surviving documents, and are used on a daily basis. At the same time an unobtrusive but helpful relationship between archivist and Brunel scholar began to grow up, whereby unidentified sketches were recognised, calculations interpreted, and Brunel's difficult handwriting deciphered, and this continues.

In 1972 came the first of two important loan deposits by the Clifton Suspension Bridge Trust of the minute books, letter books, registers, large working drawings and other documents in their possession. This was a welcome gesture of confidence in the library's ability to look after its manuscripts in a professional way and provide a helpful

Bristol hotel, piers for the Clifton Suspension Bridge, 1835 and Saltash calculations from sketchbooks (University of Bristol).

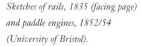

*Sketches of rails, 1835 (facing page)
and paddle engines, 1852/54
(University of Bristol).*

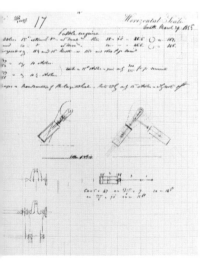

service and congenial environment for those wishing to consult them. It was easier to do this after 1976 when the library moved into a new building, taking with it most of the archives and special collections of books. This has a suite of rooms more conducive to study, and the archivist became more accessible. The Brunel Collection enjoyed a second spring. It was at about this time that an important new book on Brunel was published jointly by the University and the Institution of Civil Engineers, *The Works of Isambard Kingdom Brunel: an Engineering Appreciation* edited by Sir Alfred Pugsley.[6] This was based on the Bristol Collection and provided a real attempt to understand the engineering problems and achievements of Brunel.

In 1984 the library bought a presentation copy of the survey plan and sections of the Bristol & Exeter Railway, 1836. Brunel was the engineer of this, although the day-to-day responsibility fell on his old colleague on the Thames Tunnel, William Gravatt, with whom he later picked a quarrel. Brunel wrote:

> This survey was done in grand style – it's a good line, too – and I feel an interest in connection with Bristol to which I really owe much – they have stuck well to me... Gravatt served me well in this B & E survey.[7]

In 1987 the library was able to purchase sets of large working drawings for the Bristol Joint Station and the South Wales Union Railway Doubling, with the contract and drawings for a section of the Great Western Railway between Bristol and Bath. The last-mentioned are all signed by Brunel and by the contractor, David McIntosh, and served as evidence in a lawsuit that was not settled until long after both men were dead. Letters of a much earlier date, about Brunel's apprenticeship in France, were also bought, together with fascinating inventories of Brunel's home and office in 1858 and 1871. Most of these purchases were only made possible with the generous aid of outside bodies, in particular the Elton Memorial Fund and the Science Museum Purchase Fund as well as the University of Bristol Alumni Foundation. Meanwhile, other GWR material had arrived by a less orthodox route: a South London car boot sale. This included three bundles of correspondence, one concerning a troublesome section of the railway through Acton and Ealing (the Hanwell Embankment), just the sort of day-to-day business that gets thrown out at the end of the job and lost to engineering history.

Then, in 1990, there came an opportunity that could not be missed – to purchase two trunks full of papers that had remained in the possession of the Brunel family. They had been made available to Rolt in the 1950s but had disappeared from public view in the intervening years. Now, through the good offices of Bertram Rota and Elton Engineering Books, these papers were once more available. Serious money had to be found, and there were nail-biting moments ahead for Norman Higham and the Deputy Director of the National Heritage Memorial Fund. Eventually, with the help of generous grants from the Fund and from the J Paul Getty Jr Charitable Trust, the

Wolfson Foundation, the Pilgrim Trust and the Dulverton Trust, the University was able to purchase this great additional series of papers.

I can still remember the feeling of mounting excitement when, in the vault of a bank in Piccadilly, with the help of Julia Elton, I checked the content of the trunks. There was material from the beginning of Brunel's career which I recognised from the pages of Celia Noble's and Rolt's biographies – the personal and private diaries and some family letters – but there were other things that were unfamiliar: sketch books, two very full general notebooks, letters from Robert Stephenson and William Armstrong, and others from Brunel to Guppy, some about the *Great Britain* and looking more like a design consultation than was at all usual with him.

It was instructive to see Brunel emerging, not as superhero but as an engineer among other engineers. The letters from Robert Stephenson were particularly interesting in this regard, revealing the assiduously maintained friendship between the two, their professional rivalry, their differences over railway gauge and each man's attempt to be present at anxious and critical moments faced by the other.

There was also what could be the fullest account of any major nineteenth-century engineering project – a mass of papers on the *Great Eastern*, in some respects Brunel's greatest achievement and certainly viewed as such in his memoir.[8] GWR material included the two parliamentary bills with Brunel's manuscript notes on the summing up of the first and comments on the printed application for the second. The contents

Contractor's drawing used as evidence in lawsuit (University of Bristol).

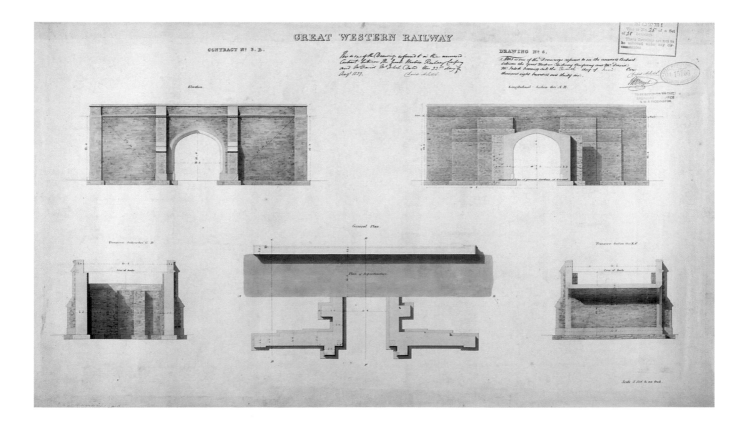

also included a bundle of Henry Marc Brunel records and an important collection of material on the Thames Tunnel containing a fascinating 'general remarks' volume to which both Sir Marc Isambard Brunel and his son contributed.

Sydney Gardens, Bath from GWR sketchbook (University of Bristol).

At about this time we were able to make a number of smaller purchases of individual items to fill in some of the gaps in the collection, perhaps the most interesting of these being three engineer's reports on the controversial South Devon Railway (the 'Atmospheric caper'), revealing growing difficulties with supply and with the airtightness of the valve which sealed the vacuum tube. One of these is heavily altered in Brunel's hand. In 1992 we also bought an album of beautiful working drawings from the GWR Drawing Office entitled Tools &c. Some of the drawings bear notes or captions by Brunel, while others have a breakdown of manufacturing cost. The dates range from 1837 to 1840. We thought of the album as a chance survivor of a neglected aspect of railway history, but a dilapidated volume of similar appearance has just turned up at an auction where it attracted little interest apart from mine. It contains 29 drawings for the construction of the broad gauge railway: switches, rails, points, traversing frames, crossings and permanent way, one of them initialled 'IKB' and dated 1839.

What I took to be the last major cache of documents owned by the Brunel family was auctioned at Christie's in 1996. Many of these went to private collectors, and the fact that we were able to buy as much as we did is attributable to funding by the Friends of the National Libraries (from the Heritage Lottery Fund), the Clifton Suspension Bridge Trustees and the University's Campaign for Resource. Geoffrey Ford was the librarian responsible for these purchases and he did much to make them possible.

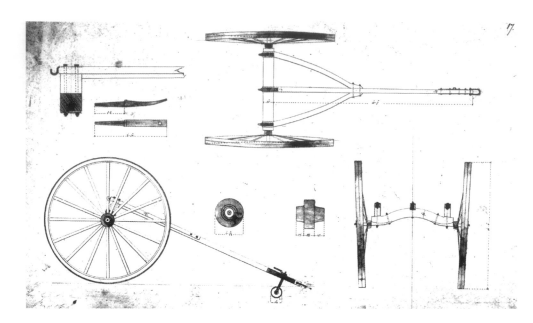

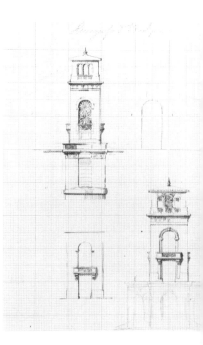

Drawing from the GWR tools volume, 1837-40 (University of Bristol).

One of the lots in which he was most interested, a set of four large drawings submitted by Brunel in the first competition for the Clifton Bridge in 1829, went to another buyer, but fortunately we were offered it subsequently, much to the delight of the chairman of the Suspension Bridge Trust, who felt that the drawings and other manuscripts had come back to where they belonged. They certainly have an iconic significance. Although Brunel did not live to see his bridge completed, it marked, according to his son Isambard:

> … a very important passage in the history of his life. Doubtless, if he had never heard of the proposed competition in 1829, or if he had been one of the disappointed competitors, he would have found some other opportunity of making a name in his profession; but, as a matter of fact, the Clifton bridge competition did give him the opportunity he desired, and all his subsequent success was traced by him to this victory, which he fought hard for, and gained only by persevering struggles. He never forgot the debt he owed to Bristol, and to the friends who helped him there; and he would have greatly rejoiced to see the completion of his earliest and favourite work.[9]

Among the autograph letters being sold were 14 from Robert Stephenson to Brunel, and we were fortunate to obtain 12 of these. As Bristol has the only collection of Henry Marc Brunel papers, it was important for us to acquire the journals of his tour to Egypt to accompany his terminally ill father in the winter of 1858/9. But the most significant lot was undoubtedly a large collection of 29 desk diaries, 20 pocket notebooks and 13 bank passbooks, which provide a great deal of information on Brunel's journeys, meetings, observations, income and expenditure during his busiest and most creative years. They take up the story at the point where the private diaries leave off and reveal the range of Brunel's undertakings, the extent of his contacts and the complexity of his finances. Notice has already been taken of them in Angus Buchanan's *Brunel: the Life and Times of Isambard Kingdom Brunel*.[10] The 976 pages of notes and sketches in the pocket notebooks contain a great deal of information on bridges, docks and railways, not all of which have yet been identified.

In 2003, unexpectedly, a few further manuscripts from the later part of Brunel's life turned up at auction. One of these, the Watcombe Garden Book, was already well known from Celia Noble's *The Brunels, Father and Son*. It offers a rather different picture of the great engineer enjoying planning the gardens of his proposed country house overlooking Babbacombe Bay, though with as much parade of style and rigorous attention to detail as ever. Alas, the garden was never to be enjoyed by him other than in anticipation.

This acquisition must bring to a close our brief history of the Brunel Collection at Bristol, described by Angus Buchanan as 'the fullest archive of Brunel material in the world' and 'a scholar's delight'.[11] But Buchanan also quite properly lists some of the other important archives, especially the two that (like us) have benefited from the Brunel family's instinct to preserve the memory of its most famous members: the National Archives and the Library of the Institution of Civil Engineers. No serious work could be done on the history of the Great Western Railway without consulting the former, and none on the Thames Tunnel or the atmospheric railway without visiting the latter. As examples of the mutual interdependence of the three collections, the National Archives hold six volumes labelled Facts (collections of notes and observations made by Brunel throughout his life), but there is a further volume at Bristol; the Institution of Civil Engineers holds all the surviving volumes of Sir Marc Isambard Brunel's Journal except two, which are at Bristol.

Hungerford piers sketch (facing page) from 1836 (University of Bristol).

Sketch of steamship to use in laying transatlantic cable (University of Bristol).

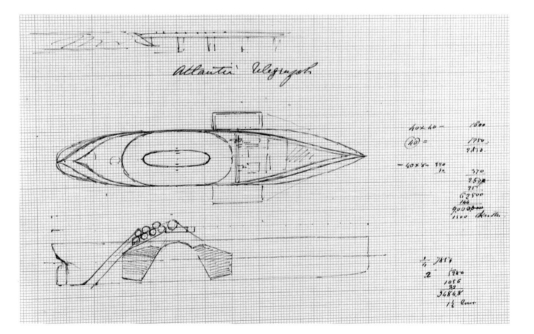

Apart from what is available in these collections and what is still in use within
Network Rail, we do now appear to have at Bristol most of the documents on which
Isambard Brunel drew for *The Life of Isambard Kingdom Brunel, Civil Engineer*, with the
exception of those used in chapters five (on the broad gauge), six (the atmospheric
system), seven (bridges) and 14 (docks), which were written, at least in part, by
William Pole, William Froude and William Bell, respectively. One intriguing
exception to this is to be found on p. 485, which quotes an entry from Mr Brunel's
private journals dated 5 May 1846: 'I am just returned from spending an evening with
R. Stephenson. It is very delightful, in the midst of our incessant personal professional
contests, carried to the extreme limit of fair opposition, to meet him on a perfectly
friendly footing, and discuss engineering points… Again I cannot help recording the
great pleasure I derive from these occasional though rare meetings.' This was the year
in which the Gauge Commissioners reported to Parliament and the Act for
Regulating the Gauge of Railways was passed, when Brunel and Stephenson were of
necessity on opposite sides in the dispute, but careful, as their correspondence shows,
to remain on good personal terms. Nothing has been heard of that journal volume
since 1870. The actual series of private journals comes to a stop after many gaps in

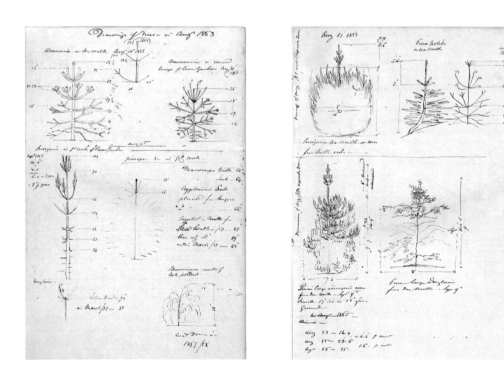

*Sketches of designs for shrubs
at Watcombe (University of Bristol).*

1840, with nothing apart from the desk diaries to record the remaining, frantically
busy 19 years. We do have a volume entitled 'Journal' which contains a few dated
entries beyond 1846, but it is not at all journal-like. Its title, with Isambard Brunel's
quotation, shows that Brunel *did* actually continue to keep such a record, as his father
had done before him, and it is a pity that it has not survived. One would like to know

Page from letter written by Brunel
in a draughty hotel room in Wootton
Bassett to his wife, Mary
(University of Bristol).

more of what passed between the old adversaries, and what it was, beyond professional etiquette, that kept them supporting each other until their too-early deaths in 1859.

Over the past 55 years many additions have been made to the Brunel Collection, several of them important, most of them unexpected. It will almost certainly continue to grow and be useful to more and more people. At present, the use made of it (about 600 consultations a year) keeps the archivist busy enough, but it stays at a manageable level because of the collection's relative physical inaccessibility. This could change if a recent proposal that it should be housed in a new Brunel study centre adjacent to the ss *Great Britain* goes ahead. The lack of an online catalogue has been another inhibiting factor. At present there is only a typescript catalogue, with a microfiche version produced for the *National Inventory of Documentary Sources: UK and Ireland*, though the collection is also described in outline on the Archives Hub and the University of Bristol Library website, and there is the ever-useful National Register of Archives which you *can* search online.

A promising recent development has been the advent of digital imaging, though this is still at an early stage. It is hard to predict how the existence of a virtual Brunel archive consultable online will affect the use made of the actual collection. Will it make consultation of the original documents redundant, or will it alert more people to their existence and lead to an increase in inquiries? During the past year digital images have been created from the sketchbooks and the individual letters, and we have applied for funding to extend this process to the rest of the collection. The images so far produced are of an impressively high quality, but they have not yet been made available for consultation. A sample sketchbook and letter have, however, been put on the web (www.brunelarchive.org). This is an important development, and one which should make the collection much more accessible.

Brunel, standing rear right, with Robert Stephenson (seated left) and others at attempted launch of ss Great Eastern, 1857 (National Railway Museum/ Science and Society Picture Library).

PROFESSIONAL COLLEAGUES

Michael R Bailey

Introduction

Isambard Kingdom Brunel was one of a generation of engineers whose innovatory talents, tenacity, and managerial and communication skills realised extraordinary engineering achievements in the mid-nineteenth century. The talent and skills of these pioneering engineers saw the development of railways, bridges, harbours, water supply, drainage and other public works in Britain and overseas. Their legacies were both tangible, in the form of extensive structures built in Britain and overseas, and enduring, with their development of the British civil engineering profession, which has been the foremost in the world since that time.

Civil engineering, which has served the world since pre-history, has evolved through the growing aspirations of each generation for social enhancements, including greater mobility and the economic stimulation of cheaper freight and mineral transport. The profession has developed to meet these aspirations and the challenges they bring. In the eighteenth century the expansion of goods and passenger transport was made possible by the construction of canals, harbours and lighthouses for water-borne movements; by road improvements, with better surfaces and bridges; and by rail wagonways for the transport of minerals. Major river improvement schemes also provided better land drainage and land reclamation in low-lying areas. Such projects were made possible by a growing understanding of soil mechanics and structural materials, matched by innovative construction techniques.

Such ambitious schemes were developed by a growing number of talented engineers whose leaders, such as John Smeaton, Thomas Telford and James Brindley, became household names. The end of the Napoleonic Wars in 1815 prompted a number of harbour and river improvement schemes to fulfil new trading opportunities, together with road improvements and new bridges to improve internal communications. Telford's major upgrading of the London to Holyhead Road, to reduce travel time between the British and Irish capitals, set new standards of understanding in civil engineering and was described at that time as 'a model of the most perfect road making that has ever been attempted in any country'.[1]

Institution of Civil Engineers

By 1818 the profession had developed to a stage where it was ready to gain from corporate organisation. The Institution of Civil Engineers (ICE) was founded, under the leadership of its first president, Thomas Telford, to develop contacts and exchange knowledge of best practice among its members. The receipt of its Royal Charter ten years later began a history of professional development that has continued to the present day. ICE's international standing remains high in the twenty-first century, with a worldwide membership of 75,000.

The earliest senior members of the Institution, such as James Walker, its second president, John Urpeth Rastrick, James Rendel and Charles Blacker Vignoles, would attend meetings at its premises in Great George Street, Westminster, and deliver papers setting out their experience in a wide range of schemes. The discussions after the papers, and the frequent 'conversaziones' that were held there, demonstrated a close dialogue between members, particularly with the incoming and innovative new generation of young engineers.

As early as 1830, the young and ambitious Brunel joined ICE in the footsteps of his father, Marc, who had himself joined seven years earlier, and went on to become its vice-president. Two other young men, Robert Stephenson and Joseph Locke, also joined the Institution in 1830. The membership of these three talented engineers coincided with the beginning of the great main line railway-building era. They were to rise to extraordinary renown through their leadership of the massive railway-building programmes from the 1830s to the 1850s, and the innovative bridges and other structures that made them achievable.

Joint portrait of George and Robert Stephenson by John Lucas, 1851 (Institution of Civil Engineers).

Steam engine and railway development

Steam engine development during the eighteenth century had been stimulated by the need of the mining industry to pump water out of mines being sunk to ever-greater depths in pursuit of mineral deposits. By the end of the century, the early work of the pioneering steam engineers, Thomas Newcomen and James Watt, had progressed to 'high-pressure' steam engines through the endeavours of the Cornish mining engineer, Richard Trevithick. His 'puffer' engines were also adapted for other industrial applications, increasing the competitiveness of ironworks, foundries and machine shops. Its adaptation to steam locomotion, first achieved at the beginning of the nineteenth century, heralded the evolution of railway traction that stimulated major reductions in mineral transport costs.

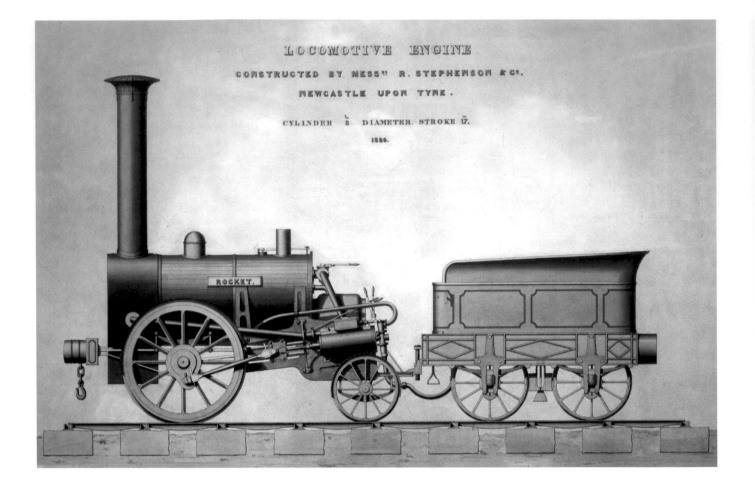

Building on the work of his fellow steam colliery engineers, George Stephenson understood the need for a co-ordinated approach to railways through route selection that was compatible with evolving track and motive power technology. The building of the Stockton & Darlington Railway in 1825, for mineral transport, presaged the introduction of main line railways, for the bi-directional movement of goods and passengers, just five years later. Stephenson pioneered the building of the first inter-urban route between Liverpool and Manchester, and his son, Robert, came to prominence with the successful development of steam locomotion demonstrated at the Rainhill locomotive trials in 1829, which were won convincingly by his *Rocket* locomotive. Thereafter, improving steam locomotive technology made possible the extraordinary development of railways in the early Victorian era.

The building of the Liverpool & Manchester Railway gave rise to new knowledge of soil mechanics and the innovative development of structures. The railway's nine-arch, 60-foot-high Sankey Viaduct, the 70-foot-deep Olive Mount rock cutting and the four-mile-long crossing of the infamous Chat Moss were just three major milestones in the development of route construction technology. George Stephenson was assisted by several young engineers, on whom he relied for innovation, energy and sheer hard work to complete these formidable tasks. Foremost was Joseph Locke, who quickly rose to become one of the nation's premier railway engineers. It was, however, Robert Stephenson who took on the responsibility for several other early railway schemes, demonstrating the talent that would make him the leading civil and mechanical engineer of his generation.

The success of the Liverpool & Manchester Railway led directly to the planning of more ambitious trunk routes between the nation's major centres of population, commerce and industry. These pioneering trunk routes were between London and Birmingham, for which Robert Stephenson was appointed as engineer-in-chief, and the Grand Junction Railway, between Birmingham and the northwest of England, for which George Stephenson, and later Joseph Locke, were engineers-in-chief.

Preparation of plans, levels and estimates for parliamentary scrutiny, against tight time constraints, gave rise to a dedication and tenacity not hitherto required in the profession. To undertake such large projects, the Stephensons and Locke recruited talented associates to assist with the extensive surveys and determine the cheapest route formations in accordance with the growing understanding of soil excavations and deposits. The associates who rose to this challenge included Thomas E Harrison, John Brunton and John Errington, each of whom went on to become a prolific builder of railways. They achieved extraordinary success in overcoming the parliamentary hurdles for these rail schemes before turning their attentions to the construction work itself.

Sankey Viaduct from an album of coloured views of the Liverpool & Manchester Railway by T T Bury, published by R Ackerman, 1831 (Institution of Civil Engineers).

The building of trunk railways was like a military operation, with its delegated responsibilities through a chain of assistant and resident engineers. They were required to stake out the ground, provide detailed drawings of bridges, viaducts, tunnels, stations and goods depots. They also supervised the contractors to ensure that work was undertaken in accordance with contracts, and that soil and rock excavations and deposits were in accordance with stated volumes. This was the recruiting ground for young engineers who became senior members of the profession in later years. They included George Parker Bidder, who became a leading consulting engineer and senior associate of Robert Stephenson, building many railways and other structures at home and overseas. Other engineers included Robert Dockray, who was appointed engineer of the London & Birmingham Railway and its associate companies, and Charles (later Sir Charles) Fox, who went on to become one of the country's leading structural and bridge engineers.

The alignments of the London & Birmingham and Grand Junction Railways were adopted in accordance with gradients and curvatures determined by the improving capabilities of steam locomotives. The resulting embankments, cuttings and tunnels were formed through a variety of soil and rock materials. Knowledge of the behaviour of these materials was gained as the engineers proceeded and, in spite of some early

J C Bourne's painting of Tring Cutting under construction in 1837 (Institution of Civil Engineers).

slippages with unstable materials such as London Clay, some embankments and cuttings were formed with remarkably steep slopes. The building work resulted in some of the country's epic formations and structures, many of which remain in use as part of the nation's West Coast main line. On the London & Birmingham route these include the two-and-a-half-mile Tring Cutting, the 45-foot-high Wolverton

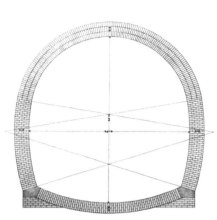

Cross section of Primrose Hill Tunnel drawn by F W Sims, 1838, with view by J C Bourne showing the tunnel's portal under construction, 1837 (Institution of Civil Engineers).

Embankment, the 2,400-yard Kilsby Tunnel and the 1,100-yard Primrose Hill Tunnel. The Grand Junction line included several fine masonry viaducts, including the 12-arch Warrington Viaduct that crossed the River Mersey with two elliptical arches of 75 feet span.

The experience being gained on these early projects allowed the two Stephensons to assist Brunel with parliamentary evidence in support of the Great Western Railway. They were questioned, in particular, about Brunel's proposal for the long Box Tunnel, and the Stephensons' evidence aided the passage of the railway's second bill. Thus began a warm and respectful association between Brunel and Robert Stephenson. Together with Joseph Locke, the trio were professional colleagues throughout their careers, consulting each other on a wide range of engineering, arbitration and national issues.

With the successful completion of the London & Birmingham Railway in 1838, Robert Stephenson was acclaimed by many businessmen who were influential in promoting new railway projects. He was appointed engineer-in-chief of several of the largest schemes, and developed an engineering practice based in Great George Street, Westminster, alongside the Institution's premises. Initially in association with his father, these railways included the North Midland Railway, between Derby and Leeds, with its several challenging features and structures, the York & North Midland Railway extending the route to that city, and the Birmingham & Derby Junction Railway linking those routes to Birmingham.

The rate of railway expansion in the mid-to-late 1830s created the opportunity for other experienced engineers to diversify their interests into railway construction. Notable among these was the Irish-born Charles B Vignoles, who built railways in the northwest of England, including the first crossing of one railway over another by an iron viaduct near St Helens, the first Irish railway between Dublin and Kingstown,

and the Midland Counties Railway linking the East Midlands with the London &
Birmingham line. John Rastrick, who had much experience in iron manufacture
including steam locomotives, also undertook the construction of several early rail
routes, particularly that between London and Brighton, with its spectacular Ouse
Valley Viaduct.

Robert Stephenson was noteworthy in selecting talented young engineers as junior
associates who undertook much of the survey, design and fieldwork for his many
projects. His Westminster practice was more akin to a barristers' chambers than to a
consulting practice with senior and junior partners. The engineers who benefited from
this association included Thomas Longridge Gooch who, under George Stephenson,
was responsible for building the Manchester & Leeds Railway including the one-and-
two-thirds-mile-long Summit Tunnel under the Pennines. He later built the North
Staffordshire and Trent Valley Railways, which form important stretches of today's
West Coast main line. Other associates included Frederick Swanwick, who built
several routes for the Midland Railway Company, and the Berkley brothers, James and
George (later Sir George), who later became senior engineers for railway projects
around the world.

Portrait of Joseph Locke by John Lucas
(Institution of Civil Engineers).

Unlike Stephenson's policy of delegation, Brunel supervised closely all the projects for which he was responsible. His assistant engineers included his chief of staff, Robert P Brereton, and William Bell.

Joseph Locke was closely associated with John Errington, with whom he formed a partnership to share their wide-ranging responsibilities. Errington undertook the building of several of the early Scottish railways, and shared the responsibility with Locke for building the Lancaster & Carlisle Railway with its notorious climb to Shap Summit that remains such a feature of the West Coast main line. Locke also engaged Albert Jee to assist him with several major rail projects, most notably the construction of the Sheffield, Ashton-under-Lyne & Manchester Railway with its daunting three-mile-long Woodhead Tunnel, the longest to be built in Britain in the mid-Victorian era.

When railway bills had been approved by Parliament, chief engineers invited tenders for building sections of line from contractors who were required to demonstrate not only experience in undertaking major civil works but had the financial backing to provide sufficient working capital to fund the construction work.

In addition to his extensive civil engineering consultancy work, Robert Stephenson was the senior partner in the successful locomotive and marine engine manufacturing business in Newcastle upon Tyne that bore his name. Together with other manufacturers, particularly Edward Bury of Liverpool, the Sharp Brothers of Manchester and E B Wilson of Leeds, he oversaw the rapid development of locomotive power and speed in the 30 years following the Rainhill locomotive trials.

Railway mania

Engineering development in the 1840s was dominated by the extraordinary 'railway mania' that saw a host of schemes promoted for the construction of many hundreds of miles of new route. Investors scrambled to promote the schemes, many of which were ill-founded, resulting in company failures and lost investments. Many others were approved and constructed to form part of the nation's rapidly growing railway network. All required the services of surveyors and engineers to undertake land surveys, determine the most appropriate alignments and best 'levels' for minimising earthworks, and prepare plans and estimates for parliamentary scrutiny. The extraordinary demand for new railways at this time saw railway mileage approved by Parliament increase from 90 in the 1843 session to 810 in 1844, 2,700 in 1845 and 4,538 in 1846.

Opportunities abounded in these hectic years for both the leading consulting engineers, particularly Stephenson, Brunel and Locke, and their junior associates. Stephenson represented proprietors for many of the larger schemes, and the sheer breadth of his work required him to delegate detailed responsibility to his several associates, particularly James and George Berkley, George P Bidder, John Birkinshaw, Michael

Transferring between broad and standard gauge at Gloucester, The Illustrated London News, *6 June 1846 (University of Bristol).*

Borthwick, Thomas Cabry, Edwin and Latimer Clark, Frank Forster, Thomas L Gooch, Thomas E Harrison, Alexander Ross, Frederick Swanwick, Henry Swinburne and George Robert Stephenson.

Other experienced and junior engineers also enhanced their reputation through both the successful representation of railway schemes through Parliament on behalf of proprietors, and the subsequent building of railways for the incorporated companies. These included Peter and William Barlow, William Cubitt, Joseph Gibbs, John Hawkshaw, John MacNeill, John Rastrick, James Rendel, Sir John Rennie and Charles B Vignoles. After the mania subsided, many of these talented engineers went on to pursue careers not only in railway building in Britain and around the world but also in other engineering disciplines.

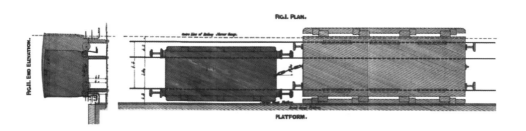

Drawing of broad and narrow gauge carriages together in a three-rail siding from Robert Stephenson's Report to the London & North West Railway of 1847 about the unsuitability of mixed gauge tracks.

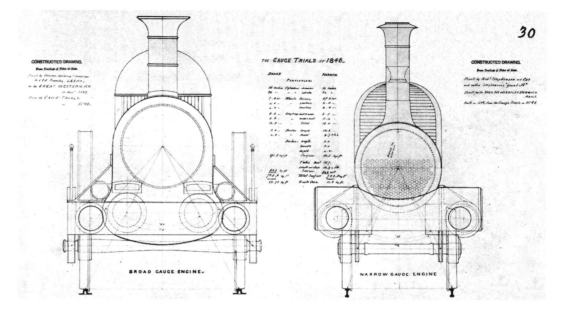

Drawings prepared by David Joy illustrating the principal contending locomotives in the gauge trials of 1845/6 (Institution of Mechanical Engineers).

In the free-market procedures of the day, and in the absence of any coherent national policy on rail network development, parliamentary consideration of rival schemes encouraged many adversarial debates. While the engineers maintained a strict adherence to their clients' best interests, parliamentary scrutiny for railways included large amounts of the engineers' time being spent defending or opposing competing schemes. The many territorial battles waged by competing boards of proprietors included the London to Birmingham corridor, where the associate companies of the Great Western Railway sought to build a competing route through Oxford. The route was to be built under Brunel's supervision and was, consequently, to be to his seven-foot track gauge.

While Brunel believed strongly in the 'broad' gauge, Stephenson believed equally strongly in the adoption of the 'standard' gauge of four feet 8½ inches throughout the country. He believed that the interface between the two gauges, in towns such as Gloucester, would give rise to considerable transhipment of passengers and goods, with consequent delays, damage, theft and costs. Other main lines that deviated from the standard gauge were those from London to Colchester and Cambridge, built to a five-foot gauge under John Braithwaite's supervision, and the Dundee & Arbroath and its associate lines, built to a gauge of five feet six inches under the supervision of the Scottish engineers, Thomas Grainger and John Miller. A Parliamentary Commission, set up to consider the gauge issues, gave rise to a strong public statement in favour of the standard gauge signed by the two Stephensons, John Hawkshaw, Joseph Locke, Sir John MacNeill and James Rendel, together with a number of mechanical engineers and railway company directors.

Brunel concentrated his endeavours on railway building and his wider engineering pursuits, while the responsibility for locomotive design and development on the Great Western Railway was undertaken by Daniel (later Sir Daniel) Gooch, with whom he had a close working relationship. The comparative trials between Gooch's broad-gauge locomotives and Stephenson's standard-gauge locomotives, which demonstrated no advantages due to gauge alone, were therefore a major event in the turbulent years of the railway mania.

In the early 1840s, Brunel and Stephenson had similar debates about the use of the 'atmospheric' form of motive power for hilly routes. Brunel was joined by other engineers, including William Cubitt and Charles Vignoles, in believing that atmospheric railways offered an economic alternative to the steam locomotive for certain routes. However, in a report to the directors of the Chester & Holyhead Railway, Stephenson highlighted the inherent inefficiencies of the system, which became evident after less than a year's experience with this form of motive power on Brunel's South Devon Railway.

Until 1847 mechanical engineering was considered to be an integral part of the civil engineering profession. However, the extraordinary growth of so many railway projects during the mania years, accompanied by the major expansion of the nation's manufacturing capabilities, brought about the establishment of the Institution of Mechanical Engineers. Today, the Institution remains the professional body for that discipline, with a worldwide membership of over 75,000.

View of Water Street Bridge by T T Bury, published by R Ackerman, 1831 (Institution of Civil Engineers).

High Level Bridge, Newcastle upon Tyne, *hand-coloured lithograph by George Hawkins after James Wilson Carmichael, published 1849 (Elton Collection: Ironbridge Gorge Museum Trust).*

Bridge designs

Abraham Darby III's innovative cast-iron bridge built at Ironbridge in the late eighteenth century led to further applications of this material, including Thomas Telford's trough-form canal aqueduct at Pontcysyllte. Some engineers, including John Rastrick and William Cubitt, developed cast-iron arched road bridges in the early years of the nineteenth century, but the growth of the railway network gave rise to more ambitious schemes for bridges and viaducts. Whereas masonry and brick bridges showed continuity of design development from road and canal bridges, the minimum clearance requirements for some under-bridges necessitated the development of beam rather than arch designs. To achieve this the innovative Manchester physicist, Eaton Hodgkinson, in conjunction with the ironmaster, William Fairbairn, developed the earliest cast-iron bridge beams. The first such bridge, of $24^{1}/_{2}$-feet span, carrying the Liverpool & Manchester Railway across Water Street into its Manchester terminus, was completed in 1830.

Brunel largely avoided the use of cast iron for his bridges. However, Robert Stephenson and his associates gained considerable experience with the material. The London & Birmingham Railway required several bridges with long spans, giving rise to designs for both cast-iron arches, such as the 66-foot span Nash Mill bridge over the Grand Junction Canal in Buckinghamshire, and the bowstring girder bridge design, the first example being of 50-foot span across the Regent's Canal in London, attributed to Charles Fox. The bowstring girder type was significantly enlarged for later applications, including the Gauxholme Bridge on the Manchester & Leeds Railway, while the finest example of this type was the High Level Bridge in Newcastle upon Tyne. Its six spans, each of 124 feet, carried the Newcastle & Darlington Junction Railway 100 feet above the River Tyne.

In addition, trussed compound girder bridges were developed by Stephenson and his team for many applications on their early railway projects, including that carrying the Northern & Eastern Railway across the River Lea. However, the collapse of one of the 98-foot spans of this type across the River Dee at Chester in 1847 dramatically ended the cast-iron bridge era.

All engineers adopted masonry and brick bridges for railways, perpetuating the types used over many years for roads and canals. Brunel, in particular, built many masonry and brick arch bridges on the Great Western Railway and its associated lines, notably the graceful Maidenhead Viaduct with its elongated main arches. Other major masonry viaducts included Stephenson's and T E Harrison's 15-arch Royal Border Bridge, which rose 130 feet above the bed of the River Tweed at Berwick-on-Tweed, and the 31-arch, 110-foot-high Crimple Viaduct between York and Harrogate, the building of which was supervised by John Cass Birkinshaw.

Endeavouring to keep the costs of railway construction within limits laid down by Parliament and company directors, however, meant that many railway bridges and viaducts were initially constructed of timber. While Brunel's Cornish timber viaducts came to be well known in later years, other engineers also adopted them, notably Stephenson who used them for a number of applications, including the 670-foot-long Sherburn Viaduct, on the Newcastle & Darlington Junction Railway in County Durham, and several on the Chester & Holyhead Railway.

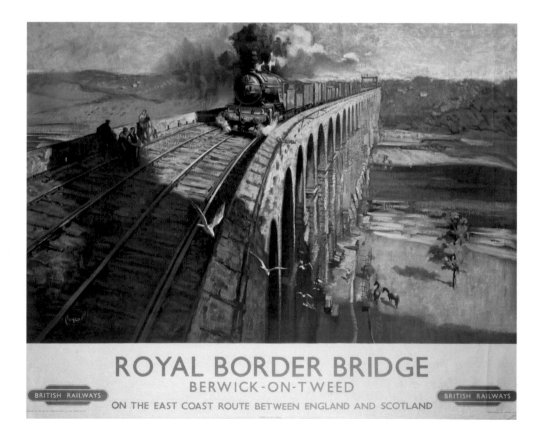

The Royal Border Bridge, *celebrated in a British Rail promotional poster, 1948-1965, artwork by Terence Cuneo (National Railway Museum/Science and Society Picture Library).*

The Wonders of the Menai, *hand-coloured lithograph by J Fagan after S Hughes, 1850, showing Telford's 1826 suspension bridge and Stephenson's 1850 tubular railway bridge (Elton Collection: Ironbridge Gorge Museum Trust).*

Conway Tubular Bridge, *1848, (right) coloured lithograph by Newman & Co after a drawing by John Lister Jnr (Science Museum/Science and Society Picture Library).*

Piece of the tube from Brunel's tubular suspension bridge at Chepstow (photograph by John Evans).

Other engineers also evolved innovative bridge designs for railways. These included Thomas E Harrison, whose Victoria Bridge across the River Wear was based on designs initiated by James Walker, and Captain William Scarth Moorsom, whose River Avon bridge near Tewksbury employed caissons sunk into the river bed by the weight of the masonry and concrete piers.

Although the further application of cast-iron bridges had concluded in 1847, thoughts had already turned to the use of wrought iron for the longest-span bridges. Stephenson and Brunel both initiated design innovations for large, wrought-iron bridges that became iconic symbols of the mid-Victorian advances in structural engineering. Stephenson embarked on a design programme for his major Britannia and Conway tubular bridges on the Chester & Holyhead Railway, with the assistance of Eaton Hodgkinson and William Fairbairn. Notwithstanding a subsequent public controversy surrounding the originality of the tubular design, their development was accompanied by the difficult design and preparation work for the erection of the piers and tubes undertaken by Stephenson's associates, Alexander Ross, Edwin Clark and Frank Forster. In the 1850s, Brunel and his chief of staff, Robert Brereton, went on to develop designs for the increasingly challenging Windsor, Chepstow and Saltash truss bridges. The surviving Conway tubular bridge and Saltash truss bridge continue in daily use and are enduring structures of mid-Victorian engineering excellence.

Overseas projects

The major progress made by the engineering profession up to the 1830s led to many approaches for engineers' services for overseas projects. Several European schemes were initiated from the mid-1830s, the first being the Belgian state railway, about

Views of Victoria Bridge (above and opposite) during construction and after completion, from J Hodges Construction of the Great Victoria Bridge in Canada *published by John Weale, London, 1860. (Institution of Civil Engineers).*

which George and Robert Stephenson were consulted. The largest involvement in Europe, however, was undertaken by Joseph Locke, who built first the routes between Paris, Rouen and Le Havre with their succession of tunnels and spectacular bridges and viaducts, such as that at Barentin. Locke was later responsible for the extension of the line to Caen and Cherbourg. He also built the first railway in Holland, between Amsterdam, Rotterdam and the German border at Emmerik. Robert Stephenson was the chief engineer for the construction of the first railway in Italy, between Florence and Livorno, requiring a substantial bridge over the River Arno. Charles Vignoles was also responsible for many railway and other engineering projects in Europe and further afield, particularly in Russia where he built the longest suspension bridge in the world across the River Dnieper at Kiev.

Robert Stephenson, George P Bidder and George Robert Stephenson expanded their European interests from 1850, particularly in Norway and Denmark. In Egypt, Stephenson was the engineer-in-chief of the Alexandria to Cairo railway with its dramatic tubular bridges across the waterways of the Nile Delta, designed by G R Stephenson and Charles Wild, and built to withstand the major challenges of the river's annual flood.

Several British consulting engineers pioneered the planning of the first railway routes in India in the 1850s. The first, with Stephenson as consulting engineer, and James Berkley as chief engineer, was the Great Indian Peninsula Railway, with its dramatic and technically challenging inclines up the steep sides of the Ghat mountains east of Bombay (Mumbai) towards both Nagpur and the east, and Poona, Bangalore and Madras to the south. Stephenson was also consulted on other railway schemes in India and Ceylon (Sri Lanka), while other early Indian schemes were in the charge of other British engineers, including James Rendel, George (later Sir George) Barclay Bruce and John Brunton.

Brunel was himself involved with overseas projects in Italy, India and Australia. Following a frustrating and inconclusive project in the Italian state of Piedmont, he undertook, through his assistant engineer, Benjamin Babbage, a short line from Florence. In India, his assistant engineer, Wellington Purdon, initiated the survey and building of the Eastern Bengal Railway, while Robert Brereton had charge of the first railway contracts in Victoria, Australia.

The 1850s also saw the building of railways, bridges, harbours and other civil works under British engineers throughout the empire, and in South America. Of particular note was the Grand Trunk Railway of Canada, with its major, 1.3-mile-long Victoria

tubular bridge over the St Lawrence River. Alexander Ross was the chief engineer for the railway and Stephenson was consulted on the best design for the bridge. The Victoria Bridge, the world's longest, was opened in 1859 just a few weeks after Stephenson's death.

Water projects

In the mid-Victorian years, the engineering profession was engaged with many improvements to urban water supply and sewage disposal. This was necessary to cope with the rising urban population and to deal with the severe problems of water-borne disease and smell, which caused such havoc to the populace. Some engineers, including John Bateman, William Tierney Clark, James Simpson and Thomas Hawksley, specialised in water and drainage schemes. In addition, Robert Stephenson was consulting engineer to both the Metropolitan Water Company and to the city of

Liverpool. He delegated the detailed study for the latter city to George Berkley and George Phipps. He was also consulted regarding schemes in Glasgow and Manchester in conjunction with Brunel. Stephenson reported on the particular problems of wastewater disposal for London and the River Thames, and Frank Forster was appointed as engineer to the Metropolitan Commission for Sewers. However, following considerable debate on the requirements, Joseph Bazalgette was subsequently appointed to undertake the major main drainage works.

Several engineers were engaged on major port and harbour projects to accommodate the growing export needs of Britain's manufacturing industry and the import needs of its rapidly growing population. The major expansion of the Ports of London and Liverpool was accompanied by the growth of new ports, harbours and harbours of refuge around the British coasts. The engineers engaged on this work included those who were dedicated to specific ports, such as Jesse Hartley in Liverpool, and those whose consulting services were wide ranging, such as James Walker.

While Brunel was so involved with his three large marine engineering ventures, undertaken with his assistants, Robert Brereton and William Bell, several engineers, including Stephenson, George P Bidder and George Robert Stephenson, diversified their engineering involvement to include harbour, river and sea-wall projects. They included major land reclamation schemes in Britain, particularly in East Anglia, and in Holland and Denmark.

Influence and achievement

Stephenson, Brunel and Locke maintained much respect for each other's experience, and consulted one another from time to time on matters of engineering development. The difficulties each faced with the execution of innovative design programmes received the practical and sympathetic assistance of the others. The paintings and photographs of the engineers with each other and their associates are enduring records of these special events in engineering history. John Lucas captures this spirit of co-operation in his 'Conference at Britannia Bridge' painting, showing both Brunel and Locke in support of Stephenson as the bridge is being built in the background. Stephenson's later, reciprocal support for Brunel as he struggled to launch the *Great Eastern* steamship was similarly captured, in a series of epic photographs.

By the 1850s, Stephenson, Brunel and Locke were seen as a 'triumvirate' of senior engineers, and were consulted by Parliament and the Board of Trade on matters of national policy. In addition, both Stephenson and Locke were Members of Parliament for the last 12 years of their lives, and played important roles on several select committees. These considered technical issues ranging from the design of the new Westminster Bridge to the ventilation of the Houses of Parliament, then being rebuilt after the disastrous fire of 1834.

It was Robert Stephenson – the senior and most influential engineer of the generation – who was called upon to undertake the widest range of major public works in Britain and overseas. He was consulted by company directors, politicians and government bodies alike, and was central to the planning of the Great Exhibition of 1851, firstly as a member of its executive committee, before being promoted to the Royal Commission itself. Such was his standing in the country that, on his death in 1859 just a few weeks after that of Brunel, he became the only engineer, after Thomas Telford, to be interred in Westminster Abbey.

Conference of Engineers at Britannia Bridge, *oil painting by John Lucas, 1851-53 (Institution of Civil Engineers). Brunel is seated at extreme right.*

THE ROEBLINGS

2006 marks the 200th anniversary of another great nineteenth-century engineer, John A Roebling, designer of the Brooklyn Bridge.

Like Brunel, Roebling was a polymath: an inventor, a European-educated professional engineer, a reader of philosophers and liberal thinkers, an artist and musician and, following the death of his wife, a spiritualist. He was never known to take a day off work and was noted for his energetic, hands-on approach to the construction of his projects.

Roebling was born on 12 June 1806 in Muhlhausen in the province of Saxony. He studied bridge construction and architecture at the distinguished Polytechnic Institute in Berlin and emigrated to the USA in 1831, establishing the town of Saxonburg in Western Pennsylvania. Six years later, having attained his US citizenship, he turned his back on farming to become a state surveyor and, later, Pennsylvania's chief engineer.

In the summer of 1841 Roebling began experimenting with the manufacture of twisted-wire rope in Saxonburg, based on a process developed in Germany. In September, he moved the manufacturing works to Trenton, New Jersey. His engineering projects, which included the first suspension aqueduct, built to carry the Pennsylvania Canal across the Allegheny River, and the Great International Railway Suspension Bridge at Niagara, made him famous. It was his wire rope, twisted into cables capable of supporting heavy loads and long spans, which made his fortune.

Roebling had first begun to consider how to build a bridge across the turbulent salt water strait that was the East River in 1857. The strait was too busy to allow for drawbridges or off-shore piers, which would have

Portraits of John and Washington Roebling
(Institution of Civil Engineers).

Bird's eye view of the Brooklyn Bridge, New York City, 1883 (Science Museum/Science and Society Picture Library).

interrupted the flow of maritime traffic. A charter for building a bridge to link the cities of New York and Brooklyn had been approved in April 1867 and Roebling was named its engineer the following month. In September he presented his masterplan to what was then called the New York Bridge Company. Preparations began in earnest in 1869 when Roebling had a series of meetings with a panel of consultants to examine his plans in detail and to silence any possible criticism from shareholders.

Roebling wrote:

> As a great work of art, and as a successful specimen of advanced bridge engineering, this structure will forever testify to the energy, enterprise and wealth of that community which shall secure its erection.[1]

His would be the largest suspension bridge built to date, measuring over a mile from end to end, including the approaches, with an uninterrupted central span of 1,595 feet. Two towers were to be built in the river close to the shore, one on the Brooklyn side, the other in New York, rising 276 feet above high tide. Roebling's early designs for the towers showed the influence of the Egyptian style used by Brunel, but he later turned to the Gothic. Below the waterline, the towers were constructed of limestone set on huge, pneumatic timber foundations resting on bedrock; above, they were of granite. The towers bore the weight of four steel cables, over 15 inches in diameter, carrying the roadway. Down the centre of the roadway ran a double pair of rail tracks

for specially built trains. A pedestrian boardwalk was set 18 feet above this. The cables also supported the land spans and were secured in two enormous anchorages on either side of the river. A system of stays at the top of the towers provided additional strength.

On Monday 28 June 1869 Roebling's foot was accidentally crushed while he was fixing the location of the Brooklyn Tower. This seemingly minor accident resulted in his death of tetanus poisoning on 22 July. His eldest son, the Civil War hero, Colonel Washington A Roebling, was named as his successor as the project's chief engineer. Washington had been educated at Rensselaer Polytechnic Institute, the first American school specialising in theoretical and practical science.

The ground for the Brooklyn Tower was broken on 3 January 1870 and what was hailed as the eighth wonder of the world was opened on 24 May 1883. Sadly, during the final years of construction, Washington had become a recluse, suffering from the after-effects of two severe attacks of the 'bends', caused by too-rapid decompression following inspections of the caissons below the river bed. The strain resulting from mismanagement and corruption on the part of some of the bridge's contractors and trustees added to his health problems. His wife, Emily, made a significant contribution to the project, ensuring it was successfully completed despite her husband's incapacity, and plaques honouring her can now be seen on each of the towers alongside those to the Roeblings, father and son.

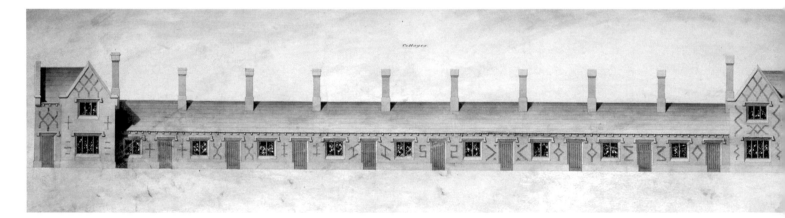

Contractor's drawing (with detail) to Brunel's specifications for workmen's cottages at Steventon, c 1835 (Adrian Vaughan Collection).

WHO BUILT BRUNEL?

Adrian Vaughan

Introduction

The mighty works of Brunel include the Great Western and other railways from Paddington to Bristol, Falmouth, Weymouth, Milford Haven and Wolverhampton, and ships that extended his Great Western Railway to New York. All this was the product and labour of Brunel's mind. The conversion of Brunel's vision into iron, stone, brick and earth embankment was the labour of body and mind of many thousands of people. Most of the people who helped him achieve his fame and glory as a railway engineer have been forgotten, or were never known in the first place. Patient searches through ancient records bring some of them back to life.

Brunel laid down an approximate route for any railway and then a small body of surveyors, no more than a dozen, made a meticulous, yard-by-yard survey of the topography of the route. The surveyors' work was precise because it constituted evidence in the future arguments in Parliament as to whether the railway was required and feasible to construct. Partly because of the need for the compulsory purchase of land on which to build the railway, an Act of Parliament was required to incorporate any railway company and give it the force of law.

While the survey was taking place, a huge amount of money had to be collected from investors. Then, thousands of labouring men representing many trades, but most famously navvies, were required. They were recruited by the civil engineering companies working as contractors. Under the direction of these contractors, the workmen descended upon the ground to carve out the route from the earth: fill in valleys, cut through hills and construct the great viaducts. The contractors' efforts were supervised and measured on behalf of the company by a few resident engineers, reporting to Brunel.

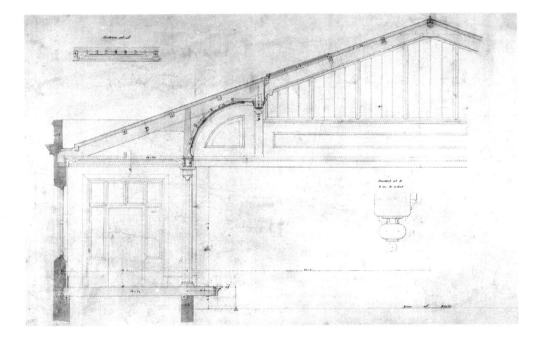

Contractor's drawing of Bath station roof, c 1835 (Adrian Vaughan Collection).

The surveyors

Brunel's private journal for 21 February 1833 records:

> [Roch] came to Mr. Osborne's [office] and informed me that a sub-committee was appointed [of which Roch was a member] for the purpose of receiving offers from me in conjunction with Townsend… as to what terms and in what time we would make a survey and lay down a line and on what terms we would afterwards undertake to lay out a line for Parliamentary plans.

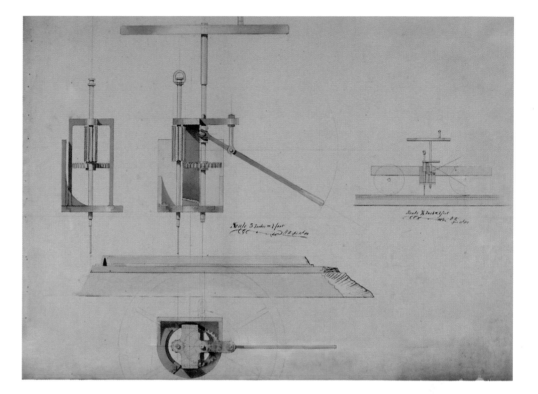

Drawing to Brunel's specification of timber bob to be used on GWR, possibly to drill holes for the attachment of rails to sleepers, c 1837 (University of Bristol).

This is the first time Townsend is mentioned by Brunel. He is coupled to Brunel on equal terms, and this seems to have been imposed on Brunel by the Bristol Railway Committee. W H Townsend was a Bristolian, a land agent and surveyor and engineer of the Bristol & Gloucestershire Railway (B&GR), a horse-drawn tramway carrying coal from Coalpit Heath to the Avon Street wharf on the Floating Harbour. Three directors of the B&GR were on the committee of the Bristol Railway. Perhaps they were hoping Townsend could convince Brunel to buy the track bed of the B&GR for the Bristol Railway. On the evening of 22 February, Brunel explained to Townsend that he was not his equal but could only be his assistant. Townsend accepted his position and Brunel's diary entry for 22 February concludes with this reference to Townsend: 'How the devil I am to get on with him tied to my neck I know not.'

Brunel and Townsend offered to carry out a survey for a railway route from Bristol to London for £500. The committee accepted their offer and Brunel wrote in his diary: 'We are undertaking a survey at a sum by which I shall be considerably a loser but succeeding in being appointed engineer – nous verrons.' The pair set out on their survey on 7 March 1833. They were, together, the first two employees of what was known then as the Bristol Railway and which was given its proper title – the Great Western Railway – in August of that year. On that first day Townsend led Brunel out of Bristol along the line of the B&GR tramway, up the steep hills through Lawrence Hill to Mangotsfield and eventually down into Bath. They came back to Bristol along the Avon valley, the obvious route. Brunel's distrust of Townsend's talents was well founded, and thereafter he confined Townsend to the work of an ordinary surveyor along the Avon valley from Bristol to Bath under the resident engineer, George Frere and also, in 1835-6, in surveying the Bristol to Exeter line under Brunel's assistant, William Gravatt. The last reference to Townsend is a letter to him from Brunel dated 2 February 1836.

Brunel had Ordnance Survey maps to show him the lie of the land between Bristol and London and he explored the entire area on horse back over a period of a month. He finally settled on the general line of the railway, which included two or three routes into London from Reading. He drew a pencil line on the maps and then employed surveyors, men like Mr Thomas Hughes and Mr Bell, to carry out the detailed survey.

View of the Maidenhead Bridge crossing the Thames, illustrated by J C Bourne, published 1846.

Contractor's drawing of Dorchester Road main building and elevation to rails, c 1835 (Adrian Vaughan Collection).

The surveyors negotiated their fee and expenses with Brunel and so they had to work with the lowest possible costs because, certainly in the early weeks of the survey, the surveyors' fees were coming out of his £500. Hughes and Bell had their sons as clerks and 'go-fers' – engaged on transcribing pencilled notes into fair copy or carrying reports to Brunel, or the luggage to the town ahead.

Hughes and Son set out from Thameside at Lambeth – Brunel's first choice of terminus – and walked to Swindon by way of Kingston-on-Thames, Chobham, Wokingham, south of Reading to Englefield and Pangbourne. Hughes and Bell even gave Brunel advice, politely, on how to carry the route through or around an obstacle – a village or a church – and generally performed a most important task of showing up vital details for consideration. Brunel would undoubtedly have accepted any well-founded objections because that is what he employed them to find out. They would report to him on the temper of the various landowners they came across, briefing him in readiness for the day when he might have to persuade them to consent to the railway crossing their land.

The survey equipment consisted of the tall, graduated staff, the very heavy, 22-yards-long 'Gunther's' chain made of steel or iron and the theodolite. The chain enabled them to measure precisely the distance from the theodolite to the staff. The theodolite was like a telescope which tilted up or down and gave an exact reading of the number of degrees it was pointing above or below horizontal. Thus two sides of a triangle were created and the height of the third side could then be calculated to show the rise or fall of the ground. To this day, the 'chain' of 22 yards is a measurement of distance

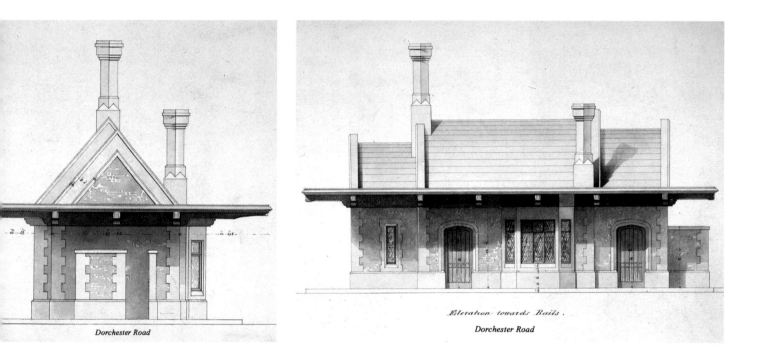

Dorchester Road

Elevation towards Rails.

Dorchester Road

specific to railways. The surveyor required a man to hold the staff while he took the angle to it on his theodolite. The same man could carry the staff and chain, leaving the surveyor to shoulder the precious theodolite. But there was also, sometimes, a need for someone to hold an umbrella over the surveyor if it was raining while he was trying to make his pencilled notes of each theodolite reading. They would pick up a labouring man as assistant in a town or village, using him for a few miles or a few days depending on how far the man was prepared to walk. They also had a box containing pen, ink, a strong ledger in which to make a fair copy of their pencilled notes, and a change of clothes. This they could send ahead of them, by stagecoach, addressed to an inn 'to be called for'.

These surveyors, out in almost any weather, had to produce precise work because their details would go before Parliament as the company's formal proposals for the line. Some evenings they arrived at their lodgings, in some village inn, soaked to the skin – and sometimes without dry clothes to change into. Sometimes they could find nowhere to stay at the end of the day and would walk on several miles to another place where they knew there was accommodation. All this is clear from the letters that Hughes wrote to Brunel.

Brunel expected to receive a letter from his surveyors every day, reporting on the previous day's work. Hughes would also say where he intended to be over the next day or two so that Brunel could fit in a visit to him if he was in the area. Letters sent off on the evening coach were in Brunel's hands the following day and letters from him to his surveyors were frequently delivered to them out in the fields. Hughes' letters to the great man would often be cut short with such a sentence as: 'I hear the mail coach coming and I must stop now to catch the post.'

The investors

The initial survey of the Great Western line took ten weeks and on 30 July 1833 Brunel presented his proposals at a public meeting in Bristol, called to publicise the railway and ask for financial support. The Act of Parliament authorising the construction of the GWR, gained in August 1835, authorised the company to raise £2.5 million, which was Brunel's estimate of the cost of construction. The capital was to be raised by selling £100 shares. The shares were allocated to an investor when he or she paid a deposit of £5 per share. The balance of the £100 was paid by 12 installments as the company 'called' for the money. There were 1,443 investors listed in the register of shareholders dated 25 October 1835.[1] Between them they held 22,911 shares – the full share issue of 25,000 was not at that date fully taken up but would have been subscribed later. Andre Gren says that the majority of shares were purchased by people living in London (6,209 shares) and Bristol (5,112 shares).[2] Those engaged in 'commerce, trade and manufacture' constituted 33.9 per cent of shareholders and 35.9 per cent were 'persons of independent means'. Other significant categories included 'other educated persons', holding nine per cent, and doctors, clergymen and lawyers, with 6.1 per cent. The residents of Clifton held 792 shares. There were 766 shares held in Bath. Cheltenham had 408 shares and little Cirencester held 129 shares.

The Liverpool party of the GWR has loomed large in the company's history. It is surprising to find that this supposedly formidable group consisted of only 49 men – 3.39 per cent of the total of shareholders – holding 4.92 per cent of the total shares.[3] Other people holding GWR shares in 1835 ranged from those with one share (equal to £6,000 in 2003) such as the widow, Susan Hall, of Kensington, and Thomas Hall, innkeeper of Broad Street, Reading, to Thomas Guppy, sugar refiner, of Bristol and

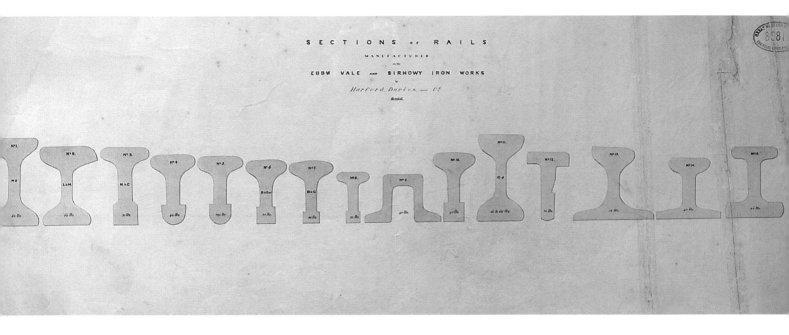

Drawing showing rail sections used on different railways across the country, c late 1830s (University of Bristol). No 9 is similar to the design used on the GWR.

Robert Bright, merchant, both members of the original Bristol Railway committee who took, respectively, 243 and 259 shares.[4] By October 1838 Brunel had increased his share holding from 50 to 80. On the 27th he was unable to pay the 'call' on these 80 and his close friend, G H Gibbs, purchased 50 of them for £113 each, which gave Brunel enough cash to pay the call on the remaining 30 shares he held.[5]

Brunel's staff

Brunel could not have been the engineer of so many projects without the untiring support of his engineering and clerical team. From March 1833 until 1835 they worked from an office at 53 Parliament Square and then – until Brunel's death in 1859 – at 18 Duke Street, Westminster. No-one came into Brunel's personal service who was not self-effacing, scrupulously honest and considered by Brunel to be 'a gentleman'. If Brunel liked a man then, by definition, that man must be a gentleman: a subjective judgement. At first his staff was small but as Brunel's railway empire extended over the years, so the team grew until by 1851 there were 33 assistants.

Brunel's personal assistant was Joseph Bennett. Brunel hired him on 5 March 1833.[6] He remained as Brunel's confidential secretary until Brunel's death. John Hammond was Brunel's first engineering assistant. Hammond had been introduced into Marc Brunel's drawing office by Marc's chief assistant, Richard Beamish. Isambard 'borrowed' him occasionally until, by 16 March 1833, Hammond was permanently on his staff. Hammond was resident engineer for the GWR's London Division and supervised the work of construction from Paddington to Steventon. He had to assist him at least two young men, W G Owen and Robert Archibald. Hammond answered to Brunel on a daily basis, receiving scant praise and always liable to get a critical letter if his performance dropped even momentarily below 100 per cent. Brunel was too busy to notice signs of illness in an assistant. During November 1846 Brunel wrote three angry letters to the loyal and long-suffering Hammond. In one he wrote: 'When I refer to our "angry discussion" I meant simply that I was angry and if I failed to make you sensible that I was angry I must be a much milder and gentler being than I thought.'[7] Hammond died a few months later.

John Gandell was Brunel's resident engineer for the line from Reading to Swindon. He came to Brunel from Robert Stephenson's staff with Stephenson's personal recommendation. Gandell had got on well with Stephenson but found Brunel's way of dealing with subordinates 'tyrannical and dictatorial'.[8] Friction quickly followed. Gandell was an architect by training, with offices at Wolverton, near Bletchley. He designed the executive village Brunel required at Steventon but objected to Brunel's interference with his designs. Brunel objected to the prices Gandell charged for his work and blamed the size of the staff Gandell employed, going so far as to order him to reduce it. Gandell refused to reduce the staffing at his private office at Wolverton. Brunel decided that since he did not like him, he was not a gentleman. When Gandell was found to have purchased some land near the site of the proposed Farringdon Road station, Brunel decided he was a speculator and dismissed him.

From Swindon to Bristol – excepting Box Tunnel – the resident engineer was George Frere, a gentlemanly person, the son of an ancient landowning family from Diss in Norfolk. He was assisted by Herschel Babbage, son of Charles Babbage, inventor of the Difference Engine, the world's first computer. It can be assumed that Frere did a

Contractor's drawing for the superintendent's house at Steventon, c 1835 (Adrian Vaughan Collection).

fine job as there are no irate letters in the record from Brunel to him. For some unknown reason, Brunel was reluctant to give him further work once the GWR had opened. Brunel did eventually offer him a job but Frere turned it down and, reading between the lines of a letter Brunel wrote to him after this refusal, one gets the feeling that Brunel did not offer him a sufficient salary. Herschel Babbage went to Italy for Brunel, to work on the Florence-Pisa line and the Genoa-Alessandria line. Babbage did surveying work on the line between Florence and Pistoia. At Brunel's request he made and sent to Brunel drawings of Tuscan churches, which he used as the basis for his designs of the pump houses for the atmospheric traction system on the South

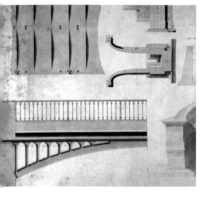

Contractor's drawings of cast-iron bridge in Sydney Gardens, Bath, c 1835 (Adrian Vaughan Collection).

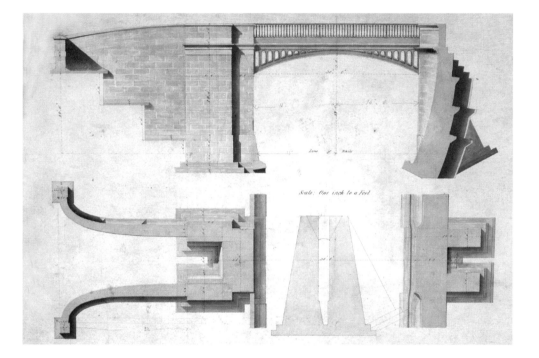

Devon Railway. Herschel Babbage deserted Brunel when he was offered more money to work for a rival engineer and, without his agent in Italy, Brunel lost the contracts.

The construction of Box Tunnel was a terrible struggle for all those directly involved, including Brunel's resident engineer, the young William Glennie. Very little is known about this man. He was well liked by Brunel and was still on his staff in 1851.

Two highly capable men from Brunel's Thames Tunnel days were working for him by 1836: Michael Lane and William Gravatt. Lane had been a foreman bricklayer in the Thames Tunnel when Brunel was resident engineer to his father, Marc Brunel. Brunel had noted Lane's qualities of craftsmanship and leadership. In 1836 Brunel was able to accept the post – and the salary – of engineer of the Monkwearmouth Docks because of the remarkable competency of Michael Lane, acting as Brunel's resident. Lane remained employed by Brunel until 1859, when, on Brunel's death, this remarkable man took over his master's post and became chief engineer of the Great Western Railway.

William Gravatt had shared the vile conditions and the dangers of the tunnel with Brunel. He and Isambard were each awarded the Royal Humane Society's silver medal on 5 March 1828 for saving the life of a workman during an irruption of the river into the tunnel on 18 May 1827.[9] Gravatt was devoted to Brunel. When Brunel became ill from conditions in the tunnel, Gravatt worked 38-hour shifts to cover for him and spent his rest periods sitting beside his friend in bed.[10]

Gravatt should have been a gentleman because he was the son of the Inspector-General of the Woolwich Military Academy, the British Army training school for artillery officers. Gravatt did not join the army. He was a born scientist and engineer, Brunel's intellectual equal. Gravatt enjoyed playing with equations involving the Differential Calculus and, on meeting Professor de Morgan, who had done a lot of

Contractor's drawing of tunnel at Saltford, c 1835 (Adrian Vaughan Collection).

development work with this mathematical tool, Gravatt insisted on writing out a cheque to the Professor's favourite charity. Charles Babbage said of Gravatt in 1855: 'Without him there would have been no calculating machine.'[11] He was a human calculating machine and, possibly because of this, he found human relationships difficult. He did have friends who enjoyed his company and admired him, but he was easily offended. He would say nothing, simmer and fester for days or weeks and then, to the astonishment of everybody, erupt. With the major collapse of the Thames Tunnel in January 1828, Gravatt went away to pursue an independent career. In 1832 he was appointed engineer of the Calder & Hebble Navigation, and was elected a Fellow of the Royal Society and the Royal Astronomic Society.

Gravatt had designed several improvements to land survey equipment when Brunel put him in charge of the Bristol & Exeter Railway (B&ER) survey in 1836 and then made him resident engineer of the railway – but with an unprecedented amount of freedom to design. There were few engineering features between Bristol and Taunton, but south of Bridgwater the line had to cross the River Parret. Gravatt designed a brick arch with a 100-foot span and only a 12-foot rise – twice as flat as Brunel's arches over the Thames. There is no record in Brunel's papers or letter books that Brunel designed this bridge, so it seems fairly certain that it was Gravatt's design, although Brunel, as chief engineer, would have approved it. There are two sets of contract drawings for

Drawing to Brunel's specification for GWR from the Paddington office, c late 1830s (University of Bristol).

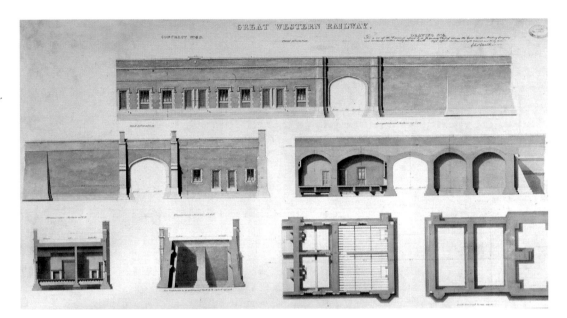

this bridge, one showing brick foundations resting on timber piles, the other showing a wide-area brick foundation to spread the load on the soft ground. Gravatt's downfall, as far as Brunel was concerned, was brought about by Gravatt's lack of tact when handling his relations with others. He had incurred Brunel's anger in 1838 for speaking harshly to a contractor, Mr Hemming, who then complained to Brunel. Brunel, who had done and would continue to do his share of raging at contractors, wrote to Gravatt on 16 April and ordered him to London to apologise to Hemming.

In July 1840 Gravatt committed the unforgivable sin for an assistant. He expressed engineering opinions contrary to those Brunel had already laid down for the construction of the B&ER. Brunel was very angry and on the 23rd he wrote a strong letter to Gravatt:

> It appears that you entertain views and opinion differing very much from my own on important engineering questions…connected as we have been as intimate friends of long standing, acting as my assistant…constantly at my side… that you

have never hinted to me that you differed and that I should hear it now, indirectly, is extraordinary…Is this the conduct of a friend, of a gentleman, of a subaltern trusted and confided in by the man above him?

Gravatt was sacked by Brunel in June or July 1841.

The Parret, or 'Somerset', Bridge, came into use when the railway was opened to Taunton on 1 July 1842. The flat arch, unfortunately, had nothing substantial to oppose its severe outward thrust: no heavy brick viaduct with conventional arches as at Maidenhead, for example. There was a very slight, but continuing outward movement in the Somerset Bridge foundations, which required the timber centring to remain in place. This obstructed the waterway and brought complaints from traders. Brunel designed laminated timber arches to rest on Gravatt's foundations and, without stopping the trains, the bridge was re-built in 1843.

If ever there was a name that ought to be closely coupled with Brunel, it is Robert Pearson Brereton, who was articled to Brunel in 1835. He was the son of a merchant family from Blakeney in Norfolk. Brereton was extremely capable, self-effacing and entirely devoted to his master. Brunel gave him important work continuously from 1839 onwards. This included the East Bengal Railway and the Florence-Pistoia route. Brereton supervised the entire construction of the Royal Albert Bridge, including the floating of the eastern or Devonshire span and the raising of both spans 100 feet above the river. Brereton belatedly became a member of the Institution of Civil Engineers (ICE) in 1860, a few months after the death of Brunel, who had been a vice-president of ICE since 1845. In 1861 Brereton read a paper on the Royal Albert Bridge to ICE members.[12] He could hardly avoid mentioning the major part he had played in its construction. This was regarded as an impertinence by Brunel's family and brought down upon him the wrath of Brunel's sons.

The navvies

Many thousands of labouring men built Brunel's railway. He cared very little for them, but in this he was no more than typical of his time. Navvies were not 'gentlemen' and they were not Brunel's concern. Brunel thought it would be nice if all contractors behaved towards their men as well as did the civil engineering contractor Sir Morton Peto but he opposed any law to oblige a contractor or company to look after its men. He said: 'The more interference [ie laws] you apply the more mischief you will produce.'[13] He felt sure that any law to lay down minimum conditions of service would 'injure the skill, activity and independence of the labourer', as if a poor labouring man had any independence to lose.[14] He was, however, sure that they had to be paid in cash – about once a fortnight, he thought, certainly not once a week – but if they were not

Selection of scenes from the building of the London & Birmingham Railway by J C Bourne, published 1839. Unlike his depictions of the GWR, made after the route was completed, Bourne's drawings of the London & Birmingham show a railway under construction (Institution of Civil Engineers).

paid, it was not his business; they had to fend for themselves. So long as the navvies could be induced to dig their way through hills and wheel out the spoil across the valleys, that was really all that mattered.

The best navvy was a man in his twenties, unmarried and muscular. A clerk of one of the great civil engineering contractors, Thomas Brassey, is recorded as saying of navvies at work:

> As fine a spectacle as any could witness is to see a cutting in full operation, every man at his post, and every man with his shirt open, working in the heat of the day, the gangers looking about and everything going like clockwork.[15]

Thomas Brassey or Morton Peto calculated their price for removing earth on the basis of each of their navvies digging out 20 tons of earth per day and throwing it above his head into a truck. Hekekyan Bey, an Armenian pupil to a British engineer, wrote in 1829:

These dissolute men exert themselves so violently in their work that I have seen many powerful, muscular, men with their blood oozing out of their eyes and nostrils.[16]

Interior of Box Tunnel, illustrated by J C Bourne, published 1846.

Interior of Long Tunnel Fox's Wood, near Bristol, illustrated by J C Bourne, published 1846.

In the Box Tunnel, the work resembled a gothic nightmare and achieving 20 tons of 'muck' per day per man was impossible. In the early stages of construction the tunnel was nothing but a set of enclosed chambers accessed only by the shaft from the ground above. When the gunpowder was exploded to bring down some more rock, the navvies received in their lungs much of the force of the blast because the shock wave was enclosed, like the men, in a dead-end chamber. In these man-made caves the navvies, breathing gunpowder fumes, floundered by candle-light in a cavern floored with stodgy mud which pulled the soles off new boots (at 15 shillings a pair).[17] The same mud was almost impossible to dig into, and having got a shovel-full it was heavy labour to then get it off the shovel and into the skip to be hoisted to the surface. On the ground above, horses attached to a capstan walked around and around, interminably, hoisting and lowering the tubs. Then, at the surface, some poor devil had to try to empty the tub. Given the circumstances, it seems unsurprising that the navvies drank heavily and fought each other for amusement.

They were not provided with lodgings by the contractor but slept and ate where they could. Some had lodgings in houses in Box or Corsham but there were not enough to go round. Half a dozen men, who perhaps worked as a sub-contracting gang, might build a bit of a shanty for themselves or men might sleep in a barn, if they could find a farmer who allowed it. Failing all else they would sleep under a hedge. Some of them married local women or came to the works with a woman. These women cooked for the men: some of them fought and drank with the men too – often they had to defend themselves. A young navvy called Chimley Charlie married a 16-year-old girl from Corsham and beat her with his leather belt whenever he was drunk. But one day she rebelled and knocked him down with a mighty blow to his head with a poker. 'I thought I'd killed 'un', she told Anna Tregelles, 'an' arter that he only hit me wi' 'is fists and we lived main comfortable.'[18]

During the final year of the excavation of the tunnel, 4,000 navvies were employed on the work. In the five years it took to complete, 'Just one hundred navvies died' digging 'Brunel's Great Box Tunnel' for him.[19] Brunel believed that this was an underestimate.[20] The GWR is said to have kept a record of deaths and injuries but it has been lost. There is no memorial to 'The Unknown One Hundred' as there should be.

Contractors

Among the civil engineering contractors who worked for Brunel were Grissell & Peto, (Paddington, Brent viaduct, Pangbourne cutting and more), William Ranger (Bristol-Bath and Sonning, failed), Hugh & David McIntosh (Earthworks at London end and Bristol-Bath), George Burge (Box Tunnel), Lewis & Brewer (Box Tunnel), Paxton & Orton (Box Tunnel shafts), Mr Knowles (Sonning cutting), Thomas Bedborough (Maidenhead Bridge), Tredwell (Worcester-Wolverhampton), Hemmings (parts of Bristol & Exeter Railway), Oldham (earthworks), Peto & Betts (Birmingham & Oxford and Oxford-Worcester) and Thomas Brassey (Gloucester & Hereford). The contractor had engineering expertise himself and hired to the GWR the brawn and brainpower of his organisation. A good contractor saved the GWR money by

Contractor's drawings for the Royal Albert Bridge, c 1835 (Adrian Vaughan Collection).

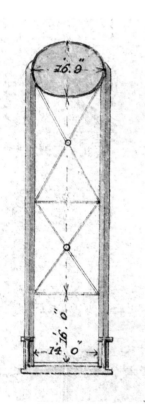
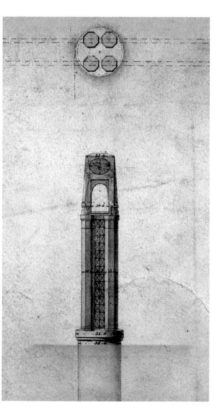

doing the work on time and within budget, while the weak contractor cost the GWR more by taking too long.

Grissell & Peto was a partnership of cousins. Peto was the junior partner and undertook the railway side of the business. He was well organised and made a point of treating his navvies well. He always carried out his work for Brunel according to his promise, but Brunel still managed to avoid paying him the agreed price. As a result, Brunel lost the services of Peto until 1847 when he carried out work between Banbury and Leamington and on the Oxford, Worcester & Wolverhampton Railway (OW&WR).

William Ranger, born in Ringmer, Sussex, had built bridges locally and completed military coastal defence contracts on the Sussex coast between 1823 and 1833. He then re-invested his earnings in undertaking heavy work for Brunel: the 100-foot span over the Avon at Bristol and all of the line, including the five tunnels, to Bath. (Bristol No.1 Tunnel was opened out into a sheer-sided sandstone cutting in 1886). In February 1836 he took on the contract for the Sonning cutting, two miles long and up to 60 feet deep. By summer 1837 he was falling behind schedule. That winter was appalling and the works became a quagmire. Brunel withheld payment and in February 1838 took away the contract. This meant that the GWR also confiscated Ranger's operational equipment because that is what the contract allowed. Ranger demanded payment for the work he had done. The court case lasted until 1865, ending in defeat for Ranger.

The Hugh & David McIntosh partnership of father and son was also an extremely large and capable organisation. They carried out heavy earthworks near Hanwell and, against his son's wishes, Hugh took over William Ranger's contracts between Bath and Bristol. Brunel and David McIntosh disliked each other. David McIntosh was an

View of Sydney Gardens, Bath (detail),
illustrated by J C Bourne, published 1846.

Daniel Gooch, photograph taken in 1845 showing the engineer standing in front of a model of his Firefly *class of express locomotive (Elton Collection: Ironbridge Gorge Museum Trust).*

extremely capable engineer and did not like being treated as a servant by Brunel. He attempted to maintain his independence and his intellectual equality and Brunel hated that. The fact is that Brunel actually went out of his way to pick a fight with the partnership, to the dismay of the GWR's solicitors who warned Brunel that he was acting illegally. The ensuing court case lasted from 1842 until 1865 and resulted in an expensive defeat for the GWR, pushing the company to the brink of insolvency.

George Burge was a solvent, competent contractor who had made St Katherine's Docks, Pool of London, opened 1828, Herne Bay Pier and Esplanade completed in 1837, and tunnels at Dover harbour. In 1837 he was working on Southampton Docks and in February 1838 he daringly signed a contract with Brunel to tunnel for 2,332 yards through the bowels of Box Hill from west to east and to form a junction with the 800-yard east-west tunnel of Lewis and Brewer in 30 months. This he did in spite of Brunel's insulting letters and his withholding of money. Even after the tunnel was complete, Burge still could not get all his pay – Brunel wanted to argue about what he considered to be Burge's high expenses. Financially, Burge survived this ordeal and went on to become involved with Peto on the East Kent/London Chatham & Dover Railway.

Brunel's contracts were one sided. If he decided that he disapproved of the contractor's methods or his rate of progress, he had the power to dismiss the contractor, confiscate his equipment and refuse to hand over any pay that was outstanding at the time the dispute arose. No discussion was allowed. Brunel was, by virtue of the contract, the arbitrator. No better illustration of what sort of personality Brunel possessed can be found than his contracts.

Friends

Of course, Brunel had a large circle of friends: 300 people – lords, viscounts, railway engineers, contractors and the permanent-way department of the Cornwall Railway – subscribed to his memorial, the Clifton Bridge, 150 of them putting in five or ten guineas. However, close friends, people he really looked to for help, professional or personal, were probably no more than ten in number.

Daniel Gooch, Sir Daniel from December 1866, counted himself a dear and close friend of Brunel's. Without Gooch, Brunel's superbly engineered route would not have shown its superiority so quickly. Gooch was the best locomotive designer of the period and his locomotives would have run just as fast on axles four feet 8½ inches long. Gooch tried to save Brunel from himself over the atmospheric system of propulsion, but without success.

Other close friends included Charles Babbage, Charles Saunders, Robert Stephenson, Benjamin Hawes, a civil servant, and the mathematician, William Froude.

Perhaps the two friends to whom Brunel most often unburdened himself and from whom he received the re-assurance he needed were T R Guppy and Charles Saunders.

Brunel (second left) playing cards with Thomas, Grace and Sarah Guppy, 28-29 September 1836, from the album of Grace Guppy (kind permission of Nicholas Guppy).

Portrait of Thomas Guppy (University of Bristol).

Thomas Richard Guppy was the son of an eminent Bristol merchant. Thomas wanted to be an engineer and applied to Maudslay, Son & Field in 1815 but Maudslay's business was so bad that he turned Guppy away. Between 1815 and 1824 Guppy set out on a nine-year educational tour, travelling to New York, Leipzig, Dresden, Munich and Paris to study engineering, architecture and painting. On his return to Bristol he took over the family sugar-refining business and in 1833 became involved with Brunel and the GWR. He was an engineer, artist and gentleman: the perfect 'older brother' for Brunel. He comforted Brunel when he was feeling 'blue-devilish' and he was one of only a handful of men from whom Brunel would take advice. Guppy contributed substantially to the design of the ss *Great Western* and the ss *Great Britain*.

Charles Saunders was educated at Winchester College – the motto is 'Manners maketh Man' – and was a merchant in Mauritius before becoming secretary to the London committee of the GWR in 1833 and, later, secretary to the GWR company, a post he held until 1863. When he died, three new posts were created to cover the duties Saunders had discharged for 30 years. Brunel abandoned his Lambeth terminus for a route into London suggested by Saunders. Brunel unloaded his tensions on Saunders as he did on Guppy. Opening his heart to Saunders in a long letter that revealed some of the private fears he kept hidden behind his public persona, Brunel wrote on 3 December 1837:

> If I ever go mad I shall have the ghost of the opening of the railway… standing in front of me… and when it steps forward a little swarm of devils in the shape of leaky pickle tanks, half finished stations, sinking embankments, broken screws, missing guard plate, unfinished drawings will quietly lift up my ghost and put him off a little further than before.

Frontispiece of J C Bourne's
The History and Description
of the Great Western
Railway, *published 1846.*

BATTLE OF MICKLETON TUNNEL

The Cotswold village of Mickleton near Chipping Campden in Gloucestershire was the scene of a series of disputes between Brunel and the contractor Robert Mudge-Marchant during work on the Oxford, Worcester & Wolverhampton Railway.

Brunel tried unsuccessfully to pass the contract for the Mickleton Tunnel to the company of Peto & Betts in June 1851, but Mudge-Marchant refused to give up control of the site, posting his navvies as guards. Attempts to evict him and reclaim the works and the equipment led to skirmishes which culminated in what is claimed by some to be the last battle fought by private armies on British soil.

On Friday 20 July, Brunel and his assistant Robert Varden led a group of navvies to the tunnel, aiming to regain possession, but they were met by local magistrates, summoned by Mudge-Marchant, who ordered Brunel and his men not to disturb the peace. The magistrates were still there when Brunel returned on Saturday and this time they were backed by armed police. The Riot Act was read and once more Brunel withdrew.

Having tricked the magistrates into believing he had given up the fight, Brunel used the Sunday to gather an army of men from other parts of the line along with some then working on the Birmingham & Oxford and GWR. There are thought to have been around 2,000 in total. Displaying 'the navvies' lawless loyalty'[1] to their employer, they moved on to the tunnel in the early hours of Monday 23 July armed with pickaxes and shovels. They charged upon Mudge-Marchant's men and fighting broke out. Mudge-Marchant threatened to shoot the attackers but in the event held his fire. He retreated, returning later supported by armed policemen, members of the Gloucestershire Artillery and two magistrates.

The two 'armies' stood face to face. Isolated pockets of fighting broke out along the line. Mudge-Marchant tried to carry on working but found it impossible to continue while outnumbered by Brunel's men, so at last he agreed to meet to discuss arbitration. L T C Rolt,

carried away with the imagery of battle, wrote in his biography of Brunel: 'the two generals retired from the stricken field (where there were many broken heads and limbs but happily no fatal casualties)'.[2] Terms were agreed by four o'clock that afternoon and Brunel resumed control of the site, just as additional troops, summoned by the police, arrived from Coventry.

Punch cartoon entitled 'Navvy in Heavy Marching Order', 1855 (Punch Cartoon Library & Archive). This illustration was originally used to show a navvy embarking for manual labour in the Crimean War.

Balaklava, (facing page) GWR broad-gauge locomotive, 1892 (Science Museum/Science and Society Picture Library).

SWINDON: A RAILWAY TOWN

Before the coming of the Great Western Railway (GWR), Swindon was a small market town of around 2,500 inhabitants. Daniel Gooch had recommended Swindon as the base for the GWR's main engine house and workshops.

He wrote to Brunel on 13 September 1840 enclosing a sketch plan and setting out his reasons for choosing the site. These were:

- Swindon was the point where the easy gradient from London changed to the heavier one to Bristol, requiring a switch to a more powerful class of engine for the Bristol end of the line.
- It was a convenient location for keeping the engines needed for the Wootton Bassett incline.
- It was close to the junction of the Cheltenham & Gloucester line so could be the principal station for both railways in that area.
- It was alongside the canal system along which coal and coke could be transported.

On Brunel's recommendation, the GWR board approved Gooch's proposal on 6 October 1840. Brunel, with assistance from the architect Sir Matthew Digby Wyatt, designed a settlement on a greenfield site to the north of the existing town – soon known as New Swindon – that, in addition to workspace, included

Swindon-built broad-gauge locomotive Amazon, *1851 (STEAM).*

Swindon steam hammer, c 1920s (STEAM).

company housing, churches, schools and other public buildings for a new community of railway employees and their families. As Swindon was a largely agricultural economy and therefore lacked the necessary skills base for the works, most of the employees were new to the area and needed accommodation and access to social amenities.[1] Many of the buildings were constructed from stone excavated during the digging of Box Tunnel.

The company of J D & C Rigby received the contract for constructing Swindon station and the workers' village, but had to do so initially at its own expense because of financial difficulties faced by GWR. As an inducement the company was offered the rental income on the cottages and a share of the profits of the station refreshment rooms and hotel, with an added sweetener of the GWR agreeing to stop all trains at the station for ten minutes. A stop was necessary for changing engines to meet the change in gradient but also meant passengers would have the opportunity to buy something to eat and drink. There were four

refreshment rooms in total, described by the architectural writer J C Loudon as being 'of noble proportions, and finished in the most exquisite style'.[2] However splendid the setting, the quality of the

Interior of Swindon engine house, illustrated by J C Bourne published 1846.

refreshments received numerous complaints, prompting Brunel to write in a letter to Mr Griffiths who held the monopoly:

> I assure you Mr Player was wrong in supposing that I thought you purchased inferior coffee. I thought I said to him I was surprised you should buy such bad roasted corn. I did not believe you had such a thing as coffee in the place; I am certain I never tasted any. I have long ceased to make complaints at Swindon, I avoid taking anything there when I can help it.[3]

Company records show that when the GWR works opened in 1843, the 131 railway cottages then built in New Swindon were occupied by 663 people, of whom 216 were men of working age. The works soon established themselves as a successful operation, meeting all of GWR's locomotive construction and maintenance needs as well as undertaking other company engineering work. From 1846, Gooch's huge eight-foot wheeled,

Group of Swindon works' office staff, c 1890s (STEAM).

Workshop during World War Two (STEAM).

broad-gauge, 4-2-2 series locomotives were being built here, including the famous *Lord of the Isles*, which was exhibited at the Great Exhibition in 1851. Swindon produced its first standard gauge locomotive in 1855 and continued to produce both types for nearly 40 years.

By the 1870s the works had developed into three large-scale 'Establishments': the rails mills, the locomotive works, and the carriage and wagon works. The workforce at this time was nearly 4,000 in number and the accommodation provision had expanded accordingly to meet demand. The 1871 census showed that the population of New Swindon had reached 7,628 – an increase of 83 per cent since 1861 – while the old town's population stood at only 4,092.

After years of wrangling, the two towns were incorporated into the municipal Borough of Swindon on 9 November 1900. By the time of the 1901 census, the combined population was just over 45,000, making Swindon the largest town in Wiltshire. Although, with the amalgamation, the GWR was no longer directly

Clerks in the Mileage Office, 1936 (STEAM).

View of the Swindon works, 1860 (STEAM).

Swindon iron foundry, 1924 (STEAM).

Assembling locomotive in Swindon workshop, c 1927 (STEAM).

involved in managing the town's affairs, it remained the borough's largest employer and the first 20 years of the century were marked by further expansion.

At its peak of production, the Swindon works employed over 14,000 people. By 1972 this had fallen to around 2,000 and, although the figures briefly rose again to 3,800 in 1980, the works were no longer tenable, finally closing on 27 March 1986, a victim of Richard Beeching's cuts to the national railway system and the formation of Bristol Rail Engineering Ltd which favoured workshops in other towns. Together the original Railway Village and the GWR works' site now constitute what is possibly Europe's main railway-engineering heritage showpiece, containing 81 listed buildings in two conservation areas.

FROM SLAVERY TO INDUSTRIALISATION

In 1698 the London-based Royal African Company, which had previously controlled all British trade with Africa, dealing in slaves, gold, ivory, dyewood and spices, lost its monopoly following lobbying by the Bristol Society of Merchant Venturers and other provincial groups. As a result of this change in policy, Bristol's merchants were free to engage legally in the highly profitable slave trade.

A View of the *Blandford* Frigate *by Nicholas Pocock, c 1760, showing trading on the coast of Africa and the passage to the West Indies (Bristol's Museums, Galleries and Archives).*

By 1720 Bristol briefly became Britain's main slaving port. The trade, not only in slaves but also in slave-produced goods, generated considerable wealth for Bristol, helping to make it Britain's second city for much of the eighteenth century.

Bristol was the starting point of over 2,000 slaving voyages until the abolition of the trade in 1807. Ships left port for the coast of West Africa, laden with textiles, firearms, luxury items, metal ware and other goods that were exchanged for African slaves, captured by locally based slave traders. The slaves were then transported in horrific conditions to British colonies in the Caribbean and North America. On arrival they were sold at auction like livestock and faced a lifetime's labour as the personal property of their owners. To complete the triangle, goods and produce dependent on slave labour were brought back to Bristol, ensuring that each leg of the three-way passage had a profitable load. There was also direct trade between Bristol and the slave colonies in America and the Caribbean.

Most of Bristol's residents were connected in some way with this traffic. It was not just the slave-ship owners, slave agents and those merchants who had invested in slaving voyages who made money out of the slaves' suffering; those involved in ancillary industries also profited. These would have included the owners of yards where slave-ships were built; ships' carpenters, riggers and labourers; the makers of sails, chains, ropes, anchors and navigational instruments; owners of brass, iron and copper works; ship-chandlers; dealers in and producers of trading goods; sugar refiners; chocolate manufacturers; port employees; and bankers and insurance companies. Indeed, anyone who smoked a pipe of tobacco, added sugar to their tea, wore a shirt of cotton, visited a coffee house or drank a tot of Jamaican rum could be said to have benefited from the trade.

The wealth brought to Bristol was reflected in the fine Georgian and Regency houses and impressive civic and commercial buildings that were built in the city in the late 1700s and early nineteenth century. It was also reflected in philanthropic endowments and bequests made by merchants which were used to support Bristol hospitals, schools and almshouses for the elderly.

While many Christians who profited from slavery, like the Bristol merchants Edward Colston and John Pinney, squared their conscience by arguing that God had provided slaves for their use, an important anti-slavery group grew up around non-conformist Protestants and some Evangelical members of the Church of England. Foremost among these were the Quakers, a movement with strong links to Bristol, who believed slavery was profoundly wrong as all people were equal spiritually. Thomas Clarkson, a friend of the Quakers and a leading abolitionist, campaigned in Bristol, Liverpool (which had overtaken Bristol as the country's leading slaving port by the end of the eighteenth century), Manchester, Birmingham and other cities. Like many others he was influenced by new Enlightenment ideals in which all people had a natural right to freedom. John Wesley, who had preached his first Methodist sermon in Bristol in 1739, declared himself against the trade in the 1770s. Others to speak out included the poet, Samuel Taylor Coleridge, who gave an anti-slavery lecture in Bristol in 1795.

The act ending the slave trade in Britain was passed by Parliament in 1807. The end of the British slave trade resulted from the damning evidence provided by some disaffected merchants and seamen, the support of leading politicians like William Wilberforce, the general shift to a waged economy, growing grass-roots support in Britain for abolition and the continuous resistance in the Caribbean by the Africans themselves. The relative weight of these factors is still a matter for heated debate. What is beyond debate is that it was another 27 years before the slaves themselves were freed from the colonial plantations. The Emancipation Act became law

Broad Quay, Bristol, *attributed to Philip Vandyke, c 1760 (Bristol's Museums, Galleries and Archives).*

'The gradual abolition of the slave trade: or leaving of sugar by degrees', by Isaac Cruikshank, 1792 (British Empire and Commonwealth Museum, www.imagesofempire.com).

in 1834 and, even then, the slaves were expected to serve a four-year apprenticeship with their former masters as part of their compulsory 'induction' into independent life.

Following emancipation, the British government paid out £20 million by way of compensation to plantation owners. Around £500,000 of this money came to Bristol. Among those to receive compensation were Evan Baillie, a slave trader, merchant and Member of Parliament for Bristol who submitted a claim of £91,632 for the 3,100 slaves his family had 'lost', and the Pinney family, plantation owners and Bristol sugar traders, who received compensation of about £36,000. Where plantation owners had mortgaged their estates to Bristol merchants, it was the merchants who were then in a position to make a claim. Some owners and merchants became quite canny in buying up estates in anticipation of emancipation, knowing that compensation would be paid. Rather than reinvesting in the plantations to ensure their continuing survival, most of the money was spent on new ventures in British engineering and manufacturing, including canals and railways. The investors thereby made a smooth transition from an economy based upon goods produced through chattel slavery to one dependent upon steam-driven production

methods which were more efficient and reliable and, therefore, more profitable.

Parliament voted for the end of the apprenticeship system for former slaves with effect from 1 August 1838. In Jamaica, many were able to leave the plantations, taking over wasteland on the island to develop for their own use. Elsewhere in the Caribbean, few had this option and had to continue working on plantations for low wages. No compensation was paid to them for their hardships and loss.

Brunel does not seem to have made any recorded comments that would indicate his own attitude to slavery and his only known partisan political act was to support the election campaign of his brother-in-law, the Radical candidate for Lambeth, in 1832. However, as he was not unduly concerned by the welfare of his own workforce, he is unlikely to have given any consideration to the conditions in which slaves worked. He was also closely associated with those who had links to the West Indian trade.

Brunel's friend and business partner, Sir Christopher Claxton, had been a slaving agent in Antigua and came from a well-established Bristol family with plantation interests around St Kitts. Claxton was also the agent of

Evan Baillie, writing inflammatory campaign pamphlets denigrating Baillie's anti-slavery opponent using scurrilous and anti-Semitic language. In 1830 he circulated a petition in Bristol in which he argued that black people were unfit for emancipation because they were a barbarous race, citing their cruelty to animals and their poisoning of their masters. He stated that their enslavement was ordained in the Bible. Historian Madge Dresser describes Claxton as someone who made no attempt to hide his racialist attitudes under the cover of modern urbanity, unlike some of his colleagues.[1]

Among the initial promoters for the Great Western Railway (GWR) were Robert Bright, a member of a family influential in the West Indian trade, George Gibbs, a West Indian merchant (the company of Gibbs, Bright & Co received compensation for the mortgages it held on slave plantations) and Thomas Guppy (another rare close friend to Brunel), whose family ran a sugar refinery. The GWR's first directors included members of the Society of Merchant Venturers, a body that had explicitly acted to protect its members' economic interests, firstly by breaking the Royal African Company monopoly on slave trading and then by campaigning against abolition and emancipation.

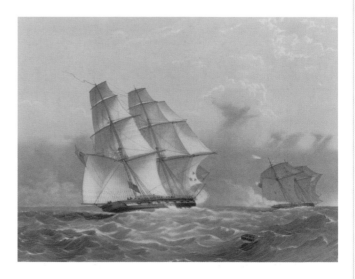

Banner calling for abolition of apprenticeships, c 1832 (British Empire and Commonwealth Museum, www.imagesofempire.com).

HM Brig Acorn, *1841 by Nicholas Matthew Condy. The* Acorn *was launched on 15 November 1838. In 1839 the ship sailed from Plymouth as part of the navy's anti-slavery operations. The* Acorn *captured a Spanish slaver,* Big Gabriel, *on 6 July 1841 (British Empire and Commonwealth Museum, www.imagesofempire.com).*

Rain, Steam and Speed – the Great Western Railway
*by Joseph Mallord William Turner, exhibited at the Royal
Academy in 1844 (National Gallery, London).*

TECHNOLOGICAL AND SOCIAL CHANGE:
THE IMPACT ON SOCIETY OF THE WORK OF BRUNEL AND HIS CONTEMPORARIES

Denis Smith

The railway was the leading technology in the nineteenth century and, together with technical developments in other fields, had unprecedented potential to change society. At the beginning of the century the fastest mode of travel over the earth's surface was on horseback on land, or on a tea clipper at sea. The railways and marine steam power changed all this. The transformation was fundamental. Technical changes gave rise to increased personal mobility, the ability to live in one place and work at a distance, the accelerated application of machine tools and the improved communication of information.

Technology provided a more varied diet, better water supply and sanitation, improved lighting, heating and ventilation of dwellings and, as a corollary to activity in all these fields, a gradual levelling of land prices across the country. By looking at changes in mobility, communications and health, the impact of Brunel and his contemporaries on society can be assessed.

Mobility

The emergence of the embryonic locomotive in the first two decades of the nineteenth century foreshadowed unimaginable changes worldwide. The steam locomotive was born in Cornwall and by-passed both Bristol and London in its development, and was brought to fruition on Tyneside in Northumberland.[1] It was destined to become *the* most important traction device in the nineteenth century. Even the earliest locomotives were heavier, and more powerful, than a horse, which led rapidly to the evolution of new design methods and the adoption of new materials in civil, mechanical and electrical engineering, rendering the old rule-of-thumb methods of design redundant. The most significant aspects of the railway were the increase in both speed and power – journey times were greatly reduced and the loads that could be hauled were dramatically increased.

The railway network necessitated huge numbers of earthworks to produce cuttings, embankments and tunnels. Brunel encountered unforeseen problems on his first contract on the Great Western Railway (GWR) at the Hanwell viaduct in the Brent Valley. The canal engineers had previously undertaken similar works, but the railway civil engineer met much greater problems. For example, an accident occurred in the Sonning cutting on the GWR on 24 December 1841 in which eight people were killed. Viscount Melbourne wrote to Queen Victoria, saying: 'The railway smash is awful and tremendous, as all railway accidents are.' His perception of the nature of the problem was sound:

> These slips and falls of earth from the banks are the greatest danger... the cuttings have been recently and hastily made, the banks are very steep, and the season has been peculiarly wet, interrupted by severe frosts.[2]

The understanding of earthwork problems was still incomplete at the end of the century, with one eminent engineer saying: 'After intervals of years, those slopes had been a perpetual torment to the engineer.'[3] All this earthmoving was undertaken by manual labour in the first half of the nineteenth century. The railway navvies lived and worked in arduous conditions, as Adrian Vaughan highlighted.[4]

The development of the steam locomotive was dramatic. For example, George Stephenson's *Locomotion* (1824) weighed just 6.5 tons, whereas by 1886 locomotives weighing 42 tons were being built. Locomotive traction relied on the friction between the wheel and the rail, so the increased weight of the locomotive reduced the

Three classes of train traveller going to the Derby in 1842, The Illustrated London News, *14 May 1892 (University of Bristol).*

Cover of The Excursion Train, *Victorian songsheet, c 1860s (Elton Collection: Ironbridge Gorge Museum Trust).*

likelihood of the driving wheels slipping when starting. The three main elements of the locomotive were the boiler, the firebox and the smokebox. A great deal of ingenuity and empirical knowledge was brought to bear on the development of these and other parts of the locomotive. Several major problems arose affecting railway safety. These included the design of iron bridges, axle-breakage, the seizure of wheel bearings and the design of brakes.

Traditional masonry bridges were used on the early railways and presented no particular problems. But when cast-iron bridges were introduced, there were problems with the greatly increased load of the locomotive and the 'hammer-blow' effect induced by the unbalanced reciprocating piston movement, which could cause a failure in the relatively brittle cast iron. However, during the decade from 1840 to 1850 cast iron was replaced by wrought iron in railway structures.

Wrought-iron axles tended to fail after running for some miles at stresses that the material tests had not shown to be critical. As there were some 200,000 axles on the railways in mid-nineteenth century Britain, there was cause for alarm. This led to the study of vibrations in France and Britain and eventually added a new technical meaning to the word 'fatigue'. This work resulted in the registering of each axle as it left the workshops and the recording of its service life.[5]

The efficiency of brakes on a train of wagons or carriages became an important issue in the 1870s and Captain Douglas Galton, who had been involved with the Board of Trade Railway Inspectorate, investigated the subject thoroughly. His conclusion was that if the railway wheels were locked and skidding, the train slowed less rapidly than if the wheels continued to rotate.[6] The study of the breakdown of lubricants was more complicated and axle bearings only began to be studied seriously in the 1880s.

The social consequences of rail transport were far reaching and affected many aspects of life, including speed of travel and the concept of time. Most of these changes were directly related to the speed of the locomotive. An eminent engineer, speaking in 1893 but recalling rail travel 50 years earlier, said:

> I well remember on two occasions travelling from Leeds to London with the mail four-horse coach, before the railway system was fully developed… the journey of 186 miles was completed in twenty-six hours… while the journey is now performed by means of the locomotive in four hours.[7]

Another advantage was the ability to live in one place and work at increasing distances, with suburban development and the commuter the result. The excursion train opened up access to the coast with a day-return ticket, and the growth of seaside pier construction was a direct consequence of this mobility. One British civil engineer, Eugenius Birch, has a unique claim to recognition in that he designed almost all the seaside piers on the English coast. He completed his first screw-pile pier at Margate

Jetty in 1853. He, and his brother John, completed piers at Aberystwyth, Blackpool, Bournemouth, West Brighton, Deal, Eastbourne, Hastings, Hornsea, Lytham, New Brighton, Plymouth, Scarborough and elsewhere. Some of these piers were parts of larger seafront improvement works. Birch was also the first to build large sea-water aquariums.[8]

Again, it was the railway, and the speed of transit, that rendered local town or village time inadequate. Railways which were aligned mainly east-west were where time differences mattered most. The mail coach traveller from London to Plymouth took 22 hours to cover the 220 miles and found the time 20 minutes later than London time – an inconsequential difference. However, the speed of the train on the GWR made such discrepancies of greater concern. Attempts to make Greenwich Mean Time (GMT) the standard were tried in the 1840s but failed, and in 1844 it was said that 'London time is kept at all the stations on the railway, which is about 4 minutes earlier than Reading time... and 15 minutes before Taunton time'.[9] But by 1852 GMT was adopted on the GWR from London to Plymouth and soon generally adopted for all clocks. In 1844 it was noted that the observatory at Clifton in Bristol contained an 'astronomical clock... accurately kept, by which Greenwich mean time may be truly known'.[10] GMT was a great boon to society generally, to timetabling on the railway and to lawyers who could arrive on time at remote law courts.[11]

The railway traveller's first contact with the railway itself was through the railway station or, more spectacularly, the terminus. It was here that the railway company wished to welcome and impress the passenger, although the technology was largely subsumed in the opulence of the architecture. The entrance façade was usually

Blackpool: Health & Pleasure, Glorious Sea, *Midland Railway poster, c 1893 (National Railway Museum/ Science and Society Picture Library)*.

flamboyant and behind it was the train shed – a word which would seem designed to keep the engineering aspects in their place. The third element in the ensemble was often the railway hotel at the terminus, which would usually be grand and comfortable. The alternative was to go to a coaching inn to climb aboard a stage-coach for an arduous road journey.

The engineering of the railway terminus often included a technologically advanced cast-iron and glass roof to the trainshed. The mid-Victorian terminus would often have its own gasworks on site for lighting the interiors, hydraulic power for lifts and turntables, and heating and ventilation systems for the offices and hotel. There was initially some reluctance on the part of the public to travel by rail, but this was overcome when Queen Victoria first used the train to travel to Balmoral. For the dead, some railway companies developed special funeral facilities on trains leading to suburban company cemeteries with their own termini, often described as *Necropolis* stations.

The nineteenth century also saw corresponding developments in maritime affairs: wooden hulls were replaced by iron and, later, steel; steam replaced sail; and screw propulsion replaced paddle-wheels. Most of these changes were in widespread use by the mid-nineteenth century. Brunel played an important role in these developments, which expanded cross-channel links with Europe, the British Empire and transatlantic commerce both for trade and leisure. There was an intriguing relationship between Brunel's engineering practice and the development of a new machine tool. Following the successful Atlantic crossings of Brunel's previous *Great Western* steamship, the

The Great Eastern *screw-engine room,*
The Illustrated London News,
10 September 1859 (University of Bristol).

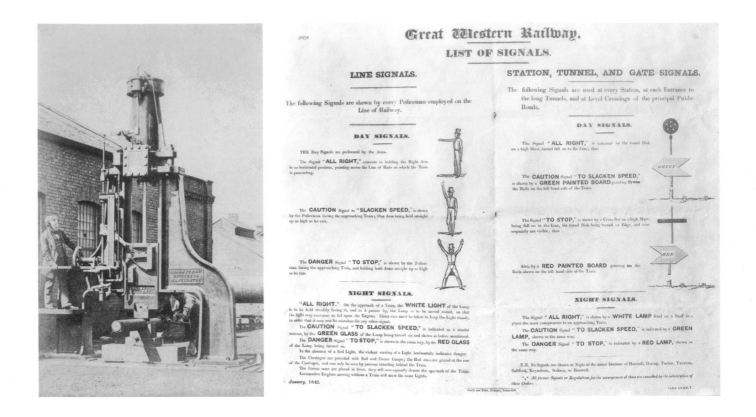

company directors commissioned a larger vessel. She was the ss *Great Britain*, whose keel plates were laid in the Bristol yard on 19 July 1839 and whose engine design required the forging of an intermediate paddle-shaft 30 inches in diameter. Francis Humphreys, who worked with Brunel, toured the ironworks of Britain only to find 'that there is not a forge hammer in England and Scotland powerful enough… what am I to do?'.[12] This led James Nasmyth to invent his now well-known steam hammer in 1839. As events turned out, the timely invention of the screw propeller led Brunel to adopt it for the *Great Britain* instead of the paddle-shaft and the steam hammer was not required by him. However, Nasmyth did build it and it was manufactured in great numbers. It was widely used in British and overseas ironworks, as well as in the civil engineering industry for driving piles for railway bridge and harbour foundation works.

Communications

The increasing speed of the railway locomotive made it necessary to have a device capable of sending a signal, or a message, which could overtake a train. Visual signalling was of great antiquity. However, the Admiralty used the shutter-type telegraph from 1796 to 1816 and the semaphore form from 1812 to 1848. Signals could be relayed from London to the Naval Dockyards on the south coast from a series of towers fitted with telescopes. The electric telegraph had its origins in scientific work in mainland Europe, but it was Britain that was first to exploit its potential and unique association with railways. There are three essential elements of the telegraph system: the battery, the conducting wire, and the send-or-receive telegraph instrument. The development of the battery was crucial to the system.

The two promoters of the English telegraph were Sir Charles Wheatstone and Sir William Fothergill Cooke, who took out a joint patent in 1837. The electric telegraph quickly became a hard-wired copper network – a Victorian internet – with enormous potential for industrial, commercial and social use. On 25 July 1837 Cooke and Wheatstone took their instrument to Euston and demonstrated their ability to communicate with each other over the wire between Euston station and Camden Town on the London & Birmingham (L&B) railway. In 1839 Cooke installed the telegraph on the GWR between Paddington and West Drayton, and by January 1843 the telegraph was extended to Slough and became the first public telegraph. The public would pay to use the GWR telegraph line on payment of one shilling. As a result, the telegraph system developed into an important part of Victorian society and played a useful role in breaking down the isolation of country districts.

In 1846 the Electric Telegraph Company was formed with six directors, including Cooke and Wheatstone and three railway men, namely Robert Stephenson, George Parker Bidder and Thomas Brassey. The telegraph was used to reduce railway accidents, relay stock-exchange prices, report home and foreign news items to the press, deliver racing results from Newmarket and accurately distribute GMT. Queen Victoria used the telegraph system to announce the birth of a son at Windsor on 6 August 1844. The electric telegraph provided a great opportunity for the employment of young men, and of women who might previously have gone into domestic service.[13] A feature of the telegraph system was the establishment in the principal towns of telegraph news-rooms. Admission was by a small annual payment and gave access to all the important and interesting news of the day.[14]

The next development in this field was the ambitious desire to cable the world by means of land lines and submarine cables. Britain had a special incentive to develop telegraphy as a means of improving communications with her Empire. Submarine telegraphy would have been technically impossible in 1840 as three essential pre-conditions were not in place at that time. These were long-distance electric telegraphy, steam engines in large ships, and the discovery of a suitable material for insulating the conductor cable. But by 1850 terrestrial telegraphy had been demonstrated successfully over 1,000 miles or more, and the necessarily large steamship was uniquely available in the shape of Brunel's *Great Eastern* with the advantages of her cable-carrying capacity and her superior forms of propulsion, comprising sail, paddle and screw. The paddle-wheels in particular were excellent at maintaining a station in mid-ocean. The insulating material, Gutta Percha, was discovered in the Malay Peninsula by a British surgeon employed by the East India

A photograph of James Nasmyth (facing page) with his steam hammer, taken by Captain Lewis, probably of the Royal Engineers, in about 1845 (NMPFT/Science and Society Picture Library).

Illustrated list of signals to be used on the GWR, dated January 1842 (National Trust).

Explanatory poster about the GWR magnetic needle telegraph (National Trust).

Company, who introduced it into Britain in 1843. Everything was then in place. Gutta Percha is contained in the sap and milky juice of the Gutta Percha tree, which coagulates on exposure to the air and 'For collecting the sap, the trees are felled, and left to dry'.[15]

The first successful underwater cable was laid across the English Channel from Dover to Calais in 1851, designed and supervised by a railway engineer, T R Crampton.[16] But it was the proposed transatlantic cable which captured the world's attention. Reassurance came when Professor Samuel Morse said in 1843 that 'a telegraphic communication on the electro-magnetic plan, may with certainty be established across the Atlantic Ocean'.[17] After a failed attempt in 1857, a transatlantic connection was made in 1858 by means of a Royal Naval ship (*Agamemnon*) and a US Navy ship (*Niagara*) meeting in mid-Atlantic, splicing both cable ends and steaming in opposite directions. In August Queen Victoria and the American president exchanged messages but the cable failed in October. A new company was formed in 1864, after a lapse of seven years, and the *Great Eastern* was converted to a cable-ship. Daniel Gooch records in his diary:

One of Robert Dudley's illustrations of the Great Eastern *during the laying of the transatlantic cable, published in W H Russell's* The Atlantic Telegraph, *1865 (Institution of Civil Engineers).*

> Our Telegraph Construction Company was formed in April 1864 and we at once set to work in making preparations for the manufacture of the cable. I spent a great deal of time & took much interest in it. We brought the ship from Liverpool to Shearness [sic] in July to fit her for the work.[18]

After a failed attempt in July 1865, a further attempt was made in 1866. The *Great Eastern* set out in July and the connection was made on 8 September. Celebrations took place on both sides of the Atlantic and The Times said:

> Since the discovery of Columbus nothing has been done in any degree comparable to the vast enlargement which has thus been given to the sphere of human activity.[19]

As a social corollary to the story of the availability of Gutta Percha, this natural hydrocarbon material was put to some unusual uses. It was used in book-binding, for the soles of shoes, in statuary and, in a flexible pipe form, as a speaking-tube in railway carriages and buildings, where it was said that 'the lowest whisper is distinctly heard;… whereby a minister may address the deaf among his congregation'.[20] Telegraphy had developed considerably between 1851 and 1872, in June of which year W H Preece of the General Post Office gave a lecture at the Royal Albert Hall. He presented the following statistics illustrating the scale of progress:

Year	Miles of line	Number of stations	Number of messages
1851	1,752	198	48,490
1862	12,711	1,616	2,676,354
1872	25,000	5,179	15,500,000

The Daily Telegraph of 19 June 1872 described the occasion as 'A telegraphic evening with a little music… Altogether the evening was one of the most instructive ever devised in connection with relaxation and amusement'.[21] The electric telegraph met competition, to some extent, from the arrival of the telephone in the 1870s.

On a different time-scale, the effect on society of 500 years of printing has been unequalled. It has been the most potent means of conveying ideas and has touched every sphere of human activity. But the most important period in the development of high-production printing was the nineteenth century. The first revolutionary invention was the mechanical manufacture of paper, which occurred at the turn of the eighteenth and nineteenth centuries. The first effective paper-making machine was introduced into Britain by the Fourdrinier brothers at Frogmore Mill in Hertfordshire in 1803. Manually operated printing machinery was replaced by the steam engine in November 1814, when it was described by John Walker in *The Times* as 'the greatest improvement connected with printing since the discovery of the art itself'.[22] In 1828 *The Times* could produce 4,000 sheets per hour, and 8,000 per hour by 1848. In 1848 the newspaper was printing on 30 acres of paper per night, 'the weight of the fount of type in constant use was 7 tons, and 110 compositors and 25 pressmen were constantly employed'.[23] Journalism and newspaper printing grew rapidly in the UK: it was said that in 1782 one newspaper was published for every 110,000 inhabitants; in 1821, one for every 90,000; and, in 1832, one for every 55,000.[24] *The Times* was the first newspaper to be printed by steam power, the first to send special correspondents 'to lie abroad' and the first to commission a war correspondent, W H Russell, who distinguished himself by reporting, by telegraphy, from the Crimean War. This technical and managerial progress was in response to the growing literacy of an ever-widening class of society.

Another agent of social change in the nineteenth century was the invention of photography. Certain techniques originated in France, but much of the development work was pioneered by W H Fox Talbot at Lacock Abbey in Wiltshire. Talbot took out a patent in 1841 for his calotype process, in which an image could be produced on iodized paper. He created early photographs of Brunel's Hungerford suspension bridge in about 1844 and the construction of Nelson's Column in Trafalgar Square in 1845. And Brunel himself appears in Howlett's iconic photograph standing in front of the stud-link chains of the *Great Eastern* on the Isle of Dogs. As the techniques of photography improved, the camera found widening uses from archaeology and engineering to family snapshots. Giving an early paper (1840) at the Institution of Civil Engineers, the lecturer prophetically suggested that:

> photographic delineations having been subjected during their formation to the rules of geometry, we may be enabled by the aid of a few simple data to ascertain the exact dimensions of the most elevated parts of the most inaccessible edifices.[25]

This anticipated photogrammetry by a century.[26]

One of a series of photographs by Robert Howlett of the Great Eastern *during construction, 1857 (Institution of Civil Engineers).*

William Henry Fox Talbot at his Reading photographic establishment, c 1845 (National Museum of Photography, Film and Television/ Science and Society Picture Library).

Health and sanitary reform

During the nineteenth century, great advances were made in the scientific study of common diseases, the development of diagnostic techniques and possible cures. These studies were increasingly undertaken in university laboratories. The ravages of such diseases as cholera, with its irregular pattern of attacks and high mortality, made the methodical gathering of statistics crucial to the emergence of the discipline of epidemiology.

One of earliest nineteenth-century experimenters in anaesthetics was Humphry Davy, who was appointed to the staff of the Royal Institution in London in 1801. Davy is known for his discovery of the anaesthetic properties of nitrous oxide, or 'laughing gas'. He had already experimented with the gas on himself, at some risk, in the Pneumatic Institution of Dr Thomas Beddoes in Bristol.

Five great scientists working in the medical field made unique contributions to the understanding of disease and the relief of suffering. They were, in order of date of birth, Edward Jenner, James Simpson, Louis Pasteur, Joseph Lister and Robert Koch.

Edward Jenner, a country doctor in Gloucestershire, used his acute sense of observation in identifying cow-pox, which when communicated to humans by vaccination gives immunity from small-pox. James Simpson in Edinburgh was a pioneer in anaesthetics, which not only greatly reduced the suffering of those undergoing surgery but also made longer surgical procedures possible. Louis Pasteur worked in Paris and Lille on fermentation in brewing and discovered that airborne micro-organisms could infect and sour beer, wine or milk. His discoveries made bacteriology an important area of research. Joseph Lister, Professor of Surgery at Glasgow University, knew that a patient who had survived surgery might die later of 'ward fever'. Conscious of the need for sterile conditions in both the operating theatre and the ward, he is often given the title of 'Father of Antiseptic Surgery'. Following the work of Dr John Snow in London, it was known that cholera was spread only by ingesting a micro-organism from polluted drinking water. But that micro-organism was not identified until 1883 by Robert Koch, in Germany. Although this work would appear to be wholly scientific, a great deal of technological development was required in building the necessary laboratory and clinical equipment and in delivering the subsequent pharmaceuticals to a waiting public.

An interesting relationship exists between the development of railways and improvements in public health associated with dietary changes. Before the railway, the transportation of perishable foodstuffs was, of necessity, a local matter and milk, fish, meat and some vegetables would either be consumed locally or taken to a nearby market by horse-drawn vehicles. The speed of rail transit made possible the delivery of fresh agricultural produce over longer distances from the country to the town. In addition, the markets themselves would often be connected to a rail-head. Before the

Royal Jennerian Society Honorary Diploma, designed by G Oben and granted to the Reverend M F Routh in 1826 (Science Museum/Science and Society Picture Library). On the left is a statue of Jenner holding a dead snake representing his conquering of smallpox. Through the arch is a view of the City of London.

railways, cattle, sheep and pigs would be driven to the market on the hoof. The railways saw an opportunity here for the movement of livestock by rail, and by the middle of the nineteenth century this traffic had become a large operation. In 1853 a reporter described the cattle department of Camden Station on the London & North Western (L&NW) railway in London, saying:

> provision has to be made for their transport by a supply of engines, trucks, pens, and landing-places at which they may be embarked or disembarked… extensive arrangements have been made on this account, and fifty waggon-loads of bullocks, sheep, calves, or pigs, may be unloaded.[27]

During 1865 records show that the L&NW railway carried 402,042 cattle, 1,260,574 sheep and 425,763 pigs, while the Great Western Railway carried 145,881 cattle, 577,534 sheep and 188,849 pigs.[28] Samuel Butler encapsulated all this activity by saying that the ulterior ends of the railways 'are the bringing or helping to bring meat or dairy produce into contact with man's inside, or wool to his back'.[29]

Working conditions in the textile mills and engineering works of the late eighteenth and early nineteenth centuries became a matter of public concern and, eventually, of government intervention. The change from cottage industry, where handloom weavers had worked from home in unregulated conditions, to textile mill produced great social problems and became a trigger for unrest. The factory system had four main characteristics: manufacture in a building provided by the owner; work supervised by the owner, his manager or supervisor; quality and quantity no longer being in the hands of the workers; and mechanical power driving specialised machinery.[30] Conditions improved gradually with the introduction of the 1833 Factory Act, under which the first factory inspectors were appointed.

The growth of Britain's population, and the marked change from rural to urban living during the Victorian period, posed many social problems. These included the lamentable public health conditions in large towns such as London and Bristol. These conditions included inadequate water supply, ineffective drainage and overcrowding in ill-ventilated and poorly heated housing. Solving these problems took the combined skills of scientists, doctors, engineers, architects, lawyers and politicians. The rising population in large cities and towns meant increased dependence on engineering and required a continuous, behind-the-scenes input of technology. Municipal authorities found it increasingly necessary to employ their own engineers to design, implement

Swainson Birley Cotton Mill near Preston, Lancashire, one of a set of drawings in pencil, pen and wash by Thomas Allom, 1834 (Science Museum/Science and Society Picture Library).

and subsequently manage a complex network of services. These could include water supply and drainage, inner-city transport systems, refuse collection and disposal, street lighting, energy distribution systems, planning controls, new buildings and dealing with dangerous structures. Most of this work would require government legislation.

The site of any large city is determined primarily by the need for a supply of potable water. Both London and Bristol began to engineer their supplies in the seventeenth century. In London it was the New River (1609) running from Ware in Hertfordshire to the City of London,[31] and in 1698 the Bristol Waterworks Company brought water in elm pipes from Hanham Mills to the city.[32] By 1850 London was supplied by nine water companies, eight of which survived in private hands until the Metropolitan Water Board was formed in 1903. In Bristol, reservoirs and a 'line of works' aqueduct had been constructed by 1850 to augment the supply of water to the city. All of this benefited the health of the local communities.

Whereas the populations of London and Bristol were prepared to pay the private companies a water rate for the delivery of clean water, it was a different story when it came to paying for improved sewage and industrial effluent disposal. With regard to such disposal, it was considered that public health should be a matter of public wealth. Outbreaks of cholera from the 1830s shocked the public. The sanitary reform movement developed strategies to raise awareness of the issues both locally and nationally. In London three figures played a particularly important role in this: Sir Edwin Chadwick, Lord Shaftesbury and Dr Southwood Smith.

It became clear that three essential elements had to be in place before the problems could be tackled: legislation, finance and engineering on a large scale. The parishes

Engraving from London: A Pilgrimage *by Gustave Doré and Blanchard Jerrold, published 1872 (Elton Collection: Ironbridge Gorge Museum Trust).*

Salt's Textile Mill, Saltaire, West Yorkshire, steel engraving by William Home Lizars, 1869 (Science Museum/ Science and Society Picture Library). As well as a textile mill opened in 1853, philanthropic mill owner Sir Titus Salt built a model village for his workers on 25 acres of land between 1851 and 1876.

SOUTHERN PUMPING ESTABLISHMENT
AT CROSSNESS.
(14 MILES BELOW LONDON BRIDGE.)
THE METROPOLITAN MAIN DRAINAGE WORKS,
will be opened at Crossness on the 4th day of April 1865
BY
IS ROYAL HIGHNESS THE PRINCE OF WALES
MEMBERS OF THE METROPOLITAN BOARD OF WORKS
JOHN THWAITES, ESQ. CHAIRMAN

*Invitation to the formal opening of
Crossness Pumping Station, 1865
(Science Museum/Science and
Society Picture Library).*

and vestries could only tinker with the problem within their boundaries. In 1848 an Act of Parliament was passed establishing the General Board of Health. It is interesting to note that during the 1848 outbreak of cholera, the General Board of Health communicated with Medical Officers of Health by electric telegraph. The Government's initial response for London was to establish a series of six *ad hoc* Royal Commissions of Sewers from 1848 to 1855. The commissioners occasionally employed consulting engineers including Brunel, Robert Stephenson, Sir William Cubitt, Thomas Hawksley and G P Bidder.

In 1849 Sir Joseph William Bazalgette was appointed assistant surveyor to the second commission – the start of a 40-year career in local government engineering. On the death of Frank Forster in 1852, Bazalgette was appointed engineer to the commission.[33] Frank Forster had already proposed an interceptor system of main drainage on both sides of the Thames. But it was Bazalgette who developed and implemented the scheme on the formation of the Metropolitan Board of Works in 1855. Of the 1,403 recorded deaths in Limehouse in 1857, 690 were of children under five years of age. Allan Clelland, the Medical Officer of Health, said:

> When the habitation of such children is an overcrowded dilapidated tenement in some close ill-ventilated court or alley, furnished with an undrained closet surrounded by untrapped sinks and festering heaps of filth, we find ourselves astonished not that so many die, but that so many survive.[34]

Bazalgette's main drainage system, covering 117 square miles of the capital, comprised separate systems north and south of the Thames, 80 miles of trunk sewers, northern and southern outfall sewers running west to east, and four lift pumping stations.[35]

In 1873 it was said that 'It is the function of the Sanitary Engineer to prevent that which the Medical Officer is called upon to detect',[36] and that during Victoria's reign 'the sanitary improvements of our cities, towns, and villages, may be said, with almost literal accuracy, to have been called into existence'.[37] It was felt that engineers were the real sanitary reformers and that 'No profession has contributed to the saving of life and preservation of health more than the municipal engineer'.[38] Describing environmental health issues and the railways in the urban landscape at the end of the century, a civil engineer said that 'it behoves the municipal engineer… to protect the inhabitants of large towns, by preserving the light and air from being obstructed by high embankments and viaducts', but he admitted that in terms of slum demolition, 'railways have been the means of clearing away the worst class of rookeries and filth'.[39]

*'The welfare of the people is the supreme law', (facing page, bottom left) a George Cruikshank caricature illustrating
a satirical poem about the pollution of the Thames, 1832 (Science Museum/Science and Society Picture Library).*

*'Monster Soup', an engraving by William Heath showing a woman discovering the quality of the Thames Water,
1828 (Science Museum/Science and Society Picture Library).*

EMIGRATION

The British Empire spread across the world in the wake of trading activity, with raw materials such as cotton, tea and rubber being imported into Britain and manufactured goods exported out. Competition for trading markets, particularly with the French and Germans, led to intensified efforts to establish British influence abroad through the scramble for Africa in the latter half of the nineteenth century and the consolidation of British rule in India. By 1901, the year of her death, Queen Victoria ruled nearly one-quarter of the world's population. Wherever possible, the British ruled through persuasion and local collaboration rather than outright occupation, as this was more cost-effective.

For British people looking to emigrate, the self-governing 'white' dominions of Australia, New Zealand and Canada were the most attractive destinations. Many more looked beyond the Empire to the USA. Over 20 million people emigrated from the British Isles between 1815 and 1914. Most of the poorer emigrants were fleeing unemployment, food shortages, clearances, deprivation and a dearth of opportunities, and were attracted by the potential for making a fresh start in a new country with higher wages and cheap land. Some had their fares paid by local charities or landlords (a cheaper option in the long term than sending them to the workhouse), while around two million were given assisted passages by colonial governments that were looking to attract skilled workers and tradesmen. Others were being sent away in disgrace by their families and would only receive their regular allowance for as long as they remained in the colonies. Reports of gold strikes added to the fervour. The emigrants soon outnumbered the indigenous populations.

On the ss *Great Britain*'s first Australian run in 1853, a steerage passage cost £16. This was 'more than half a year's wages for a farm labourer'.[1] Other passengers travelled First, Second or Third Class. The voyage was expected to last eight weeks, but the captain was forced to turn back to St Helena to buy more coal and it took 83 days to reach Melbourne. After this, the average journey for the 12,000-mile trip was around 60 days non-stop, following the eastward trade winds around Cape Horn and spending as much time as possible under sail to save coal. It is thought that the ship carried the forebears of around 250,000 modern-day Australians.

Ford Madox Brown's painting, *The Last of England*, was inspired by the emigration of the Pre-Raphaelite sculptor, Thomas Woolner, who left for Australia in 1852. He became a gold prospector but returned to England in 1854, having calculated that the £50 of gold he had dug had cost him £80 to retrieve. Brown used himself as the model for the central male figure with his wife, Emma, holding their child, posing as his companion on the voyage from Dover.

The Last of England *by Ford Madox Brown, 1855 (Birmingham Museums and Art Gallery).*

RAIN, STEAM AND SPEED

J M W Turner's dramatic painting, *Rain, Steam and Speed – the Great Western Railway*, (see page 256), can be read as a homage to technical advances. Curator James Hamilton says that by depicting 'a steam engine passing both along Brunel's line *and* over his Maidenhead Bridge *and* in a violent storm, Turner is allying himself directly with the engineer, and publicly applauding his triumph'.[1]

It is also a painting of contrasts, showing 'steam harnessed with rail by man… in the natural circumstances of a showery day… [While] below, in Old Masterish calm, some female figures in Claudean costume stand at the river's edge'.[2] The speed of the train is emphasised by the narrowness of the rail (which appears to be single track, though this is ambiguous) and the blurred effect of its rushing by, the hare running for its life ahead of it.

The painting was inspired by a train journey Turner took between Swindon and Exeter in a storm. Lady Simon, a fellow passenger, recalled the event:

> When I had taken off my cloak and smoothed my plumes, and generally settled myself, I looked up to see the most wonderful eyes I ever saw, steadily, luminously, clairvoyantly, kindly, paternally looking at me… I looked at him, wondering if my grandfather's eyes had been like those. I should have described them as the most "seeing" eyes I had ever seen…

Well, we went on, and the storm went on more and more, until we reached Bristol; to wait ten minutes. My old gentleman rubbed the side window with his coat cuff, in vain: attacking the centre window again in vain, so blurred and blotted was it with the torrents of rain! A moment's hesitation, and then:

'Young lady, would you mind my putting down this window?'

'Oh no, not at all.'

'You may be drenched, you know.'

'Never mind, sir.'

Immediately, down goes the window, out go the old gentleman's head and shoulders, and there they stay for I suppose nearly nine minutes. Then he drew them in, and I said: 'Oh please let me look'.

'Now you *will* be drenched'; but he half opened the window for me to see. Such a sight, such a chaos of elemental and artificial lights and noises, I never saw or heard, or expect to see or hear. He drew up the window as we moved on, and then leant back with closed eyes…

The next year, I think, going to the Academy, I turned at once, as I always did, to see what Turners there were.

Imagine my feelings: -

RAIN, STEAM, AND SPEED,
Great Western Railway, June the _, 1843

I had found out who the "seeing" eyes belonged to! As I stood looking at the picture, I heard a mawkish voice behind me say:

'There now, just look at that; ain't it *just* like Turner? – whoever saw such a ridiculous conglomeration?'

I turned very quietly round and said:

'*I* did; I was in the train that night, and it is perfectly and wonderfully true'; and walked quietly away.[3]

Although 'Turner and Isambard Kingdom Brunel were well known to each other... documentary evidence of personal contact has yet to be discovered'.[4] Mutual friends included C R Leslie, Clarkson Stanfield and John Callcott Horsley, who were among those commissioned to paint Shakespearean subjects for a room at Brunel's Duke Street home. Both were members of the Athenaeum Club and Turner owned an inn near the Wapping entrance of the Thames Tunnel.

J M W Turner (University of Bristol).

'The Pleasures of the Rail-road – Showing the Inconvenience of a Blow Up', hand-coloured etching
by Hugh Hughes, published 1831 (Elton Collection: Ironbridge Gorge Trust Museum).

LINES, LANDSCAPE AND ANTI-MODERNISM
UNDERSTANDING VICTORIAN OPPOSITION TO THE RAILWAYS

Marcus Waithe

The railways transformed travel in nineteenth-century Britain. But not everyone was happy with such progress. Victorian England's most successful and prolific art critic, John Ruskin, once announced that he would happily 'destroy most of the railroads in England, and all the railroads in Wales'.[1] He made his feelings even clearer several years later, on describing the 'iron road' as 'the loathsomest form of devilry now extant, animated and deliberate earthquakes destructive of all wise social habit or possible natural beauty, carriages of damned souls on the ridges of their own graves'.[2] These sentiments are expressed with uncommon ferocity, but Ruskin was not alone in voicing concerns of this kind and his loathing took hold on the wider public imagination.

Victorian opposition to the railways

No longer the embodiment of a dirty industrialism, the rail networks frantically laid out across England in the 1840s now represent the best available alternative to the country's polluted and congested roads. It requires a special effort, then, to understand why they caused so much controversy at the time of their construction. The strength of opposition to the railways is especially difficult to comprehend in the light of renewed appreciation of the men who built them. For several decades now it has been fashionable to present the ingenuity and risk-taking of Victorian engineers like Isambard Kingdom Brunel as a model for modern entrepreneurial talent.

On what grounds, then, did the Victorian public take exception to the modern miracle of mass transport? Why did an invention transparently so *useful* engender such hostility? The most common forms of animosity were associated with the sheer novelty of the railway. The 'iron road' struck unwary onlookers as a strange, sudden and other-worldly apparition. It was in this spirit that W Cooke Taylor pointed out in 1842 that the 'steam-engine had no precedent':

> the spinning-jenny is without ancestry, the mule and the power-loom entered on no prepared heritage: they sprang into sudden existence like Minerva from the brain of Jupiter. [3]

Taylor places great emphasis on the importance of precedent in English life. Customary respect for what seemed familiar and established was widespread among all orders of society. This outlook, so deeply entrenched, goes some way towards explaining why the emergence of the railway was profoundly unsettling. Instead of building upon change in an organic way, it appeared to initiate a whole new era. The speed and the novelty of this shift combined to suggest an anarchic 'spin', a perceived loss of control over the pace of change. It is an impression conveyed eloquently and poignantly by Tennyson in his poem, 'Locksley Hall': 'Let the great world spin for ever down the ringing grooves of change.'[4]

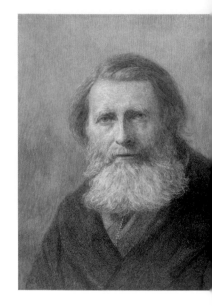

Railway accident near Dawlish, The Illustrated London News, *3 March 1855 (University of Bristol).*

Pencil drawing of an impending railway accident by Sir John Tenniel used for an engraving published in Punch, *1 August 1891 (Elton Collection: Ironbridge Gorge Trust Museum).*

John Ruskin, c 1880s (facing page) (University of Bristol).

East Coast Route to the Lake District, North Eastern Railway poster, 1898 (National Railway Museum/ Science & Society Picture Library).

Change itself is always unnerving, no matter which century or society one considers. What distinguishes the effects of the railway from the impact of other innovations is the sheer range and diversity of interests affected. It was widely feared that the human body would not be able to tolerate the speeds of which the railway was capable. Farmers were worried that the noise of passing trains would stop their cows from milking. A number of other prejudices have proven more resilient. The middle classes often opposed new lines on the grounds that the railway tarnished the social status of the residential areas it served. Snobbery of this kind was augmented by concerns that the lower orders were achieving far too much mobility as a result of cheap and fast transport. A new class of person could now reach the country's coastal retreats and spa towns. For those accustomed to the relatively intimate and exclusive environments of the Georgian stagecoach and resort, the railway's threatened democratisation of travel was hardly a welcome development.

Other sources of opposition came from those with a financial interest in the status quo. Canal owners, who feared they would lose business to this fast and cheap competitor, were in the forefront of protests. Landowners, aggrieved by the devaluation of their property caused by new lines, raised similar objections. Some of the earliest large-scale opposition was mounted by men professing a more altruistic concern with the damaging effects of railway construction. Exemplified by William Wordsworth's battle against the proposed London, Northwestern Kendal & Windermere Line, this kind of campaign had its roots in the Romantic cult of scenery and wilderness. For those involved, the idea of 'opening up' the Lake District – a landscape defined by poets so consistently in terms of its remoteness, its immaculate hills and its antique manners – seemed tantamount to vandalism.

There were, finally, those who expressed concerns directed less towards the railways themselves than towards their wider effects. New lines required the enactment of private bills proposed in Parliament by the railway companies. This procedure encouraged suspicion that the railway lobby was becoming too powerful, that it was becoming a law unto itself. Allegations of corruption and official collusion were common. Other complainants were concerned by the dangers posed by this new form of transport. Rail accidents were not rare occurrences. One famous victim was Charles Dickens. In 1865 he narrowly escaped injury when his train met a gap in the track on a bridge near Staplehurst in Kent. He, like many of those concerned by the mounting casualty rate, felt unable to overcome the sense of man's hubris in placing such trust in new technology.

Thomas Carlyle and John Ruskin: medievalist opposition to the railways

Victorian objections to the advent of rail transport ranged from selfish financial interest through to the fear that precious aspects of British life and landscape were

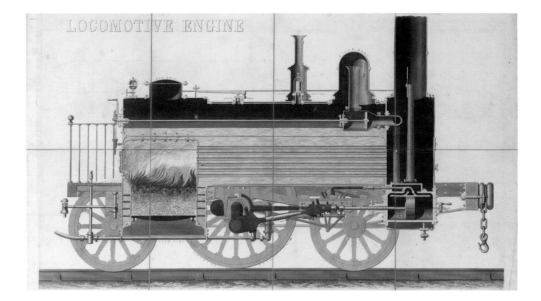

being swept away. It remains to consider a more comprehensive and multi-faceted objection to the railways than any so far treated. This comparatively ideological form of opposition was propounded by two moral commentators, the radical conservative and modern-day prophet, Thomas Carlyle, and Ruskin, his intellectual heir.

What, then, was Carlyle's contribution to the debate? Best known in his time as the author of the long essay 'On Heroes, Hero-Worship and the Heroic in History', he pioneered that brand of prophetic and doom-laden commentary that would later come so naturally to Ruskin in the face of such issues.[5] Carlyle's message in his tract on heroes was simple. The poor were dying of inanition while the rich were leading unproductive lives, bedazzled by trinkets and false idols.[6] The country required true leaders to correct this sad state of affairs. Carlyle might, perhaps, have looked to industrialists, and in particular to the new breed of engineer. In fact, he did nothing of the sort. He was not opposed to modern industry *per se*. Nor would he have found fault with the capacity of men like Brunel for hard work. But such men did not fit with what he had in mind.

Ruskin's position is similarly complex. The wilds of the Lakeland landscape inspired him with poetic feeling. But so too, on occasion, did the sublime inventions of man. Such feelings rise to the surface in *The Cestus of Aglaia*.[7] In this work, he describes 'the amazed awe, the crushed humility' with which he sometimes watched 'a locomotive take its breath at a railway station'.[8] The machine is like a monster, snorting steam and breathing fire. It is wild, and untamed. But it also represents an artifice, a kind of automaton built with 'the precision of watchmaking'. These reflections, expressed in a style distinctly resonant of Carlyle's prose, bring the makers to mind, the men who 'beat it out'. Ruskin's profound, if anxious, respect for the creations of these 'Iron-dominant Genii' can be traced back to childhood – to his first poem, in fact.

Entitled 'On the Steam Engine', it expressed similar regard for the monstrous beauty and utility of the modern locomotive.[9]

The heroic qualities that Carlyle and Ruskin associate with 'makers', whether considered as engineers or workmen, certainly seem to complicate the picture. Indeed, the untamed and Romantic poise of their creations would appear to challenge any crude distinction between natural beauty and man-made ugliness. Is it not, then, surprising that two men so well known for appreciation of heroic endeavour should become such consistent enemies of the railways? For Ruskin, we should remember, the railways were the 'loathsomest form of devilry now extant'. What prompted the apparent change of heart?

Like Dickens, Ruskin felt tremendous affection for the culture of travel that preceded the railway age. He remembered the old coaching inns and the bonhomie of the long journey. He esteemed the warmth of hospitality available at the roadside, and the gradual witnessing of changing landscape through the stage-coach window. Such memories and customs combined to produce a romantic and nostalgic vision of 'authentic' travel, far removed from the new scenery of the upholstered train compartment and the railway platform. Many of those who experienced the dying gasps of Pickwickian travel in the late Georgian and early Victorian periods were haunted in a similar way. But Ruskin, it seems, was more haunted than most. In his great multi-volume study of Gothic architecture, *The Stones of Venice*, he breaks off from the main narrative and remembers

> … the olden days of travelling, now to return no more, in which distance could not be vanquished without toil, but in which that toil was rewarded, partly by the power of deliberate survey of the countries through which the journey lay, and partly by the happiness of the evening hours… – hours of peaceful and thoughtful pleasure, for which the rush of the arrival in the railway station is perhaps not always, or to all men, an equivalent.[10]

Ruskin's conception of travel, as set out here, lays great stress on the importance of work and reward. The modern traveller, he implies, has things too easy. By seeking to cover distance at the greatest possible speed, and with as little inconvenience as possible, he loses something valuable. Such a person allows his fixation on destination to obscure the real pleasures and lessons of transit.

Most of Ruskin's comments on this subject focus on the reward of close observation, as well as on the unexpected vistas that greet the wanderer who passes through unfamiliar territory. Indeed, much of Ruskin's artistic and political philosophy hinged upon the importance of vision, a faculty that required training and hard-won experience to function effectively. He had been taken as a youth on long tours of the continent and the English Lakes by his parents. On these trips, he captured in his sketchbook the forms of wild flowers. He also sketched minerals, picturesque villages and castles. It is important to appreciate that the understanding of travel promoted here – travel considered, that is, as an activity intimate with the processes of art – was what underpinned many of his most virulent anti-railway diatribes. For it was precisely the effect of the railway to insulate the traveller from the 'responsibility' of close observation.[11] *The Seven Lamps of Architecture* contains a forceful assertion of this point:

> The railroad is in all its relations a matter of earnest business, to be got through as soon as possible. It transmutes a man from a traveller into a living parcel. For the time he has parted with the nobler characteristics of his humanity for the sake of a planetary power of locomotion.[12]

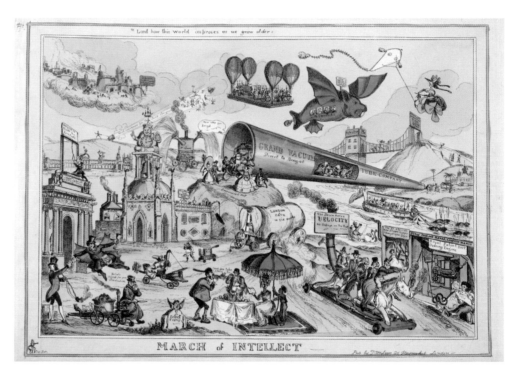

The March of Intellect, *hand-coloured etching by William Heath, c 1829, satirising the view that there was no limit to what could be achieved by harnessing science and machine (Elton Collection: Ironbridge Gorge Museum Trust).*

This description of the traveller as a 'living parcel' is clearly an extravagant outgrowth of Ruskinian rhetoric. The routes of stagecoaches, while less rigid, were also formalised and prescribed. Still, the concern outlined here – that passengers were no longer either able or inclined to take note of their surroundings – is clearly echoed in one successful painting of the era. This was Augustus Egg's *Travelling Companions* (1862). Egg's composition features two young women sitting in a carriage. They are conspicuously dressed in period finery. Outside, a dramatic and mountainous landscape looms. Yet the occupants of the upholstered interior pay no attention to what passes outside. One girl is lost in a book; the other, her companion, is fast asleep. The painting seems thus to realise Ruskin's vision of human parcels, as well as the more compelling concern that the educative value of travel was being weakened by the convenience of speedy transit.

The hostility Ruskin showed to the construction of new lines also stemmed from concern for the country's architectural heritage. In 'Samuel Prout' (1849), he wrote that 'There is not, so far as we know, one city scene in central Europe which has not suffered from some jarring point of modernisation'.[13] 'The railroad and the iron wheel,' he added, 'have done their work, and the characters of Venice, Florence, and Rouen are yielding day by day to a lifeless extension of those of Paris and Birmingham'.[14] Across Great Britain and Europe, railways encroached upon the settings of many castles and great houses, among them Furness Abbey (Cumbria), Conway Castle (Wales) and the Castle of Chillon (Switzerland).[15] Historic buildings that survived in the vicinity of such engineering projects were seen as having lost their integrity.

Ruskin's objections extended even to the state of art itself. Before considering this matter in more depth, it is worth pointing out that the railways were attracting a good deal of positive attention from other artists. In 1837 John Cooke Bourne was

Travelling Companions (above) by Augustus Egg, 1862 (Birmingham Art Gallery and Museum).

Construction of the Euston Arch, London (right), wash drawing by J C Bourne, October 1837 (National Railway Museum/ Science and Society Picture Library).

Completed Euston Arch (far right), April 1838, also by Bourne (Institution of Civil Engineers).

commissioned by the London & Birmingham Railway Company to produce a set of lithographic prints covering the Watford Tunnel, as well as work still in progress at Boxmoor, Tring and Wolverton.[16] In his *Drawings of the London & Birmingham Railway* (1839), 36 images were published. Among them, he includes a striking wash drawing of London's Euston Arch under construction. This scheme of illustrations had been suggested by the writer, topographer and antiquary, John Britton. Britton's encouragement was motivated in part by a desire to challenge widespread hostility to the railways among the public and the artistic community. Bourne's series of sketches, he hoped, would establish the railways as a fit subject matter, one worthy of artistic attention.

The popular acclaim that Bourne's work attracted laid the ground for a new genre of industrial art. It also complemented the less minute depictions of locomotives offered by Turner in *Rain, Steam and Speed* (1844), by W P Frith in *The Railway Station* (1863) and by Monet in *Gare St Lazare* (1877). So while Ruskin saw the railways as a force practically detrimental to art, other artists were embracing new opportunities. They saw aesthetic potential in the new geometric forms rising above the English landscape. In the passage from *The Cestus of Aglaia* already quoted, Ruskin makes direct allusion to the incompatibility of the artistic and industrial worlds. He wonders what the men who created the modern locomotive would think of his 'weak hand... timidly leading a little stain of water-colour'.[17] A more forthright passage in *The Seven Lamps of Architecture* leads an assault on 'artistic' attempts to decorate railway stations. Such efforts, Ruskin declares, could only be counterproductive, for a railway station is 'the very temple of discomfort'.[18] It follows that 'the only charity that the builder can extend to us is to show us, plainly as may be, how soonest to escape from it'.

There was a more profound cause of anti-railway sentiment. In simple terms, the railways symbolised a competing ideology. They were representatives of a world-view antagonistic to that brand of medievalism promoted by figures like Carlyle and

Ruskin. Understood in this context, *medievalism* implies an attraction to the relatively static social conditions of the middle ages. What, then, was the significance of this period of history? From playful beginnings in the late-eighteenth-century taste for 'Gothique' architecture, popular interest in the middle ages had matured, by the mid-Victorian period, into an historically alert appraisal of a preceding age. This form of admiration had far-reaching consequences for art, architecture, theology, private philanthropy and social policy. Although medievalism embraced a range of divergent political positions – from radical Catholic, to High Tory, to proto-Socialist – it is quite clear that the values promoted by the railway remained inconsistent with this vision, even when considered in its most general sense. Where medievalism valued the charms of rural life, the railway promoted the growth of towns; where medievalism implied revivalism, the railways symbolised defiant modernity; where medievalism relied implicitly upon fixity and loyalty, the railways facilitated movement, speed and anonymity. In simple terms, the railway seemed to reflect and promote a newly vigorous liberalism and capitalism.

For those confident in the providential merits of these forces, no problem attached to this perception. But for Ruskin, the alliance between finance, speculation, political influence and the railway's poor safety record caused serious concern. In Letter 35 of *Fors Clavigera*, he quotes from an article entitled 'Month's List of Killed and Wounded' (published in the *Pall Mall Gazette*). The author wrote of 'miserable greed, and an entire callousness of conscience on the part of railway directors, railway companies and the railway interest alike'.[19] He goes on to impute to 'the Government and the Legislature a most unworthy and unwise cowardice'.

The first idea for the picture of the Railway Station by W. Frith

Congestion in Fleet Street, an engraving from London: A Pilgrimage *by Gustave Doré and Blanchard Jerrold, published 1872 (Elton Collection: Ironbridge Gorge Museum Trust).*

The unseemly connection between the parliamentary 'railway interest' and the human cost of railway construction informed the Carlylean and the Ruskinian commentary on the environmental side-effects of urban and industrial life. Ever since constructing a sound-proof study on the top floor of his London abode, Carlyle had been an opponent of the noise pollution that disturbed his processes of composition and thought. Ruskin followed Carlyle in attributing to the noise, smoke and pollution of the railways a moral quality symptomatic of a wider, cosmic decline. In 'The Storm-Cloud of the Nineteenth Century', a dark prediction of environmental catastrophe, Ruskin drew comparison between the effects of atmospheric pollution and a darker 'plague-cloud', fed by greed and exploitation in the new industrial economy.[20] In *The Seven Lamps of Architecture*, a sense of continual unrest and excitement encourages him at points to apply a biological metaphor, suggestive of delirium and disease. The railways, he complained, aggravated 'the ceaseless fever of… life'.[21]

Before moving on to consider the more dynamic, proactive aspect of Ruskin's criticism, one last ground of disapproval needs to be dealt with. Although multi-faceted, this may be summed up broadly as a concern for the state of the landscape and for the culture supported by it. As an inheritor of classical and Romantic assumptions regarding the characteristics of idyllic locations, Ruskin placed a premium on the virtues of isolation. There is again a link here with his medievalism, especially as it informed an interest in pre-industrial economies. The places Ruskin associated with these virtues, notably the Lake District and the Swiss Alps, were not in actual fact free of trade or tourists. Still, it is clear that the arrival of the railway destroyed even the semblance of idyllic isolation. Such places were losing their geographic integrity.

They were effectively moving closer together. This is a point that Carlyle raised in his essay, 'Hudson's Statue'. The eponymous Hudson was George Hudson. Dubbed the 'Railway King', Hudson was responsible for financing huge swathes of new track in the 1830s. He gained immense political power and wealth as a result. Carlyle's essay is concerned primarily with the British public's worship of false idols, but it also contains an interesting passage on the odd spatial effects of railway construction:

> Much as we love railways, there is one thing undeniable: Railways are shifting all Towns in Britain into new places; no Town will stand where it did, and nobody can tell for a long while yet where it will stand. This is an unexpected and indeed most disastrous result. I perceive, railways have set all the Towns of Britain a-dancing.[22]

Carlyle's concern that the railways had 'set all the Towns of Britain a-dancing' communicates an amusing image. But it also expresses real anxiety that what once seemed unchallengeable in its fixity might suddenly be set loose. Ruskin's writings express fears of similarly anarchic consequences, though his concern with space is more expressive of conservationist instincts. He points out a central and abiding paradox that the railways were proposed 'to render beautiful places more accessible or habitable'[23] whilst in fact destroying that very beauty by eliminating the conditions that produced it – not least among them, inaccessibility. It is not hard to detect an anti-democratic agenda behind such sentiments, a disappointing suggestion that natural beauty should remain forever the sole preserve of the rich man and the naïve peasant.

The second element of Ruskin's attitude to landscape concerns time. Idyllic places are necessarily cut off in a temporal as well as a spatial way. Their rhythms are not those of the city, but of a place tied more closely to the passage of the seasons. Ruskin valued diversity, the kind of gentle variation experienced by the Coniston peasant who, before the arrival of the railway, would not hesitate to walk the distance to Ulverston.[24] Apart from bringing towns and their inhabitants closer together, railways had the effect of instituting a new uniformity of time. Before their arrival, time had been regulated on a local level, with considerable variance even between such places as Oxford and London. The Great Western Railway was the first to insist that all stations adhere to London time. That was in 1840. By the subsequent decade, many other railway companies had followed suit, standardising their timetables in accordance with Greenwich Mean Time. This process eventually led to an Act of Parliament in 1880 which imposed one national time. These changes naturally reduced the sense of apartness that some areas of the country had formerly enjoyed. It represented yet another far-reaching consequence of railway construction, a source of anxiety for those opposed to the centralisation and uniformity favoured by modern industrial civilisation.

*Portrait of Carlyle, 1865
(University of Bristol).*

Protests and alternative visions: Ruskin's vision of an England without railways

There is undoubtedly much to criticise in Ruskin's position, especially where his remarks imply a conception of beauty that is out of keeping with modern democratic values. Yet it is important to recognise that there was a more positive side to his engagement with this issue. This becomes clear when we grant attention to Ruskin's actions as well as his words. There was, in short, a utopian tendency underpinning his anti-railway campaigns.

By the 1870s Ruskin was living at Brantwood, a house he chose for its location, perched high above Lake Coniston. Just as Wordsworth had reacted in the 1840s to plans for a new railway in his neighbourhood with the zeal of a local resident, so Ruskin responded with deep personal feeling when plans were announced for the construction of more lines in the 1870s. His first major intervention took the form of a written warning. This was published as a supporting preface to Robert Somerville's *A Protest Against the Extension of Railways in the Lake District*.[25] Here, Ruskin voiced many of the concerns already mentioned above, together with the half-serious prediction that the railroad company would succeed only in opening 'taverns and skittle grounds around Grasmere, which will soon, then, be nothing but a pool of drainage, with a beach of broken ginger beer bottles'.[26] The minds of workmen on holiday, he declared, would be 'no more improved by contemplating the scenery of such a lake than of Blackpool'.[27]

Even more amusing than the imagery to which Ruskin resorted is the public response to his intervention. On 5 February 1876, *Punch* featured a humorous sketch of the venerable art critic and sage, entitled 'Lady of the Lake Loquitur'.[28] Ruskin the

'Lady of the Lake Loquitur',
Punch, *5th February 1876*
(Punch *Cartoon Library & Archive*).

'Amateur navvies at Oxford – undergraduates making a road as suggested by Mr Ruskin', from The Graphic, *27 June 1874, pp 612-613 (Bristol Library Service).*

medievalist is seen dressed in armour. His attire at once evokes antiquarian tastes and a liking for moral crusades. On his breastplate it is possible to make out the word, 'colour'; and in his hand, he raises not a sword but a paper knife. The aesthetic tastes symbolised by these objects, as well as by the flower trodden under foot, seem ill-suited to the task of protecting the unhappy lady in the background, or to challenging the iron machine hurtling towards them.

Around the same period, *The Saturday Review* described Ruskin as 'the Don Quixote of the nineteenth century, who makes war against chimneys and manufactories instead of windmills'.[29] The description sits well with the *Punch* cartoon, in that it suggests good intentions disabled by a tenuous grip on reality. Yet it remains the case that Ruskin's crusade did, in fact, bear fruit. It inspired younger men to take an interest in conservation. And in subsequent years, the Lake District Defence Society was established. This organisation achieved notable success in preventing further mechanised incursions into the scene of the Lakes.

Earlier the same decade, Ruskin had sponsored a separate utopian scheme. Now known as the Ferry Hinksey Road Dig, this bizarre event was occasioned by Ruskin's determination to influence a new generation of young men. While lecturing at Oxford as Slade Professor of Fine Art, he issued a warning and a call to action. Modern undergraduates, he announced, were devoting too much time to frivolous forms of exercise. They were spending their time on rowing and games, when they might instead employ their physical energies in the service of their communities. A party of his most loyal supporters, mainly members of Balliol College, decided to take the professor up on his suggestion. They laid plans to mend and drain a damaged road that ran through the outlying village of North Hinksey. Thence began the bizarre and long-running spectacle of the dig. Teams of Oxford undergraduates appeared each day to carry on the work. Dressed in sporting flannels, and lacking the muscles and staying power needed to execute this punishing task, they were laughed at by the locals and dons who came out to watch them.

The dig is interesting on several different levels. First of all, it represents a practical illustration of Ruskin's then rather unconventional belief that all classes of person should undertake physical work. More relevant is the dig's secondary objective. Ruskin was especially taken by the idea of the road as 'a Human Pathway'.[30] Such a route would work as far as possible in harmony with what he called 'a lovely country… rightly adorned'.[31] This sense of a wider context, both appropriate to the road and improved by it, was particularly important to Ruskin. He wanted to demonstrate the advantages of the scheme's human scale because he was trying to promote it as an acceptable substitute for the railroad's mechanical and intrusive presence. 'I am always growling and howling about rails,' he admitted in one letter, 'and I want them to see what I would have instead, beginning with quite a by-road through villages.'[32]

GWR promotional poster for travelling to historic Bath by rail, 1935, artwork by Dora M Batty (National Railway Museum/ Science & Society Picture Library).

The works at North Hinksey, it cannot be denied, were a failure. The road soon deteriorated into dust and mud. And in any case, it was never fully finished. Whether considered as a drainage project, as a road improvement scheme, or as an exemplary alternative to the intrusively linear and hasty methods of rail transport, Ruskin's experiment left a great deal to be desired. But the dig remains an interesting demonstration of one crucial fact: that there existed, in certain quarters, continued confidence and hope that alternatives might be found to the perceived blight of 'railway vandalism'.

Legacies

It would be easy to imagine that someone so intractably opposed to the expansion of the railways as John Ruskin would not choose to use them. This was not in fact the case. Ruskin, like William Morris, was forced to accept that the railways were here to stay. There could be no doubting their usefulness, after all. This was especially so where habitual travellers like Ruskin were concerned. In 1875 he responded to the inquiries of a female correspondent regarding the extent to which he himself relied upon the railways. Assuming a strange and haughty tone of resignation, he replied: 'I do so constantly, my dear lady; few men more.'[33] 'I use everything that comes within reach of me,' the letter unashamedly explains. It would seem, in the light of this evidence, that the dictates of utility and daily life required something quite different from the 'nobler' vision of the lectures, in which the author is 'looking always hopefully forward to the day when their railway embankments will be ploughed again, like the camps of Rome, into our English fields'.[34]

Pastel drawing of Canon Rawnsley, co-founder of the National Trust, 1895 (National Trust Picture Library).

Yet even if Ruskin's somewhat misleading distinction between utility and hope is accepted, it is apparent that he was willing at times to allow the railway to play a role within his dream of a better England. In a letter to *The Daily Telegraph* in August 1868, he asserted that neither 'the roads nor the railroads of any nation should belong to any private persons'.[35] In *Munera Pulveris*, he had indicated the need for 'quadruple rails, two for passengers and two for traffic, on every great line'.[36] These bold schemes are in many ways prescient of nationalisation, forward-thinking in so far as they demonstrate recognition that the railways were natural monopolies whose power to impose prices should be regulated. Thus despite the recurrent critical tone, Ruskin's willingness to include railways in his futurology ensured that he retained some influence over the subsequent development of the railway network. This is particularly true of the immediate post-war era. Clement Attlee, the Labour Prime Minister who presided over nationalisation in 1948, cited Ruskin as an early influence on the development of his politics.

In the 1980s these orthodoxies came under pressure. The state monopoly was broken up in 1993 and the network was split into a set of private franchises. In the intellectual arena, similar developments were occurring. Historians like Martin Wiener linked the decline in British industrial fortunes to the role Ruskin had played in popularising the idea that vigorous industrial capitalism was vulgar and distasteful. The English love of rurality, it followed, was not only misguided and sentimental but actually economically harmful.[37] Although it would clearly be inaccurate to dismiss Ruskin as an enemy of risk-taking or, indeed, of grand projects, it is clear that by the 1990s his vision of the railways as a communal asset had fallen out of favour in political circles. And yet Ruskin's concern to preserve regional diversity and to safeguard rare habitats is still very much with us. Canon Rawnsley, one of the men who took part in the dig at Hinksey, went on after his time as an 'amateur navvy' to found the National Trust, a body that now owns and protects vast swathes of coastline and countryside across Britain. In this respect, Ruskin's vision can still be felt today as a tangible influence on the country's landscape and preserved history.

Conclusion

Victorian objections to the railways, as we have seen, were wide-ranging and diverse. Some were based on a financial interest in the status quo, some on a mistaken belief that this outlandish presence would occasion harm to livestock or to people. More serious were concerns regarding accidents and the influence of the companies who built and ran the new networks. Carlyle and Ruskin interpreted the railways as the harbinger of a misguided modernity, conducive to restlessness, fever, uniformity, philistinism and environmental damage. But even these opponents of the railways were capable of seeing what was dignified, heroic and sublime in them. Looking at these arguments, and considering their environmental and social motivation, one cannot help but be struck by the fact the railways now symbolise something quite different. Since electrification, they have represented a sustainable alternative to the motorcar. Thus Ruskin's concern for conservation, and Brunel's grand engineering schemes, find themselves reconciled in the present: the railways have become symbols of the public sphere and the public spirit, an alternative to the individualism of the motorcar that is at once clean, efficient and fast.

Lord of the Isles, *GWR express passenger locomotive, built 1892 (private collection).*

BRUNEL THE CONSERVATIONIST

The spectacular Avon Gorge is the site of rare plants, limestone grasslands and ancient woodlands, and has long been a popular destination for naturalists. Brunel is credited with one of the earliest attempts to save an endangered species growing there, the autumn squill, whose pale mauve flowers appear in late summer.

In 1831 Mrs Glennie, wife of one of Brunel's assistants, told Brunel that the building of the bridge would destroy the autumn squill colony. Brunel instructed some of his workers to dig up carefully the turf that contained the bulbs and replant them further down the Gorge, out of harm's way. It has since been found that the species found in the Gorge is distinct from other colonies found in the South West, thus increasing its rarity value.

Although it would be far fetched to claim Brunel was particularly 'green', Frederick Scott has also given him some credit as a conservationist for his attempt to devise a rail route in South Devon that, as far as possible, would cause minimal disruption to its beautiful surroundings. Brunel's atmospheric system was quieter and cleaner than steam locomotion and, Scott writes, 'From his efforts can be read an intention of fitting his own work into the greater work of nature'.[1]

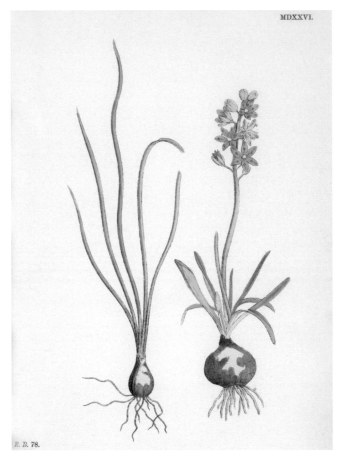

Autumn squill from Sowerby's English Botany *Vol 9, published 1869 (University of Bristol).*

Page of yew tree designs from Brunel's gardening diary (facing page) for the Watcombe estate (University of Bristol).

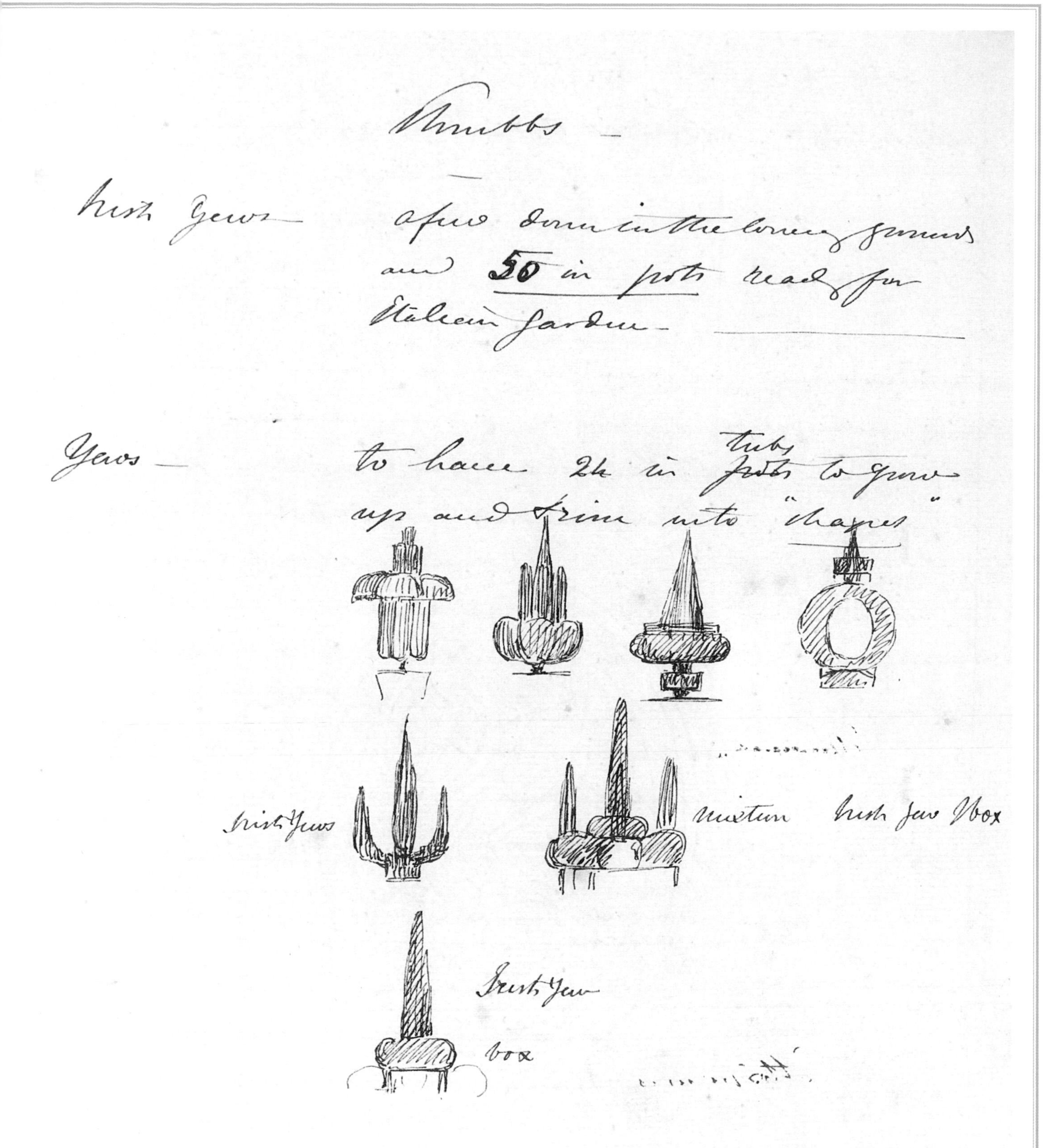

Shrubbs —

—

Irish Yews —   a few down in the lower ground
are **50** in pots ready for
Italian Garden. _____

Yews —   to have 24 in tubs to grow
up and trim into "shapes"

Irish Yews

nineteen   Irish Yew & box

Irish Yew
box

THE BRUNELS AT WATCOMBE

In 1847 Brunel purchased an estate at Watcombe overlooking Babbacombe Bay near Torquay in Devon, where he planned to build a house for his eventual retirement. Brunel had made a temporary base in Torquay while working on the South Devon Railway and the family were regular visitors to the area, involving themselves in local affairs. They hosted a garden party for the estate workers, Brunel was made patron of the annual poultry exhibition and in 1854 he helped overturn a decision to build a gas works on Babbacombe Beach by speaking in opposition to the scheme. Mary Brunel continued to visit Torquay after her husband's death.

At Barn Close in North Torquay, Brunel designed a complex of houses, schoolroom and chapel for his workers in anticipation of work commencing on his estate. It was to be paid for with money Brunel received for his contribution to the Great Exhibition of 1851, where a key theme was Model Housing. As it was, only two of the houses were built before his death.

Photograph of gardeners working at the estate (University of Bristol).

Brunel oversaw the planting of the grounds at Watcombe, with the help of the distinguished landscape gardener, William Nesfield, among others, and he devised tools for transplanting young trees to their chosen site. His son, Isambard, wrote that his father:

> had always a great love and appreciation of beautiful scenery, and in his choice of a place in which to plant and build he provided amply for his complete gratification.

> The principal view, which, if the house had been built, would have been the view from the terrace, is one of the loveliest in that part of Devonshire. On one side is the sea, and on the other the range of Dartmoor, while in front is spread undulating country, bounded by hills on the further side of Torbay, the bay itself looking like a lake, being shut in by the hills above Torquay.

> When Mr Brunel bought this property it consisted of fields divided by hedgerows; but, assisted by Mr William Nesfield, he laid it out in plantations of choice trees. The occupation of arranging them gave

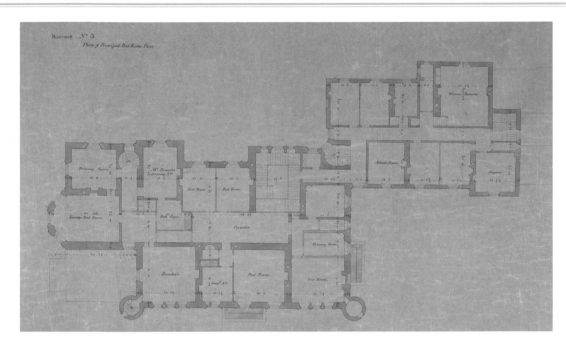

Architect's drawings of Brunel's proposed home at Watcombe (left and below left) (University of Bristol).

Photograph of Henry Marc Brunel (below) (University of Bristol).

him unfailing pleasure; and, although he could seldom spare more than a few days' holiday at a time, there can be little doubt that the happiest hours of his life were spent in walking about in the gardens with his wife and children, and discussing the condition and prospects of his favourite trees.[1]

However, the family home Brunel proposed for the time when he could at last withdraw from an active professional life 'and make room for others' was never built and he gave orders for the estate to be sold a week before he died.[2] The plantations of trees he planted now form Brunel Woods. A house, not to his design, was later built in the grounds and is now used as a conference centre.

Henry Marc, Brunel's second son, followed his father in the engineering profession and, among his

achievements, contributed to the design of Tower Bridge. Retaining the Devon connection, Henry lived for a time in Paignton at the home of William Froude, an engineer and naval architect and former assistant to Brunel. Later they moved to a house at Chelston Cross, Torquay, which served both as a home and a research centre. Henry designed an innovative 'flying' staircase for the house and in 1860 he completed a wooden bridge across the Teignmouth Road at Watcombe that had been started by his father.

Bristol passenger shed, showing decorative hammer-beam roof, illustrated by J C Bourne, published 1846 (Bristol's Museums, Galleries and Archives).

THE FUNCTION OF ORNAMENT
THE CONSOLATION OF DESIGN IN THE INDUSTRIAL AGE

Claire O'Mahony

In his classic portrait of Isambard Kingdom Brunel standing in front of the *Great Eastern*'s chains, Robert Howlett created not only a unique record of Victorian engineering innovation, but also an icon of heroic Victorian industrialism. His portrait has become associated with the interpretation of the Victorian age as a period of superlative, but also problematic, innovation. The stovepipe hat provides a thin veil of respectability over the thrusting entrepreneurialism implicit in Brunel's posture and cigar. The creased, splashed trouser legs, the disarray of his suit coat and scuffed shoes connote the 'hands-on' involvement of the inventor.

The image creates an aura of swagger, declaring the cult of individualism necessary to his achievements. He is also vulnerable, however, dwarfed by the gigantism of the chains. Who really is in control, inventor or invention?

The issue of scale is not all that dwarfs humanity in Howlett's image. The chains vie with Brunel for our attention, forcing us to contemplate the fragility of his humanity with the permanence of their silent lifelessness. The traditional, arrogant belief in the centrality and authority of humanity in the natural order is rendered fragile by his juxtaposition with unnatural machinery.

'Cropped' version of Howlett's portrait of Brunel from the collection of the Institution of Civil Engineers. For the full portrait see the section on Robert Howlett on page 37.

Steam! Steam!! Steam!!! songsheet, c 1860s (Elton Collection: Ironbridge Gorge Museum Trust) with detail, right.

Victorian art theoreticians, including John Ruskin and William Morris, argued that such destabilising effects of technology had the capacity to diminish individual, aesthetic experience. An illustrated popular song sheet of the period, *Steam! Steam!! Steam!!!*, underlines this impact of the contact with technology in a more satirical but nonetheless suggestive way. The forceful rhythmic puff of the engine repeated in the title of the song is envisioned as a locomotive bearing down upon the viewer. This prefigures the Lumière brothers' film of *The Train Arriving at La Ciotat* of 1895, which early audiences leaped out of their seats to avoid. The new technology forces aside everything in its path, threatening death and destruction. Perhaps even more worryingly, its impact is not just a single tragic incident such as a wreck; rather, tampering with human physiology, it creates a mutated, mechanised race of stick people, part machine, part abstraction, at best diagrammatically human.[1] The power of the new technology has literally dehumanised life into geometric schema.

The complex relationships between visual culture and technology in the Victorian era should not be over simplified, however. Many engineers were aesthetes; many writers and artists were not Luddites. As Herbert L Sussman has argued:

> … it would be a mistake to see the [Victorian] literary response to the machine solely in terms of opposition. The proper metaphor for culture, as Lionel Trilling suggests, is that of tension, the attempt to hold together contradictory ideas. The machine is both the unwearied servant and the sacrificial god to whom mankind has offered its soul. And as each writer criticises the destructive union of mechanization, he also admires his age for its technological skill. None are Luddites. Rather they see the machine as a servant who would be terribly useful if he would only not insist on ordering the household according to his own needs.[2]

The fascination and continued relevance of the Victorian age resides in the inextricable relationships *between* art and science, industry and aesthetics, as much as their oppositions. Indeed the most quintessentially historicist architecture of the Victorian era relied upon the new modes of transport and production created by the engineering of Brunel and his contemporaries.[3] New methods of manufacture and transport links allowed a greater variety of materials and hence designs, such as the polychromatic brick patterns, from Accrington red to Staffordshire blue brick, which are the signature style of so many Victorian buildings. Eighteenth-century innovations in cast iron allowed the possibility of load-bearing construction, transferring innovations from marine to civic architecture, as in Richard Turner's Palm House at Kew, which borrowed the ribs of a wrought-iron ship to create its majestic arches. These discoveries were used in Victorian Gothic churches as well as railway station and warehouse architecture.[4]

In their use of ornament in railway stations, Brunel and his contemporaries show how art and engineering could work together to ease the Victorian encounter with modernity. Brunel's talent typifies the interconnections in the Victorian period: inventive engineer and daring financial risk taker, he was also a knowledgeable and sociable patron of the arts. His contributions to the age were not only in creating new feats of engineering that, in turn, led to new experiences of speed and travel, but also in counteracting the depersonalising and degenerative effects of such technological innovations through the visual language of ornament – in essence creating objects that were functional *and* beautiful. Brunel's distinctive 'clothing' of the railway station undermined traditional oppositions between engineering and architecture, modernity and history, and the different classes of people that used and worked in these spaces. However, before examining Brunel's decorative interventions in detail, it would be helpful to set them and their creator in the context of contemporary debates about art and industry.

In his famous essay, 'The Signs of the Times', written in 1829, Thomas Carlyle indicated that if he were asked to characterise the nineteenth century, it would not be as 'Heroical, Devotional, Philosophical or Moral' but rather as 'the Age of Machinery, in every outward and inward sense of that word'.[5] The temptation to reduce this evocative passage, and the Victorian age it comments upon, to a simplistic conflict of culture versus technology is misleading, however. Modernity and its embodiment in 'machinery' did not simply supplant values of heroism, piety and morality; rather they transformed these conceptual modes into operating according to new sets of criteria. The impact of industrialisation was both through external, 'outward' manifestations of technological change and through the 'inward' internalisation of mechanistic modes of thought.

As the cover of *Steam! Steam!! Steam!!!* signalled, even popular illustration referred not only to the new technologies but also to the theories of social Darwinism through which their impact was critiqued. George Cruikshank's *The British Bee Hive* eloquently manifests another intersection between aesthetic and technological modes of experience and interpretation. British society is conceived as a coherent, organic whole where artistic and scientific, artisanal and professional modes of work all rely upon each other. Scientific endeavours (medicine and chemistry) sit either side of aesthetic culture (literature and art) and education (schools and colleges).[6] This visualises the inter-relatedness of concepts of industriousness and industrialisation, construed as both the moral virtue of work and the economic and cultural impact of technological innovation. As Paul Atterbury explains:

> Inextricably linked to the process of industrialisation was the increasingly stratified structure of British society, made in any case more rigid by the application of the work ethic, a reflection of a revived Protestant philosophy that had been largely dormant in the pre-Victorian period. Classification was the order of the day in a society whose values were linked to a well-defined moral code fortified by concepts such as 'improvement' and 'self-help'… Inherent within this vision, and enshrined in society was the idea that the rungs of the social ladder could be climbed only by hard work and integrity, an idea that was constantly explored by painters and writers. By this process the concept of industry achieved its own morality, with the word consciously achieving the dual meaning of hard work and industrialisation.[7]

Just as the term 'industry' can refer both to a work ethic and material realities, the intersections between artistic expression and mechanistic culture were inseparable in the Victorian age, despite the protests of writers from Ruskin to Nikolaus Pevsner.

Cruikshank's model encompasses the disparate modes of Victorian professional labour and intellectual enquiry within the ancient symbol of the beehive, rebuilt to house Victorian social theory. The image integrates vibrant, irregular nature with the forces of mechanistic and social determinism. Artisanal craftsmanship comfortably inhabits

The British Bee Hive *by George Cruikshank, 1867 (Victoria and Albert Museum).*

the structures of 'Banking', 'Free Trade' and stratified class hierarchy. For Ruskin such an orderly reconciliation of supposedly rival forces is irrevocably destructive as it rationalises and commodifies the Romantic vitality of individual expression.[8] Technology for Ruskin enslaves unless it nurtures the 'inwards' life:

> Telegraph signalling was a discovery; and conceivably, someday a useful one…You knotted a copper wire all the way to Bombay, and flashed a message along it, and back. But what was the message and what the answer? Is India the better for what you said to her? Are you the better for what she replied?[9]

The ornament in Brunel's railway stations offered a comforting message of familiarity to Victorian travellers, making the dehumanising spaces of modern travel 'better' places to be.

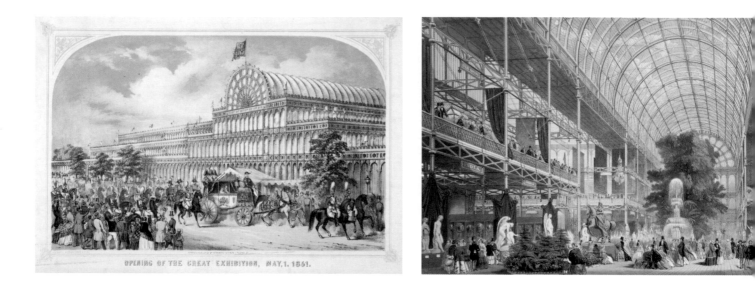

OPENING OF THE GREAT EXHIBITION, MAY 1, 1851.

Paxton's Crystal Palace, site of the Great Exhibition: exterior view showing opening on 1 May 1851 and interior view of transept, lithographs by Dickinson Brothers (Elton Collection: Ironbridge Gorge Museum Trust).

Brunel and his age of heroic industrial architectural projects have always posed challenges to art historians. Although these sites were and are praised as triumphs of engineering, many commentators deemed them incapable of aesthetic expression. The many contemporary dismissive responses to the Crystal Palace can stand for the 'dilemma' which J Mordaunt Crook has argued confronted Brunel and his peers: 'Engineering of the highest merit and excellence, but not architecture'; 'Not architecture but a packing case'.[10] Even advocates of modern materials such as J Fergusson who had hailed Paxton's achievement with vigour, said: 'Art … will not [have] been regenerated by buildings so ephemeral as Crystal palaces, or so prosaic as Manchester warehouses, nor anything so essentially utilitarian as the works of our engineers.'[11] Such spaces of modern industrial experience, from the Crystal Palace to the Bristol passenger shed, were deemed incapable of achieving a message, telegraphic or otherwise, let alone one that made the world better for its expression.

Ruskin articulates this position dramatically:

> Another of the strange and evil tendencies of the present day is the decoration of
> the railroad station. Now if there is any place in the world in which people are
> deprived of that portion of temper and discretion which are necessary to the
> contemplation of beauty, it is there. It is the very temple of discomfort, and the
> only charity that the builder can extend to us is to show us, plainly as may be, how
> soonest to escape from it… The railroad is in all its relations a matter of earnest
> business, to be got through as soon as possible. It transmutes man from a traveller
> into a living parcel. For the time he has parted with the nobler characteristics of
> his humanity for the sake of a planetary power of locomotion. Do not ask him to
> admire anything… All attempts to please him in any other way are mere mockery,
> and insults to the things by which you endeavour to do so. There was never a more
> flagrant nor impertinent folly than the smallest portion of ornament in anything
> concerned with railroads or near them. Keep them out of the way, take them
> through the ugliest country you can find, confess them the miserable things they
> are, and spend nothing on them but for safety and speed… Better bury gold in the
> embankments, than put it in ornaments on the stations.[12]

Happily, Brunel and his contemporaries did not heed Ruskin's advice. As Edward
Kaufman has argued: 'Victorian architectural theorists believed that buildings were
capable of conveying meanings in a direct and precise way, rather like books, painting
or even orators.'[13] Railway stations, as much as Augustus Pugin's churches or Morris'
Red House, are legible, purveyors of meaning. Despite their modern materials and
solutions, the industrial buildings of Brunel and his contemporaries are not

Views of two of the display areas showing the Victorian interest in both the ancient and modern: Pugin's medieval court and the machinery section, lithographs by Dickinson Brothers (Elton Collection: Ironbridge Gorge Museum Trust).

aesthetically illegible or silent in the way that Fergusson and others implied; they
deploy resonant and varied architectural idioms through which to signify the
complexities of the Victorian age. In *The Poetry of Architecture*, Ruskin draws a parallel

between portraiture and architectural ornament, arguing that the selection of particular decorative surrounds and the arrangement of windows within a villa's facade reflect both the owner's individual nature and a deeper manifestation of the character of his nation and time.[14] A consideration of the use of architectural decoration within buildings which Brunel helped to devise offers some insights into the artistic physiognomy of Brunel and his age.[15]

What were Brunel's aesthetic predilections and how did they influence him in the selection of ornamentation in his architectural projects? What is their relationship to larger aesthetic currents of the time? Tantalising glimpses of answers to such questions can be found in some of the work on Brunel, but art history remains uninterested in Brunel's projects, which appear briefly, if at all, in standard histories of Victorian architecture, and hardly ever in undergraduate survey lectures.[16] Analysis of Brunel's artistic empathies have usually been overshadowed by analysis of his technological achievements or the dramatic incidents of his life, from drowning to financial disasters.

Brunel's aesthetic formulation is signalled in Angus Buchanan's biography, most attention being focused on his formation – through the instruction of his father and the Lycée Henri-Quatre in Paris – his precocious skills as a draughtsman and the contact with creative personalities in his wife's family circle, from Callcott to

Contractor's drawing of GWR public house at Steventon, part of a complex of railway buildings designed by Brunel, c 1835 (Adrian Vaughan Collection).

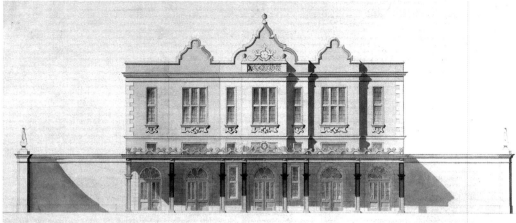

Contractor's drawing of north front of Bath station, c 1835 (Adrian Vaughan Collection) and interior view showing original roof spanning the platforms, illustrated by J C Bourne, published1846. Brunel chose a Jacobean style for Bath that contrasted with the surrounding classical influences seen in the city's Georgian architecture.

Mendelssohn, culminating in the entertainments held in his house in Duke Street, London.[17] John Binding and Steven Brindle's work on Temple Meads and Paddington respectively have outlined which aspects of each site's design and ornamentation were Brunel's contribution and which were those of his colleagues and employees, thus granting a more accurate set of evidence from which to consider Brunel's aesthetic tastes.[18]

Given Ruskin's claim that domestic fenestration reveals one's soul, providing a coded revelation of the character of the owner and his age, Brunel's garden design and his unrealised plans for an Italian villa house at Watcombe in Devon offer an intriguing, yet undefinitive portrait.[19] The 1858 inventory of the Duke Street household equally shows a man of opulent tastes in domestic interior decoration, including a large and varied collection of fine glass, china and silver services, Indian carpets and carved furniture.[20] Buchanan has implied such decorative extravagance and Brunel's painting collection undermine potential claims for his aesthetic judgement as a patron-collector: 'The choice of commissions suggests a normative, not to say conventional, aesthetic standard', reflected in Brunel's failure to purchase either the works of the Pre-Raphaelite Brotherhood or Turner's majestic evocation of the Great Western Railway *Rain, Steam and Speed*.[21]

However, Faberman and McEvansoneya argue that Brunel's project for a 'Shakespeare Room' in the enlarged Duke Street house was an ambitious and aesthetically informed project. It did reflect the widespread contemporary vogue for subjects inspired by great British authors such as Shakespeare and Milton, initially triggered by Boydell's 1786 Shakespeare Gallery, and culminating in 1,400 Shakespearean-inspired paintings being exhibited in the Royal Academy of Arts in London between 1769 and 1900, with approximately 20 each year at mid-century.[22] Rather than deriding this topicality as a sign of conventional taste, one can construe it as evidence of Brunel's awareness of the aesthetic debates and climate of his age.

Brunel planned every aspect of the Shakespeare room. He redesigned the *piano nobile* of 17 Duke Street (adjoining his original office and residence at 18). The dining room was the centrepiece, and was to be filled with Shakespearian-inspired imagery. Several biographers have commented upon the care with which Brunel decorated this space in Tudor-Elizabethan style effects, such as graining of the plaster walls to look like oak and selecting red velvet curtains and Venetian mirrors probably acquired during the family's Italian tours of 1842 or 1852.[23] Brunel commissioned the paintings for his Shakespeare room from an impressive pantheon of contemporary artists: Charles West Cope, Augustus Leopold Egg, Sir Edwin Landseer, Frederick Richard Lee, Charles Robert Leslie and Clarkson Stanfield. His sketches show how intimately Brunel planned the installation of these paintings.[24] At least three of the paintings, two scenes from *Henry VIII* by Leslie and Augustus Leopold Egg's *Launce's Substitute for Proteus' Dog* (1849) inspired by *Two Gentleman of Verona*, represent Tudor interiors like the one in which Brunel intended them to hang.

Brunel's admiration for the Tudor Gothic style, which is the keynote of Bristol Temple Meads station, was clearly not just a public statement but also a private passion. Brunel's interior designs at Temple Meads, from the specially designed newel posts for the oak grand stair and the elaborate ceiling plasterwork, to the massive carved fireplace and oak panelling in the Boardroom, deploy the Tudor Gothic style in all its vigour. The passenger shed is perhaps most audacious of all, juxtaposing magnificent timber hammer-beam ceiling rafters and an arcade of Tudor arches, all supported by declaratively cast-iron columns – tradition and modernity literally supporting each other. Pugin abhorred this bastardisation, declaring that the Great Western stations deployed 'mock castellated work, huge tracery, shields without bearings, ugly mouldings, no-meaning projections and all sorts of unaccountable breaks, to make up a design at once costly and offensive, and full of pretension'.[25]

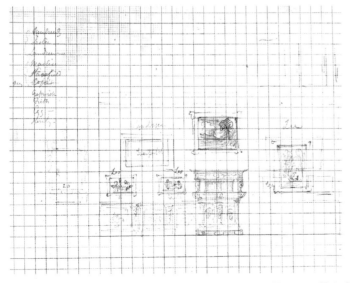

Brunel's sketches showing the proposed layout for his Shakespeare Room (University of Bristol).

The GWR boardroom (British Empire and Commonwealth Museum).

Exterior of Bristol terminus, c 1850 (Bristol's Museums, Galleries and Archives).

However, one suspects that Pugin's revivalism, rooted in Oxford Movement theology with its medieval ideals and theories of pattern and its origin in nature, was not central to Brunel's Tudor Gothic. Chris Bartram has argued for the distinct tenor of Tudor revivalism within the context of railway architecture:

> It was an architecture of action and adventure to which the vivid interplay of violence and compromise, of excitement and restraint, has given a special character utterly different from that of earlier and later buildings. The architecture of Tudor and Elizabethan England is the architecture of the new ruling class feeling its way to power.[26]

As such, Brunel circumvents Pugin's accusation of message-less eclecticism, instead creating a vibrant interplay of specific historical association.[27] The Tudor and Elizabethan period was an age of cathartic change similar to his own extraordinary era and, as such, provided a decorative mode relevant to the problematic modern sites of the railway. Thus his juxtaposition of modern and traditional modes of ornament avoids the trivial or commonplace. It instead creates a resonant dialogue emblematic of the space it adorns, a crossroads of historical change and of every echelon of society.

The Moorish Arch (left), hand-coloured aquatint by S G Hughes after Talbot Bury, published by Ackerman & Co, 1832 (Elton Collection: Ironbridge Gorge Museum Trust). This ornate architectural style seen at Edge Hill on Stephenson's London & Birmingham was less popular with railway builders than the classical Greek, Roman and Italianate, or the Gothic Revival.

One last document relating to the Shakespeare room not only offers insights into this methodology of juxtaposition but also suggests continuities between Brunel's working methods and aesthetic preferences and larger questions of private and public experience intrinsic to the railway station as cultural site. A letter to Sir Edwin Landseer about the project demonstrates Brunel's attitudes both to issues of artistic independence and the aesthetics of contemporaneity and tradition. He insists repeatedly upon the sacrosanct freedom of the artist taking precedence over himself as patron: '… the choice of subjects I leave to the artist limiting only to selections from the *Acted* and *popular* plays of the *Author*.' He hoped that 'in choice of subject and in treatment each artist would endeavour to record his own peculiar style and produce as it were a characteristic picture of himself'.[28]

Brunel thus reveals an admirable professional courtesy, a useful counterpoint to the frequent accusation that his inability to delegate reflects an unnaturally controlling persona, insisting instead upon each artist's creative authority and individualism. However, this attentiveness to individual temperament is coupled with a commitment to collective aesthetic experience and accessibility. Visual culture must be at once private and public, bearing a trace of the expressive peculiarities of an individual, but this trace must also be rooted in collective knowledge, drawn from 'popular' and 'acted' plays.[29] Such interplay between individual expression and social collectivity was deemed impossible within the context of the 'Age of Machinery' and its engineering structures. The conception and creation of the ornamental strategies at Paddington suggests another story.

In the decade between the creation of Bristol Temple Meads and Paddington, Brunel enlisted a set of ornamental idioms which achieved a greater synthesis of historical and contemporary aesthetic references. The timber hammer-beam roof of the Bristol Temple Meads passenger shed has given way to fully fledged and undisguised iron and glass construction. The Tudor Gothic mode is supplanted by a rhetoric suggestive of recognisable ornamental motifs which nonetheless avoid explicit classification. Paddington's decoration was the product of a fruitful collaboration between Brunel and Matthew Digby Wyatt.[30] Brunel recognised in Wyatt a colleague with sympathetic attitudes to metal ornament, not least because Wyatt had declared his admiration of Brunel in an 1850 article in the *Journal of Art*:

> His independence of meretricious and adventitious ornament is as great and as above prejudice as his engineering works are daring in conception and masterly in execution. From such beginnings, what glories may be in reserve, when England has systematised a scale of form and proportion, a vocabulary of its own, in which to speak to the world the language of its power.[31]

GWR Tunnel No 2 near Bristol, illustrated by J C Bourne, published 1846. When the ground slipped during construction, causing masonry to fall from the top left corner of the tunnel entrance, Brunel decided to plant it with ivy so it would look like a Romantic ruin.

In a letter of 13 January 1851 inviting Wyatt to assist in the Paddington project, Brunel recognised Wyatt as a like-minded colleague who would 'carry out strictly and fully, all those correct notions of the use of metal which I believe you and I share'.[32] The collaboration with Wyatt and its products demonstrate a remarkable parallelism of working practices and aesthetic aims. Henry Russell Hitchcock declares Paddington Station's restrained but elaborate ornament to be: '… so coherent an integration between the general handling of the interior space and the detailing of the structural elements that a true collaboration between Brunel and Wyatt on the overall design must, I think, be assumed.'[33] The purportedly incompatible professions and outlooks of the Victorian engineer and architect are denied by the synthesis of Paddington Station's ornamentation.

This collective yet harmonised conception may account for the variety and unusualness of the abstracted and yet referential ornament. The tensile grace of the geometry of iron grows organically from load-bearing function into the patterns of classical anthemia and Moorish arabesques. This ornamentation serves to create a built environment which is at once manifestly modern, and yet familiar. It creates a mood of anticipation, hinted at through allusions to the exotic, but also reassurance, in the humanising effect of an ornamental skin not laid over but integrated into the stark geometry of functional materials, amidst the potentially disorientating new spaces of travel. Ruskin was precisely wrong to declare the uselessness of decoration in the dispiriting spaces of modern travel; it is precisely in such contexts that it is most needed.

The Railway Station (detail) by William P Frith, 1862 (Royal Holloway, University of London). Full painting can be seen on page 286.

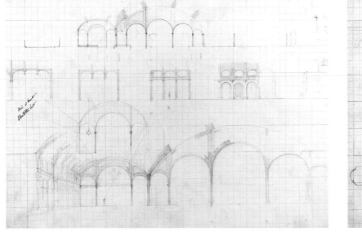

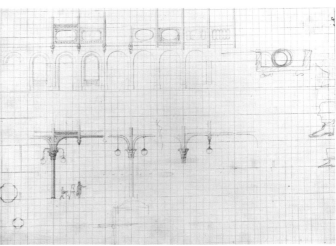

The Paddington ornamental programme helped in the negotiation of the destabilising, potentially traumatic, experiences of departure, arrival and travel. William Powell Frith's famous painting of Paddington, *The Railway Station* (1862), eloquently shows the variety of forms of psychic strain inherent in the core experiences undertaken in the built environment of the railway station. Mary Cowling

Brunel's sketches of Paddington (University of Bristol).

has persuasively demonstrated the complex discourses of social control and classification that this painting, and its enormously popular engraving, enacted upon the crowds it attracted and imaged.[34] Frith's painting, like the station it represents, not only affords frightening evidence of the perils of Victorian industrial society fermenting crime and social degeneration, but also shows the sentimental engagement with the delights of new modes of travel and communication, from honeymoons to daytrippers.

This emotional force of the space of the railway station is the final concern. The impact of the experiences and spaces of modern travel echoes across the years despite the maiming of Euston and the closure of branch lines. James Scott's 1913 account of the infinite permutations of the 'Romance' of the railway station is both wonderfully close to Frith's image in its cast of characters and timelessly evocative of this catharsis:

> What of the passengers and their friends! Who may fathom, much less portray, the thoughts and emotions surging within their minds and hearts? The grief and pain of separation, the hopes, the fears, the loving care, the prayers, the joys, the trust! How ill-concealed by some in affected gaiety of mood; how patent in others who do not attempt to conceal! Here are soldiers and sailors, with troops of acquaintances to see them off. There are some boys and girls going away to school, their fathers and mothers filling up moments of waiting with many injunctions in order to shut out their anxieties which their children must not see. At another place a wife is bidding good-bye to a husband whom duty calls hence. Elsewhere an only son and brother is setting out into the great world to win a name and place. In the corner of another carriage there sits, his face screened by a magazine, some lonely soul who has no one to bid him adieu at this end of his journey or welcome him at the other. At another compartment a happy wedding party is assembled, and amid merriment and showers of confetti the 'happy pair' are getting a good send-off. These and a hundred other scenes one may witness at the departure of such an important train as this. There may be tears or mirth, or calm demeanour, but in all the life of feeling runs high...

He continues:

> A railway station speaks of epochs of decision in life, a parting of the ways, cross-roads in conduct. Shall we embark upon this adventure; shall we definitively declare our hand; shall we make a break in habit; or a departure from principle? Are we fleeing from Nineveh and duty, or going where Love and Right beckon us? As we wait at the station are we still counting the cost, and weighing the consequences in the balance? Are we making a sacrifice in going away? Have we left a clean record behind? When the train has borne us away and we settle in to our corner, have we feelings of remorse or satisfaction? Where will the journey's end be?[35]

Modern travel and communication, and the built environment in which it transpires, is a cathartic process – at times delightful, at others traumatic. The function of ornament in these sites of memory is to provide a setting at once personal and public through which to negotiate these complexities and to make the experience of travel a good one.[36]

Railway stations and their ornament continued to serve this stabilising function in the twentieth century. In *Vision and Design*, Roger Fry echoed Ruskin's advice against decorating Victorian railway stations. However, whilst he criticises the clash of rococo brocade wall paper with classical mouldings in these Victorian spaces[37], he stumbles upon the consoling function of ornament: 'One must remember that public places of this kind merely reflect the average citizen's soul as expressed in his home.' Ornament in the modern, impersonal spaces of transport makes the disorientated traveller feel at home, saving us from moments of desperation. Resonant with the social panorama of Frith's painted crowds and Scott's dramatic description, David Lean's cinematic masterpiece *Brief Encounter* also reveals how the railway station and its decorated refreshment room can provide a space of familiarity in which to regain one's equilibrium and survive life's traumas, from coal dust in the eye to the end of an affair.[38] At the climax of the film, Lean captures Laura Jesson's despair over Alec's departure by tilting the camera as she rushes out to the platform edge just before an express train thunders past. As the camera returns to the horizontal position,

David Lean's Brief Encounter *(1945): two scenes in the station refreshment room showing the actors Celia Johnson (Laura), Trevor Howard (Alec), Joyce Carey (Myrtle) and Stanley Holloway (Albert) (British Film Institute).*

reflecting the end of her destabilisation, Laura's voice explains her decision not to commit suicide to her husband Fred. She locates her return to sanity and security explicitly in the humanised, decorated space of the refreshment room:

> I meant to do it, Fred, I really meant to do it. I stood there trembling right on the edge, but I couldn't. I wasn't brave enough. I should like to be able to say that it was the thought of you and the children that prevented me but it wasn't. I had no thoughts at all, only an overwhelming desire not to feel anything ever again. Not to be unhappy anymore. I turned. I went back into the refreshment room. That's when I nearly fainted.

Laura had become Ruskin's 'living parcel', unable to feel, risking her own destruction. The refreshment room provided a familiar, decorated haven, helping Laura to regain her humanity, fainting. Her final destination is a return to post-war domesticity, far from the adventures and traumas of modernity undertaken in and around steam trains and their stations during the Second World War.[39]

From the Euston Arch to branch line stations, many beautifully ornamental railway stations were lost in the decades of the twentieth century. The survivors are important places of memory and should be cherished. Happy or sad, sites of majestic and brutalising technology or aesthetic consolation, the railway stations of Brunel and his contemporaries capture the consoling dialogue of ornament and technology of the Victorian age, a rich repository of memory and history to be treasured, preserved and contemplated.

BRUNEL AND THE ART OF ENGINEERING DRAWING

In his biography of Brunel, L T C Rolt refers to the importance of drawing to the engineer:

> Young Isambard began to display his talent for drawing when he was only four years old, and by the time he was six he had mastered his Euclid. Such a precocious display of inherited talent obviously delighted Marc so that he determined to foster it to the limit of his means… Marc always insisted that this drawing habit was as important to an engineer as a knowledge of the alphabet, and it was undoubtedly in this way that both father and son developed such extraordinarily acute powers of observation.[1]

In studying the geometry of Euclid, Brunel would have gained an understanding of the importance of pattern, proportion and order, bringing a coherence and sense of perspective to his working drawings. He would also have learned visual techniques from studying the drawings of the architects Filippo Brunelleschi and Andrea Palladio, and Leonardo da Vinci's 'exploded' views of machines.

Brunel received his early training from his French father and at French colleges, and it was a Frenchman, the military engineer Gaspard Monge, who developed an influential drawing system for tackling complex machine forms that was introduced by the French government into the curriculum of the École Polytechnique. Other technical schools on the Continent and America took up and taught this system in the early nineteenth century, but it did not reach Britain until the 1840s. Mathematical precision, and the use of shading and colour, gave an engineer's working drawings a persuasive illusion of reality. Natural scientists of the period also needed to be able to draw accurately to record their observations and, as part of their own training, the classical curriculum for art students would have included the study of anatomy, contours and design, and life drawing, illustrating the links across diverse disciplines evident at that time.

Descriptive, geometrically accurate drawings are among the engineer's tools 'of prediction and analysis… [that] greatly expands… the capacity for sustained innovation' within design, construction and manufacture.[2] They could also be used as a form of control over the work process and, if of particularly high quality, could impress clients and promote an engineer's work. The Industrial Revolution had seen a development in the skills of draughtsmanship to meet the rapidly changing needs of the new technology. Such drawings served a practical purpose but many can also be looked on today as creations of beauty, enjoyed for their visual quality as well as for the insight they provide into engineering solutions of the time. In 1987 the Royal College of Art marked its 150th anniversary with an exhibition entitled *The Great Engineers* that, in addition to celebrating continuing developments in engineering technology, examined the close relationship between engineering, art and design. It is often overlooked that the college was formed in 1837 'to forge and maintain links with industry and to encourage an element of design within engineering by absorbing it into the process as a whole'.[3]

John Scott Russell's drawing of Brunel's Great Eastern, *published 1865.*

Brunel's own interest in art – as a collector and patron – has been mentioned earlier, as has his artistic sensibility. His portrait of Sarah Guppy, shown here, is further evidence of his artistic talent. Rolt felt that this was one of Brunel's strengths, setting him above the generation of engineers that followed. He writes:

> So long as the artist or the man of culture had been able to advance shoulder to shoulder with engineer and scientist and with them see the picture whole, he could share their sense of mastery and confidence and believe wholeheartedly in material progress. But so soon as science and the arts became divorced, so soon as they ceased to speak a common language, confidence vanished and doubts and fears came crowding in.[4]

He goes on:

> There are some few men, rare in any age, who are mysteriously endowed with such an excess of creative power that it can be truly said that they are born to greatness… That Brunel was of this company there can be no question. Creative power such as he possessed will always seek whatever outlet promises to yield it the richest satisfaction. In 15th-century Italy, he might have been [a] great painter, sculptor and architect, in 17th- or 18th-century Europe a master of the baroque, but because he was born in 19th-century England he became an engineer.[5]

Portrait of Sarah Guppy, sister of Thomas, painted by Brunel, from the album of Grace Guppy, c 1836 (Nicholas Guppy collection).

BRUNEL: THE LEGACY

Andrew Kelly and Melanie Kelly

Brunel left two legacies: a physical legacy in his ships, bridges, railway lines and buildings, and a legacy of inspiration and vision which continues to have an impact today. Brunel is, more than ever, a mentor to artists, designers, project managers and scientists, among many others. At the same time it is accepted, albeit reluctantly, that there may never be another individual like Brunel: someone who unites art and science in grand plans and who can think laterally, raise funds widely, accept risk and failure freely, be ready to work on a problem tirelessly until it is solved and, in the process of all this, capture the public imagination.

Too much stands in the way of a Brunel appearing today. Why is this? Setting aside the issue that few would wish to contemplate the kind of punishing workload that Brunel accepted, the impact of legislation and financial constraints plays its part, as does increasing educational specialisation within schools. The need to make quick returns to shareholders is another obstacle. So too is the split between art and science, something that Brunel and other engineers of that era would have found impossible to contemplate.

This has been seen as a problem for some time. In December 1989, commenting on the frustration felt at the development of British road and rail systems, *The Economist* said that engineers needed to 'pick up where they left off a century ago':[1]

> Isambard Kingdom Brunel, pictured famously with a cigar in his mouth and mud on his trousers, probably spent more time talking to parliamentary committees than designing railways. Many of the great civil engineering projects of the nineteenth century were run by engineers who also had to be designers, managers and entrepreneurs all at the same time. [T]oday's engineering industry [does not] produce Brunels. Modern engineering consultants are technicians as different from nineteenth-century entrepreneurs as a cameraman is from a film producer.

That is probably true. But many of the challenges faced by Brunel continued to arise on engineering projects long after his death and these have been met with creative approaches that would doubtless have gained his approval, resulting in such recent triumphs as the landmark performance space Sage Gateshead (2004) and the beautiful cable-stayed Viaduc de Millau (2005). Although no one individual is likely to emulate the full range of Brunel's expertise again, many Brunel-like characteristics can be seen in outstanding engineering work of today.

James Dyson, inventor of the bagless vacuum cleaner, cites Brunel as one of his inspirations, praising his can-do approach and his creativity, while criticising today's education and training, lack of imagination, the failure to invest in research and development, and the short-term nature of many businesses and governments. One of Dyson's mentors, the founder of Rotork, Jeremy Fry, also cited Brunel as an inspiration. Like Dyson, Fry was an entrepreneur who encouraged a spirit of inventiveness and creativity in his staff, as well as being a good organiser of people and projects and an outstanding marketeer. Like Brunel, he was sometimes difficult to get on with, but this was countered by his enthusiasm and optimism.

The Science Museum's 2005 exhibition *Building to the Limits* used a range of twenty-first century structures to demonstrate how innovative and eye-catching designs are being created 'that are safer, stronger, smarter, greener and more eye-catching than ever before'.[2] Such structures are 'future-proof' as they are built to withstand the demands of natural and man-made extremes (desert, flood, wind, earthquake, terrorist

Foster and Partners were the architects of Sage with consultation from Arup Acoustics, Mott Macdonald, Buro Happold and others (photograph by Mark Westerby).

attack), have a lasting iconic identity, are sustainable in their use of resources and aim to combat the problem of climate change. Although some of these concepts would have been alien to Brunel – the wider implications of energy use, for example, were of little concern to the Victorian engineer beyond seeking greater fuel efficiency for his transport systems – the combination of risk-taking, experimentation, inspiration and determination needed would have been shared by Brunel, the Stephensons, Daniel Gooch, Joseph Paxton and other nineteenth-century innovators.

And it is this approach that continues to inspire engineers today. Peter Kydd, Parsons Brinckerhoff's director of environment, safety and risk management, said:

> Brunel was a first class engineer with a far-reaching vision. He was interested in creating new markets, introducing people to travel by making it easier and less costly. He was also a prolific and ambitious engineer with an entrepreneurial flair, although like a lot of entrepreneurs, not all of his projects were financially successful. However, it is his vision that is the most appealing of his qualities and the one that we need in today's world more than ever. The term visionary implies ahead of its time, future–proofed, ambitious, state of the art and innovative.

He added:

> Meeting the engineering challenges of today in a sustainable and effective manner requires all of these qualities. However, there is also something else – and that is the emotional energy that is also implied in being visionary: the creative, imaginative and appealing nature of visionary projects will become ever more important as the pressures of global economics result in more commoditised design solutions for the great civil engineering projects of the future.[3]

Parsons Brinckerhoff were appointed in 2005 to support the London Underground Ltd Tunnel Cooling Team as engineering advisers. Kydd said:

> The project has previously been described as mission impossible because of the ever-changing tunnel air currents, heat generated from the trains themselves and possible future air conditioned trains. We are looking at sustainable and visionary methods to achieve this – and the fact that identifying any solution, let alone the correct one, is in itself a challenge, would have appealed to Brunel.

Other major projects undertaken by the company include the Boston Central Artery – the largest and most complex urban transportation project ever undertaken in the USA – and Jumeirah Palm Island off Dubai, which will be the largest man-made island in the world.

Mark Whitby, founding partner and design director of Whitbybird Ltd, said that the qualities of Brunel that his company carries forward are 'tenacity in the face of challenges that seem insurmountable and the desire to create value for our client and society at large'. He continued:

> Brunel was the nineteenth-century heroic engineer. Today's engineers work in a softer environment where they live up to their twenty-first century definition of the profession of civil engineering being the practice of maintaining and improving the built and natural environment. This means they need softer skills – such as communications, project management, teambuilding – as well as those more often associated with Brunel.

View of the Central Artery (Parsons Brinckerhoff). The project comprises 161 lane-miles of interstate highway, over half of it underground, and includes the world's widest cable-stayed bridge to date, North America's deepest underwater immersed tube tunnel connection and an unprecedented ground-freezing programme to stabilize Boston's historic soils during construction.

BBC W1 PROJECT

Langham Place, London W1

whitbybird

www.whitbybird.com

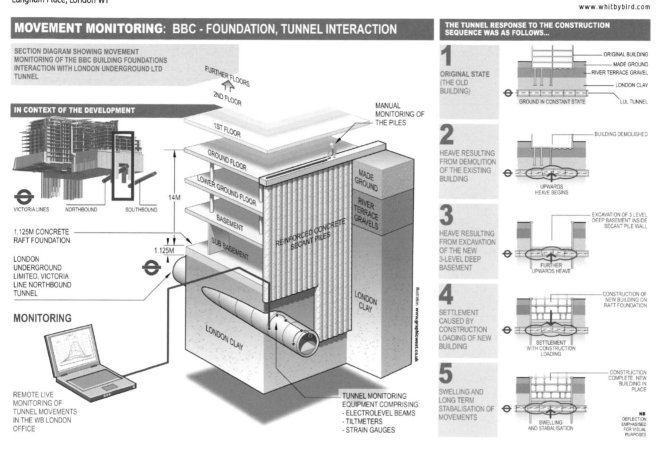

MOVEMENT MONITORING: BBC - FOUNDATION, TUNNEL INTERACTION

SECTION DIAGRAM SHOWING MOVEMENT MONITORING OF THE BBC BUILDING FOUNDATIONS INTERACTION WITH LONDON UNDERGROUND LTD TUNNEL

FURTHER FLOORS

2ND FLOOR

IN CONTEXT OF THE DEVELOPMENT

1ST FLOOR

MANUAL MONITORING OF THE PILES

GROUND FLOOR

LOWER GROUND FLOOR

MADE GROUND

VICTORIA LINES   NORTHBOUND   SOUTHBOUND

BASEMENT

RIVER TERRACE GRAVELS

14M

1.125M CONCRETE RAFT FOUNDATION

SUB BASEMENT

1.125M

REINFORCED CONCRETE SECANT PILES

LONDON UNDERGROUND LIMITED, VICTORIA LINE NORTHBOUND TUNNEL

LONDON CLAY

MONITORING

LONDON CLAY

REMOTE LIVE MONITORING OF TUNNEL MOVEMENTS IN THE WB LONDON OFFICE

TUNNEL MONITORING EQUIPMENT COMPRISING:
- ELECTROLEVEL BEAMS
- TILTMETERS
- STRAIN GAUGES

THE TUNNEL RESPONSE TO THE CONSTRUCTION SEQUENCE WAS AS FOLLOWS...

1 ORIGINAL STATE (THE OLD BUILDING)

ORIGINAL BUILDING
MADE GROUND
RIVER TERRACE GRAVEL
LONDON CLAY
GROUND IN CONSTANT STATE
LUL TUNNEL

2 HEAVE RESULTING FROM DEMOLITION OF THE EXISTING BUILDING

BUILDING DEMOLISHED
UPWARDS HEAVE BEGINS

3 HEAVE RESULTING FROM EXCAVATION OF THE NEW 3-LEVEL DEEP BASEMENT

EXCAVATION OF 3 LEVEL DEEP BASEMENT INSIDE SECANT PILE WALL
FURTHER UPWARDS HEAVE

4 SETTLEMENT CAUSED BY CONSTRUCTION LOADING OF NEW BUILDING

CONSTRUCTION OF NEW BUILDING ON RAFT FOUNDATION
SETTLEMENT WITH CONSTRUCTION LOADING

5 SWELLING AND LONG TERM STABILISATION OF MOVEMENTS

CONSTRUCTION COMPLETE. NEW BUILDING IN PLACE
SWELLING AND STABILISATION
NB DEFLECTION EMPHASISED FOR VISUAL PURPOSES

Illustration: www.graphicwest.co.uk

Diagram by Paul Weston showing movement monitoring at BBC W1 (Whitbybird Ltd). The project consists of the construction of two new, linked buildings and the refurbishment of Broadcasting House. The architecture is by MacCormac Jamieson Prichard. Whitbybird is providing structural, façade, geotechnical and fire engineering. The expected competition date is 2007.

For Whitby, the BBC West One broadcasting complex is an example of a project that demonstrates a Brunellian approach. 'This is a heroic rebuild of the heart of British broadcasting,' he said, adding:

> After it is opened the engineering will be taken for granted but the creation of this structure has all the hallmarks of Brunel's legacy of tough engineering, hard analysis, detailed negotiations – particularly with London Underground – alongside the huge constructional logistics provided by Bovis, the construction managers, and the client in the form of a private-public partnership by Land Securities Trillium. Through this effort we have been able to re-engineer the whole of the BBC's West One operation, consolidating their news-broadcasting onto the single site and creating a new structure to house a national institution that is leading the broadcasting challenges of the twenty-first century.

From Arup (formerly Ove Arup and Partners), Tristram Carfrae, Arup Fellow, said:

> The engineering challenges today are as rich and diverse as in Brunel's time. At the core of our work should be creativity and the desire to innovate. My true passion lies in designing and creating interesting structures that do more with less. While the concept of sustainability is relatively new, the qualities that are needed to see it realised, reflect those of Brunel.

He added:

> One is a determination to succeed – to stay true to the vision – while others around you may easily give up or think it too hard. The second is collaboration: the ability to enrol others and then work together towards a common purpose. In essence they are simple qualities, and yet determination and collaboration will help to achieve remarkable legacies.

Carfrae believes that the National Swimming Centre for the 2008 Beijing Olympics, a dramatic structural design known as the Water Cube, which is based on the natural formation of soap bubbles, illustrates these qualities. He said: 'It is a fine example of a creative, determined team working together. A shared vision of a box made of bubbles that will become a lasting legacy for the people of China.'

Mike Glover of Arup said he associated Brunel with 'Brilliance, courage and tenacity in dealing with bureaucracy, failing contractors and ground-breaking engineering challenges'. He described the Channel Tunnel Rail Link (CTRL) project as 'a true

Arup won the international design competition for the National Swimming Centre in July 2003 in a consortium with the architecture firm PTW, the China State Construction and Engineering Corporation (CSCEC) and the CSCEC Shenzhen Design Institute (Computer-generated image Arup + CSCEC + PTW).

View of the Severn Bridge (Neil McFadyen, Flint & Neill Partnership).

The Millennium Bridge was designed by Wilkinson and Eyre Architects, and Gifford and Partners. It won the Royal Institute of British Architects Stirling Prize 2002 (Doug Hall, I2i Photography).

Brunellian venture' because it required 'applying the above qualities in huge volumes, and overcoming enormous resistance and obstacles to create this important national asset'. He said the rail link, 'like Brunel's contributions to our infrastructure, will help the wealth of the nation for centuries'. Arup's association with the link, the first new railway in England for over 60 years and the first to operate at high speed (300kph), began in 1988. Section one, running from the tunnel near Folkstone to Fawkham Junction in North Kent, was opened in 2003. Section two, from North Kent to St Pancras, will open in 2007.

There are other projects that might be described as Brunellian in spirit, including four major suspension bridges: the Forth Road Bridge (completed 1964), Severn Bridge (1966), Bosphoros Bridge (1973) and Humber Bridge (1981). All four involved the firm of consulting engineers Freeman Fox & Partners and the designer William Brown. Charles Fox, who founded Freeman Fox in 1857, had been responsible for the design, fabrication and erection of the Crystal Palace, working with Joseph Paxton and William Cubitt. All three men were knighted for their work.

Another is the building of the Hong Kong and Shanghai Banking Corporation headquarters, Hong Kong, completed in 1986. An innovatory structure comprising two rows of steel masts linked by five levels of two-storey suspension trusses, it was designed by the British architect Norman Foster with the engineers Arup. Ove Arup himself, founder of the company, has been praised for his multi-disciplinary work and was described as 'the Robert Stephenson of the 20th century: a designer who could organise and inspire simultaneously, capable of generating loyalty, dictating quality and delegating responsibility with the certainty and infallibility of a papal edict'.[4]

More recent has been the remarkable transformation of the North East of England, starting with Anthony Gormley's *Angel of the North* in 1998 and followed three years later with the Gateshead Millennium Bridge. Both are beautiful and iconic structures that have become symbols of a confident region.

Brunel would have liked these projects. He would have liked their boldness and beauty. He would have admired their ambition. He would have liked the fact that many of the engineers involved saw him as an inspiration. There may be some misgivings expressed about current attitudes to risk-taking, concerns about the ways people are trained and developed, frustration sometimes at the cost of bureaucracy that inhibits innovation, but the spirit of Brunel remains alive and strong in contemporary engineering.

All that is good. But there is a wider potential. It is often felt that the progress that characterised the period since the Enlightenment, typified most of all in the nineteenth century, came to an end in the twentieth century. Fears now, in the twenty-first, of global warming and climate change, population growth exceeding resources available, terrorism and war, and the potential end of oil supplies, may mean that progress, at least insofar as it can be described as incremental and irreversible improvements in knowledge and living standards for the majority of people, may be at an end. Optimism, humanism and freedom from fear – the very essence of Brunel – are in short supply.

If that is so, visionaries are needed. Some may doubt that the image of a man in a stovepipe hat, with mud-spattered boots, standing in front of oily chains, has a connection with the twenty-first century engineer, or indeed society. Some may fear, as Dyson does, that engineering is seen as a relic of a bygone age. However, Dyson still believes that 'the fact that Brunel's functional and inspiring works which put Britain at the forefront of structural engineering continue to exist not only physically, but also in the psyche of the British public' shows that 'his legacy is something that should inspire us and that we should aspire to replicate'.[5]

Daniel Gooch's diary entry, on hearing of the death of his friend, was quoted in the introduction to this book. What is not quoted is the end of that entry. Gooch wrote: 'The commercial world thought him extravagant; but although he was so, great things are not done by those who sit down and count the cost of every thought and act.'[6] Brunel built the modern world in the nineteenth century. His inspiration can help to build the new world now.

The engineering behind the art of the Angel of the North was provided by Arup (Doug Hall, I2i Photography).

APPENDICES

ENDNOTES

Introduction

1 Quoted in Rolt, L T C (1957) *Isambard Kingdom Brunel* Longmans: London p 298.

2 Clark, Kenneth (1969) *Civilisation* BBC and John Murray: London p 331.

3 Lord, John and Southam, Jem (1983) *The Floating Harbour: a landscape history of Bristol City Docks* Redcliffe Press Ltd: Bristol.

4 See Channon, Geoffrey (1985) *Bristol and the Promotion of the Great Western Railway* Bristol Branch of the Historical Association: Bristol.

5 Hutton, S (1907) *Bristol and its Famous Associations* J W Arrowsmith: Bristol p 35.

Isambard Kingdom Brunel

1 For a more detailed account of events mentioned in this chapter, see Buchanan, R A (2002) *Brunel: the life and times of Isambard Kingdom Brunel* Hambledon and London: London. Also Brunel, I (1870) *Life of Isambard Kingdom Brunel, Civil Engineer* Longmans, Green and Company: London (reprinted by David & Charles Reprints, 1971) and Rolt, L T C (1957) *Isambard Kingdom Brunel* Longmans: London.

2 See, for instance, Beamish, R (1862) *Memoir of the Life of Sir Marc Isambard Brunel* London, and Noble, C B (1938) *The Brunels, Father and Son* Cobden-Sanderson: London.

3 Personal Journal, Brunel Collection, p 28.

4 The best account is still MacDermot, E T (1927) *History of the Great Western Railway*, I, London (revised edition printed 1964 by Ian Allan).

5 Corlett, E (1975) *The Iron Ship: the story of Brunel's* Great Britain Moonraker Press: Bradford-on-Avon gives a full account of the ss *Great Britain*.

6 See Binding, J (1997) *Brunel's Royal Albert Bridge* Twelvetrees Press: Truro. On Brunel's contribution to girder bridge development, see Buchanan, R A and Jones, S K (1980) 'The Balmoral Bridge of I K Brunel' in *Industrial Archaeology Review* 4 pp 214-226.

7 On Brunel's contribution to the water towers erected at Sydenham for the Crystal Palace, see Buchanan, R A, Jones, S K and Kiss, K (1994), 'Brunel and the Crystal Palace', *Industrial Archaeology Review* 17 pp 7-21.

8 The Brunel Collection has been curated with great professional skill by the Bristol University Library, and as usual I am glad to acknowledge the expert help of the current archivists, Michael Richardson and Hannah Lowery.

9 Private Diary, Brunel Collection, 5 December 1831.

10 Personal Journal, Brunel Collection, p 3.

11 Beamish, op. cit. p 234.

12 Smiles, S (1862) *The Lives of the Engineers* (3 vols) London, did much to establish this sense of an Heroic Age for British engineering, but he did not include the Brunels in his biographical studies of the leading figures because he could not get access to the documents which were being jealously guarded by the Brunel family, with ideas of their own about publication.

John Callcott Horsley

1 Buchanan, R A (2002) *Brunel: the life and times of Isambard Kingdom Brunel* Hambledon and London: London p 195.

2 Brunel, I (1971) *The Life of Isambard Kingdom Brunel, Civil Engineer (1870)* David and Charles Reprints: London p 506-507 (originally published Longmans, Green and Company).

The Howlett photograph

1 Powell, R (1985) *Brunel's Kingdom: photography and the making of history* Watershed Media Centre: Bristol p 52.

Marc Isambard Brunel

1 Smiles, S (1863) *Industry Biography: iron workers and tool makers* John Murray: London p 215.

2 Clements, P (1970) *Marc Isambard Brunel* Longmans: London pp 25, 26.

3 Through the recent scholarship of Jonathan Coad we now know much more about the relationship between Bentham and Brunel, and have an invaluable insight into Bentham's pre-existing plans for Portsmouth, which facilitated the acceptance of Brunel's system. See Coad, Jonathan (2005) *The Portsmouth Block Mills: Bentham, Brunel and the start of the Royal Navy's industrial revolution* English Heritage: London.

4 Gilbert, K R (1965) *The Portsmouth Blockmaking Machinery* Science Museum: London.

The Thames Tunnel

1 Quoted in Rolt, L T C (1957) *Isambard Kingdom Brunel* Longmans: London p 30.

2 Ibid. p 36.

Sophia Kingdom and the Reign of Terror

1 Our thanks to Nigel Overton of Plymouth Museum & Art Gallery and his colleagues for their ongoing research into the possible Kingdom connection to that city.

2 As described in Noble, C B (1938) *The Brunels, Father and Son* Cobden-Sanderson: London pp 95-97.

The Nineteenth-Century Engineer as Cultural Hero

1   'The catalogue of the Great Exhibition' in *Edinburgh Review* (Oct 1851) p 574 quoted in Mckean, John (1994) *Crystal Palace: Joseph Paxton and Charles Fox* Phaidon Press: London p 21.

2   MacLeod, Christine 'James Watt, heroic invention, and the idea of the industrial revolution' in Berg, Maxine and Bruland, Kristine (eds) (1998) *Technological Revolutions in Europe: historical perspectives* Edward Elgar: Cheltenham & Northampton Ma pp 96-7.

3   Rolt, L T C (1974) *Victorian Engineering* Penguin: Harmondsworth p 163; Rolt, L T C (1957) *Isambard Kingdom Brunel* Longmans: London pp 318-21. Rolt's views were grist to the mill of Wiener, Martin J (1981) *English Culture and the Decline of the Industrial Spirit 1850-1950* Cambridge University Press: Cambridge p 30. Cf Buchanan, R A (1989) *The Engineers: a history of the engineering profession in Britain 1750-1914* Kingsley: London.

4   Rolt, L T C (1974) ibid. pp 194-5; Smith, E C 'Memorials to engineers and men of science' *Transactions of the Newcomen Society* 28 pp 137-9; *Westminster Abbey: Official Guide* (1997) Dean and Chapter of Westminster: London pp 18-24; MacLeod, Christine (forthcoming) *Heroes of Invention: celebrating the industrial culture of nineteenth-century Britain*.

5   Celebrity status was itself a novel phenomenon in the early nineteenth century, promoted by the industrialisation of printing: Tom Mole 'Are celebrities a thing of the past?' http://www.bris.ac.uk/researchreview/2005/1115994436 [13/5/05].

6   Chrimes, M M, Elton, J, May, J, Millett, T (eds) (1993) *The Triumphant Bore: a celebration of Marc Brunel's Thames Tunnel* Science Museum: London pp 25-9; Falconer, Jonathan (1995) *What's Left of Brunel* Dale House: Shepparton pp 24-9.

7   Trench, Richard and Hillman, Ellis (1993) *London under London: a subterranean guide* 2nd edn John Murray: London, front cover.

8   Ibid. pp 104-15; Chrimes et al, op. cit. pp 21-3; Chrimes, Michael 'The engineering of the Thames Tunnel' in Kentley, Eric, Hudson, Angie and Peto, James (eds) (2000) *Isambard Kingdom Brunel: recent works* Design Museum: London pp 26-33.

9   Klingender, Francis D (1972) *Art and the Industrial Revolution* Paladin: London p 141.

10 Vaughan, Adrian (1991) *Isambard Kingdom Brunel: engineering knight-errant* John Murray: London p 175.

11 Ibid. p 197.

12 Quoted in Freeman, Michael (1999) *Railways and the Victorian Imagination* Yale University Press: New Haven & London p 202. See also Trench and Hillman, op. cit. p 113.

13 'George Stephenson' line engraving by T L Atkinson (1849) after John Lucas (1847) National Portrait Gallery London reprinted in Hart-Davis, Adam (2000) *Chain Reactions: pioneers of British science and technology and the stories that link them* National Portrait Gallery: London p 151.

14 Printed advertisement by Messrs Henry Graves & Co with obituaries and order form 4pp [1848] in the Library of the Institution of Mechanical Engineers London.

15 *The Times* 1 Aug 1844 p 6a; MacKay, Thomas (ed) (1905) *The Autobiography of Samuel Smiles* John Murray: London p 135.

16 Buchanan (1989), op.cit. pp 80-2.

17 Bailey, Michael R (ed) (2003) *Robert Stephenson – the eminent engineer* Ashgate Publishing Ltd: Aldershot pp 333-5; Chew, Kenneth and Wilson, Anthony (1993) *Victorian Science and Engineering Portrayed in* The Illustrated London News, Alan Sutton for the Science Museum: Stroud & Dover NH pp 54-5.

18 *The Times* 29 Aug 1851 pp 7a-b. In 1850 Stephenson had 'gratefully and respectfully' declined Queen Victoria's offer of a knighthood (following her visit to the new 'Royal Border Bridge' over the river Tweed): ibid. 3 Sept 1850 p 4f.

19 Bailey (ed), op. cit. pp 231-2, 235; frontispiece and plates 18, 19. For an assessment of Stephenson's achievement and a comparison with Locke and Brunel see ibid. pp 297-9.

20 Ibid. pp xxi 233-6. He was also president of the Mechanicals (1849-53) and the Civils (1856-7) and a Fellow of the Royal Society.

21 *The Times* 13 Oct 1859 p 7c-d. For Brunel's obituary see ibid. 19 Sept 1859 p 7c.

22 Chew and Wilson, op. cit. p 46.

23 Falconer, op. cit. p 18.

24 Bailey (ed), op. cit. pp xxi, 236; *The Illustrated London News* 29 Oct 1859 p 423; *The Times* 15 Oct 1859 p 12f; 20 Oct 1859 p 7f; 22 Oct 1859 p 7b; *The Builder* 29 Oct 1859.

25 Buchanan, R A (2002) *Brunel: the life and times of Isambard Kingdom Brunel* Hambledon and London: London pp 46-9; Buchanan, R A and Williams, M (1982) *Brunel's Bristol* Redcliffe Press: Bristol pp 15-23; Rolt (1957), op. cit. pp 315-16.

26 Buchanan and Williams, ibid. pp 24-7; Falconer, op. cit. pp 119-23.

27 National Archive WORK20/253 no 1: *Brunel Memorial* [printed subscription list nd].

28 *The Times* 31 Mar 1860 p 5c; 17 June 1861 p 12a.

29 Buchanan (2002), op. sit. p 227.

30 *The Times* 26 Jan 1869 p 10f; Brunel, Isambard (1870) *The Life of Isambard Kingdom Brunel, Civil Engineer* Longmans, Green and Company: London p 520n.

31 Bristol University Library [BUL] Brunel MSS Henry Marc Brunel Letter Book 8 ff 50-6. A quatrefoil is a round window composed of four equal lobes, like a four-petaled flower, and is a common pattern in Moorish and Gothic architecture.

32 *The Times* 26 Jan 1869 p 10f. The Brunel window was removed to the south side of the nave in 1952: Falconer, op. cit. p 145. The Locke window is said to have been destroyed during the Blitz: *Westminster Abbey Official Guide* (1966) p 119, but E C Smith reported that it had been taken down before the First World War: Smith E C (1951-3) 'Memorials to engineers and men of science' p 138. I have been unable to find any description of it other than this reference to its colour-scheme.

33 *The Times* 29 July 1862 p 11f; *The Builder* 20, 26 July 1862. The executors were George Robert Stephenson, his cousin, and George Parker Bidder, his friend (both of them civil engineers), in company with his solicitor Charles Parker: Jeaffreson J C (1864) *The Life of Robert Stephenson FRS* Longman, Green, Longman, Roberts and Green: London p 253.

34 A cinquefoil is a five-lobed circle or arch, usually applied in windows and panels in Gothic architecture.

35 *The Builder* 20, 26 July 1862 p 537; 2 Aug 1862 p 557; *The Times* 29 July 1862 p 11f; *Westminster Abbey: Official Guide* (1997) p 23.

36 *The Times* 13 Mar 1862 p 5f; depicted in *The Illustrated London News* 22 Feb 1862 p 195.

37 Mackay (ed), op. cit. p 255; Jeaffreson J C (1864), op. cit. For Jeaffreson see 'Jeaffreson, John Cody (1831-1901)' by G H Martin *Oxford Dictionary of National Biography* [*ODNB*] 2004-5 Oxford University Press: Oxford.

38 Smiles, Samuel (1874) *Lives of the Engineers: the locomotive: George and Robert Stephenson* new edn John Murray: London vol 5.

39 Brunel, op. cit.; Rolt (1957), op. cit. p 328.

40 Mackay (ed), op. cit. p 255. Smiles is credited with the review of Beamish's biography of Marc Isambard Brunel in *The Quarterly Review* 104 (1862): Buchanan (2002), op. cit. p xvii.

41 *The Builder* 20, 20 Dec 1862 p 901. For recent assessments of Locke see Haworth, Victoria 'Inspiration and instigation: four great railway engineers' in Smith, Denis (ed) (1994) *Perceptions of Great Engineers: facts and fantasy* London & Liverpool pp 55-83; 'Locke, Joseph (1805-1860), by G C Boase' rev. by Ralph Harrington *ODNB*.

42 Rolt's bibliography is revealing: Rolt (1957), op. cit. pp 328-9, 332-3. See also Buchanan (2002), op. cit. pp xv-xix and Bailey (ed), op. cit. pp xxi-xxii.

43 Stanton, Corin Hughes (1985) 'Stylish pictures put Brunel on top' *New Civil Engineer* 22/29 Aug 1985 p 14 (offprint in BUL). See also Buchanan (2002), op. cit. p xi and the commemorative plaque to mark the centenary of Paddington station in 1954 repr in Falconer, op. cit. p 65.

44 Rolt (1957), op. cit. p 317; repr in Dugan, Sally (2003) *Men of Iron: Brunel, Stephenson and the inventions that shaped the modern world* Channel Four Books: London Basingstoke & Oxford p 180.

45 National Archives WORK20/253: W Cowper to Lord Shelburn 17 June 1860. For the controversy over Jenner's statue and its ignominious removal from Trafalgar Square to Kensington Gardens see Empson, John (1996) 'Little honoured in his own country: statues in recognition of Edward Jenner MD FRS' *Journal of the Royal Society of Medicine* 89 pp 516-18. I am grateful to David Mullin, Director of the Jenner Museum at Berkeley (Glos), for this reference.

46 National Archives WORK20/253: same to same 3 May 1861.

47 National Archives WORK20/253: C Manby to W Cowper 1 July 1861; Devey, Joseph (1862) *The Life of Joseph Locke, civil engineer* Richard Bentley: London p 354.

48 National Archives WORK20/253: Marochetti to W Cowper 27 June 1867; W Cowper to Marochetti 3 July 1867.

49 National Archives WORK20/253: Memos dated 12 Dec 1868. It cannot have helped their case that the committees had commissioned bronze statues on red granite pedestals, when Cowper had stipulated Sicilian marble on grey granite to tone with Canning's statue: ibid. Lord Shelburn to W Cowper 15 June 1861; W Cowper to Lord Shelburn 20 June 1861.

50 BUL Brunel MSS Henry Marc Brunel to Isambard Brunel 11 July 1871 Letter Book 8 f 276.

51 National Archives WORK20/253: C Manby to Office of Works 8 May 1871; Memo dated 20 June 1871. Ward-Jackson, Philip 'Carlo Marochetti, sculptor of Robert Stephenson at Euston Station: a romantic sculptor in the railway age' unpubl paper pp 1-2, 7-8; I am grateful to Philip Ward-Jackson for kindly allowing me to read this paper. Robert Stephenson's statue is still in place although its present surroundings scarcely set it off to its best advantage; George Stephenson's is now in the National Railway Museum York.

52 National Archives WORK20/253: Memos dated 15 June 1871 5 Apr 1877; C Davenport to Office of Works 3 Apr 1877; *The Times* 3 July 1871 p 11a; 18 Nov 1871 p 6d; 3 May 1877 p 11a; 5 July 1877 p 7b; 30 July 1877 p 11b. *The Builder* was far from complimentary about the new masonry into which the statue was inserted in 1877: 13 Oct 1877 p 1035.

53 *The Times* 29 Apr 1864 p 7f; 1 Jan 1866 p 9d; 19 Jan 1866 p 12e; Devey, op. cit. pp 353-4. The statue remains in situ: my thanks to Amanda Howe for driving me to see it.

54 *The Times* 16 Apr 1861 p 10e; 19 Jan 1866 p 12e. The latter report said that £3,000 had been given to the Grammar School which provided ten scholarships.

55 *The Builder* 20, 26 Apr 1862 p 296.

56 MacLeod (forthcoming), op. cit.

57 *The Scotsman* 20 Jan 1893 p 5; *The Times* 23 Jan 1895 p 4f; Meyer V C (1930) *James Watt and the Watt Institution: Papers of the Greenock Philosophical Society 1929* Greenock pp 18-20.

58 MacLeod (forthcoming), op. cit.; Fara, Patricia (2002) *Newton: the making of genius* Macmillan: Basingstoke & Oxford.

59 *The Times* 27 Mar 1851 p 8b; Smiles, op. cit. p 354. The statue, with a book open on the knee, uses the traditional Roman iconography of the intellectual. My thanks to Gillian Clark for this observation.

60 *The Times* 17 Dec 1851 p 7f; 3 Jan 1853 p 5f; 11 Apr 1854 p 10a. For Lucas's painting see *The Times* 5 June 1849 p 5f.

61 *The Times* 11 Apr 1854 p 10a.

62 Mackay (ed), op. cit. p 162. In Smiles' estimation, as much as he admired Robert, George was the greater engineer – as his son himself modestly admitted: ibid. p 255.

63 Ibid. p 163.

64 By the end of the century, Smiles estimated this had risen to 60,000 copies in Britain alone but, to keep this in perspective we should note that his *Self Help* (1859) sold 258,000 copies by 1905: Mackay (ed), op. cit. pp 221, 223.

65 MacLeod, Christine 'Concepts of invention and the patent controversy in Victorian Britain' in Fox, Robert (ed) (1996) *Technological Change: methods and themes in the history of technology* Harwood Academic Publishers: Amsterdam pp 143-5.

66 Mackay (ed), op. cit. pp 255-6. For an assessment of the accuracy of Smiles' and other accounts of Stephenson, see Jarvis, Adrian 'The story of the story of the life of George Stephenson' in Smith, Denis (ed) (1994) *Perceptions of Great Engineers: fact and fantasy* Science Museum for the Newcomen Society, National Museums and Galleries on Merseyside, and the University of Liverpool: London & Liverpool pp 35-46.

67 *The Builder* 10 Oct 1857 p 581; 17 Oct 1857 p 598. See also the engraving of the cottage and disapproving text in *The Illustrated London News* vol 33 9 Oct 1858 pp 323-4.

68 *The Times* 14 Feb 1860 p 12f; 10 June 1881 p 7e. Robert Stephenson had been a major benefactor of the school, which he intended as a memorial to his father. George Stephenson's birthplace is now in the keeping of the National Trust which itself is indicative of his standing.

69 *The Times* 29 Oct 1858 p 8e; *The Builder* 2 Apr 1859 p 240; 7 May 1859 p 317; 10 Sept 1859 p 599; Thomas Oliver, Architect (1858) *The Stephenson Monument: what should it be? A question and answer addressed to the subscribers* 3rd edn Newcastle upon Tyne pp 5-7 and passim.

70 *The Times* 15 Sept 1862 p 10b; *The Builder* 31 Mar 1860 p 204; 6 July 1861 p 468; Smiles, op. cit. pp 355-6. An engraving of the statue may be found in *The Illustrated London News* vol 41 1 Nov 1862 p 456. For Stephenson's invention of a miners' safety lamp, the priority of which he disputed with Sir Humphry Davy see *The Times* 20 Jan 1818 p 3c-d repr. from the *Newcastle Chronicle*. A full account of Stephenson's monument in Newcastle may be found in Robert Coll's 'Remembering George Stephenson: genius and modern memory' in Coll, R and Lancaster, B (eds) (2001) *Newcastle upon Tyne: a modern history* Philimore: Chichester pp 267-92.

71 *The Times* 3 Oct 1862 p 10b; *The Builder* 11 Oct 1862 p 736.

72 Pevsner, Sir Nikolaus (1957) *Northumberland* Penguin: Harmondsworth pp 307-8.

73 *Newcastle Chronicle* 5 Nov 1858 repr. in *The Illustrated London News* vol 33 11 Dec 1858 p 555.

74 The monument was erected by Lord Ashton in 1906: 'Williamson, James Baron Ashton (1842-1930), by Eric J Evans' *ODNB*.

75 *The Builder* 17, 18 June 1859 p 401. It reported that Stephenson's statue was about to be sculpted by Thomas Woolner: *The Builder* 1 Dec 1860. Pevsner however attributes it to Joseph Durham: Sherwood, Jennifer and Pevsner, Sir Nikolaus (1974) *Oxfordshire* Penguin: Harmondsworth p 282.

76 *The Illustrated London News* vol 40 4 Jan 1862 pp 25-6. See also Dugan, op. cit. p 29.

77 *The Engineer* 15 Apr 1881 p 278.

78 Duncan, William (ed) (1881) *The Stephenson Centenary, 1881* new edn Frank Graham: Newcastle 1975 pp 8-9, 11-19, 39-45.

79 *The Engineer* 10 June 1881 p 430; 17 June 1881 p 449; *The Times* 10 June 1881 p 7e-f.

80 *The Times* 12 Feb 1881 p 12b; 15 Feb 1881 p 10e; 12 March 1881 p 7f; 10 June 1881 p 7f.

81 *The Engineer* 29 Apr 1881 p 314. Also see *The Times* 18 Oct 1877 p 4a-c; 16 July 1879 p 4f.

82 *The Times* 10 June 1881 p 7f.

83 *The Times* 4 Nov 1879 p 3c.

84 *The Times* 28 Oct 1879 p 6c. In 1925 Italian railwaymen presented a large bronze plaque 'in honour of George Stephenson' to mark the centenary of the Stockton and Darlington Railway: *Railway Magazine* July-Dec 1925 pp 130-3. It is now on display at the National Railway Museum York; my thanks to John Clarke of the NRM (NMSI) for providing information about it.

85 My thanks to William Kingston for drawing these statues to my attention and to Tim Cole who, while visiting Budapest in 2004, kindly took photographs of the statues.

86 Ironbridge Gorge Museum; for the poetry visit *Literature Online* http://lionchadwyck.com (1/7/05); MacLeod *Heroes of Invention* (forthcoming). For the legend of Watt's kettle see Robinson, Eric (1956) 'James Watt and the tea kettle: a myth justified' *History Today* 6 pp 261-5; Miller, David Philip (2004) 'True myths: James Watt's kettle his condenser and his chemistry' *History of Science* 42 pp 333-60.

87 'George Stephenson' engr. by J R G Exley for The Institution of Mechanical Engineers London: National Portrait Gallery Archives.

88 The Institution of Civil Engineers' memoir found Brunel less 'sound' and 'practical' than Robert Stephenson, succeeding rather through 'his intuitive skill and ready ingenuity'. 'The characteristic feature of his works was their size and his besetting fault was a seeking for novelty where the adoption of a well-known model would have sufficed': BUL Brunel MSS DM1289 *Memoir of Mr Isambard Kingdom Brunel, excerpt from the annual report of the Institution of Civil Engineers 1859-60* (1862).

89 Chew and Wilson, op. cit. p 15.

90 Brunel University in Uxbridge received its royal charter in 1966. The BT telephone directory for Bristol 2004/5 lists 26 firms and organisations whose names begin 'Brunel'.

91 City Museum and Art Gallery Bristol. The details of the 30 figures who are featured may be found at: http://www.discoveringbristol.org.uk/results.php?s=&categoryId[]=8&p=3 (27/6/05).

92 Buchanan (2002), op. cit. p xix.

93 BUL Brunel MSS DM1039 Brunel Society newsletter March 1979 no 27 (typescript).

94 BUL Brunel MSS DM1039 Brunel Society newsletter 1968-92 nos 1-47.

95 Falconer, op. cit. pp 80-5.

96 BUL Brunel MSS DM982 *Isambard Kingdom Brunel at Broad Quay & Paddington* [1982] Bristol and West Building Society (leaflet); DM1039 Brunel Society newsletter Nov 1982 no 33; Falconer, ibid. pp 15-40.

97 http://www.bbc.co.uk/history/programmes/greatbritons.shtml (28/6/2005).

The Brunel statues

1 Quotation provided by John Sansom who wrote this section.

2 The reference was to *Brunel's Bristol* by Angus Buchanan and Michael Williams which Redcliffe Press published for Bristol & West and republished in its own name in a revised version with new illustrations in 2005.

'Suspensa Vix Via Fit' – the saga of the building of the Clifton Suspension Bridge

1 The bridge would only have been economically viable if considerable property development had taken place on the Leigh Woods side and/or the proposed Portbury deep-water port had been constructed in the 1840s. Bridges' ambitious design reflected the anticipation of a Bristol building boom that was unrealised. In the event, it was only in the 1930s with the increase in private motor car traffic that the toll income became significant. The debts arising from the construction of the bridge were not discharged until 1952 when the Trustees completed purchase of the share issue. From the outset, the project was chronically under-funded with unrealistic estimates and poor financial control. A sequence of political and economic events led to lack of confidence within the business sector who perceived more attractive investment opportunities in railways, shipping, overseas trade and building projects. See Portman, D (2002) 'A business history of the Clifton Suspension Bridge' *Construction History* 18, pp 3-20 for further details.

2   The full wording from the will is '... I am of the opinion that the erecting of a stone bridge over the River Avon from Clifton Down in the county of Gloucester to the opposite side on Leigh Down in the county of Somerset for carriages as well as horse and foot passengers toll free would be of great publick utility.'

3   Portman, op cit, p 5. The quote is attributed by Portman – John Mathew Gutch 1776-1861 wrote 12 letters using the pseudonym Cosmo to Felix Farley's Bristol Journal concerning 'The impediments which obstruct the Trade and Commerce of the City and Port of Bristol'. They were published in 1823. The quotation is from this latter, publication VIII, pp 1-53.

4   Pugsley, A (ed) (1976) *The Works of Isambard Kingdom Brunel: an engineering appreciation*, Institution of Civil Engineers/University of Bristol: London and Bristol p 52 – quoting from Porter Goff, R F D (1974) *Brunel and the Design of the Clifton Suspension bridge*. Proc. Instn Civ. Engrs, Part 1, 56, 303-321.

5   Pugsley, ibid. p 53; Rolt, L T C (1957) *Isambard Kingdom Brunel* Longmans: London p 54: 'Brunel had allowed for the effect of wind pressure, notably by use of extremely short suspension rods at the centre of the span, thus bringing the chains almost down to the level of the platform, by transverse bracing and by the addition of inverted chains as introduced by his father...' 'Catenary' is a term used in mathematics to describe the curve created by a flexible chain or cable hanging from two points and acted upon by a uniform gravitational force.

6   Dugan, Sally (2003) 'The Great Gorge' *Men of Iron – Brunel, Stephenson and the inventions that shaped the modern world* Channel 4 Books: London p 39.

7   Brunel, I (1870) *The Life of Isambard Kingdom Brunel, Civil Engineer* Longmans, Green and Company: London p 51.

8   The 1830 Clifton Suspension Bridge Prospectus. University of Bristol Brunel Archive.

9   Brunel, op. cit. p 52.

10 Rolt, op. cit. p 55.

11 Pugsley, op. cit. p 53; quoting from: Beamish R (1862) *Memoir of the Life of Sir Marc Isambard Brunel* Longmans: London.

12 Proceedings of Trustees 1830-1900 – University of Bristol, Special Collections – Brunel Archive.

13 Body, G (1976) *The Clifton Suspension Bridge – an illustrated history* Moonraker Press: Bradford on Avon pp 16 and 19.

14 Proceedings of Trustees 1830-1900 – University of Bristol.

15 See Pugsley, op. cit. p 55-56. The chains on this bridge are not round metal hoops linked to each other like the anchor chain on a boat. They are similar to those on a bicycle chain only much, much bigger. These are called bar chains. Each section of chain is made up of ten flat metal bars 7.6m long (25ft) laid side by side. These look like spanners except that the ends are closed with holes in them. They are connected to each other by large bolts, which pass through the holes at either end. The design of the joints between the links is very important. It keeps the weight of the structure to a minimum but maintains its strength. (Modern computer analysis has proved that Brunel's design was close to the ideal.) The shoulders also allow for damaged sections of chains to be removed. This was a condition set by Gilbert in 1831. A main shape of a suspension bridge is like that of an upside-down arch. To build this curved shape the chains were first joined as single links and adjusted to form the required profile or shape. Once the correct profile had been constructed, nine parallel links were added to each section. Then the two additional sets of chains were built above the first. On this bridge three sets of chains were needed to carry the weight of the road structure and its traffic and remember that this bridge was designed before the invention of motor cars.

16 From Wright, Lewis (1866) *A Complete History of the Clifton Suspension Bridge*. The first official guide book.

17 Rolt, op. cit. p 58.

18 Wright, op. cit. p 11.

19 Ibid.

20 Vaughan, Adrian (1991) *Isambard Kingdom Brunel: engineering knight-errant* John Murray: London.

21 Body, op. cit. p 91.

22 Clifton Suspension Bridge Trust Minute Book, 31 March 1838.

23 An early forerunner of the modern cable stay suspension bridge eg. the second Severn crossing. James Dredge ('brewer of Bath') took out his patent in 1836. His system consisted 'of making the chains of sufficient magnitude and strength at the points of suspension, to support ... the greatest ... load ...; and from thence, to taper ... them gradually to the middle of the bridge, where strain becomes essentially evanescent', thus attempting to use the material in equal stress through the length of the chain... (Turnball, William, London, Weale 1841) Elton Engineering Books Catalogue No. 4 1989 p 8.

24 Regrettably, there is not space here for a thorough appraisal of Brunel's structural designs for this bridge. By far the most cogent account remains that of Sir Alfred Pugsley, op. cit. chapter 3.

25 Wright, op. cit. p 14.

26 Ibid.

27 Speech by the chairman, Captain Mark Huish at the first general meeting, August 2, 1861 – from Brunel, op. cit.

28 The road deck is constructed from two systems of girders joined at right angles to each other. The two parallel longitudinal girders are the main element suspended from the chains by the suspension rods. These girders can be seen separating the road from the footpaths. They were built in five-metre (16-foot) sections riveted together. In profile they look like a capital letter I. Under these are cross girders which are bolted at right angles to make a rigid frame. The two types are tied together by diagonal ties which are designed to brace the deck against horizontal wind pressure.

29 When cars and people cross the bridge they add extra weight, which is called a load. This changes depending on the amount of traffic crossing. These loads cause the chains to sag very slightly. The chains also expand and contract with different seasonal temperatures throughout the day or night. This means that the chains must be able to move. To cope with this need each tower has an iron saddle on top, to which the chains are bolted. The saddles move according to the load by sliding slightly back and forth on solid iron rollers. They each weigh 30 tonnes. These saddles provide a strong but flexible link between the chains in the centre with those at either end. They also transmit the load on to the towers. There are also ground saddles which perform a similar task at the point where the chains enter the anchorage tunnels at either end of the bridge.

30 Wright, op. cit. p 20.

31 Ibid. p 21.

32 Henry Marc Brunel, Letter Book, 5, p 270, 29 November 1864 (05270).

33 Henry Marc Brunel to W Froude: Henry Marc Brunel Letter Book, 5, p 270, 29 November 1864.

34 Body, op. cit. p 46. Quote ascribed to 'a local reporter'.

35 Joseph Chaley, French (1795-1861); John Roebling, American (1806-1869); Charles Ellet Jnr, American (1810-1862).

36 The Clifton Suspension Bridge Trust is a registered charity set up under an Act of Parliament of 1952. There are 12 trustees, who between them have a range of expertise, particularly in the field of engineering. The bridge master is responsible for the day to day running of the bridge.

N.B Dimensions, weights and other structural details from: *A Complete History of the Clifton Suspension Bridge* by Lewis Wright (1866), the first official guide book, and Barlow, W H (1867) Description of the Clifton Suspension Bridge, Minutes and Proceedings, Institution of Civil Engineers: London.

The authors wish to give special thanks to John Mitchell, Bridge Master, Clifton Suspension Bridge for expert opinion, guidance and editing. Also to Pat Pascoe for proof reading and suggestions.

Bristol Riots 1831

1 Parliamentary constituencies where the number of electors had declined drastically yet still retained the right to elect a member of the House of Commons.

2 Quoted in Kelly, Andrew (2003) *Queen Square* Redcliffe Press: Bristol p 38.

Hungerford Bridge

1 As described in Vaughan, Adrian (1991) *Isambard Kingdom Brunel: engineering knight-errant* John Murray: London p 147.

2 From an undated press article assumed to have been published in the spring of 1845 contained in a scrapbook of contemporary prints and articles in a private collection.

Egyptian revival

1 Wilton-Ely, John in Turner, J (ed) (1996) *The Dictionary of Art* Oxford University Press: Oxford 10 p 96.

2 Letter dated 27 March 1831 quoted in Noble, C B (1938) *The Brunels, Father and Son* Cobden-Sanderson: London p 109.

3 Greenacre, Francis and Stoddard, Sheena (1991) *W J Müller* Bristol Museums and Art Gallery: Bristol p 105.

Brunel in Bristol Docks

1 For the general background to the history of the Port of Bristol, see Wells, C A (1909) *A Short History of the Port of Bristol* Arrowsmith: Bristol; Neale, W G (1968) *At the Port of Bristol* 1, Port of Bristol Authority: Bristol; and Buchanan, R A and Cossons, Neil (1969) *The Industrial Archaeology of the Bristol Region* David & Charles: Newton Abbot. See also Minchinton, W E (1962) *The Port of Bristol in the Eighteenth Century* Historical Association Bristol Branch for the idea of Bristol as 'the Metropolis of the West'.

2 See Williams, A F 'Bristol Port Plans and Improvement Schemes of the Eighteenth Century' in *Transactions of the Bristol & Gloucestershire Archaeological Society* (TB&GAS), 81, 1962 pp 138-88.

3 See Buchanan, R A 'The Construction of the Floating Harbour in Bristol: 1804-1809' in TB&GAS, 88, 1969 pp184-204.

4 See Buchanan, R A 'I K Brunel and the Port of Bristol' in *Transactions of the Newcomen Society* (TNS) 42, 1969-70 pp 41-56. I remain grateful to those officers of the Port of Bristol Authority who assisted me in the research for this paper 40 years ago, and especially John Corin and Olive Crouch, who were responsible for public relations and supervised the extensive archives of the PBA. Most of these records were subsequently transferred to the Bristol Record Office. For a general view of the Bristol context of Brunel's career, see Buchanan, R A (2002) *Brunel: the life and times of Isambard Kingdom Brunel* Hambledon and London: London, especially Chapter 4.

5 See Brunel's report to the BDC of 31 August 1832: the first quotation is from p 6 and the second from p 12. This report is the first of a series, neatly written in different styles and handwriting, presented by Brunel in the years between 1832 and 1848.

6 Technical details are from Watkins, George 'Drag-Boat by Brunel' in *Model Engineer*, 24 May 1962, 7 June 1962, and 24 January 1963.

7 For an account of Brunel's successors in the Port of Bristol, see Buchanan, R A (1971) *Nineteenth Century Engineers in the Port of Bristol* Bristol Branch of the Historical Association, Pamphlet No. 26.

8 The phrase regarding the steam packets is from the Minutes of the Bristol Docks Company, 28 September 1835.

9 The report of 16 January 1836 is in manuscript in six large format pages: Brunel's list of options includes a fourth dealing with the Junction Lock between the Cumberland Basin and the Floating Harbour, but no action was taken on this.

10 On Brunel's Bristol steamships, see Graham Farr's pamphlets in the Bristol Historical Association series, *The Steamship Great Western* (Bristol, 1963) and *The Steamship Great Britain* (Bristol, 1965). Corlett, Ewan (1975) *The Iron Ship* Moonraker Press: Bradford-on-Avon and, more recently, Griffiths, Denis, Lambert, Andrew and Walker, Fred (1999) *Brunel's Ships* Chatham Publishing: London.

11 Brunel's letter of 3 June 1844 to Christopher Claxton is copied in Private Letter Book No 3 in the Brunel Collection at the University of Bristol Library.

12 The report dated 15 June 1844 consists of six manuscript pages in the PBA Collection.

13 See Buchanan, R A 'The Cumberland Basin, Bristol' in *Industrial Archaeology* 6: 4 1969 pp 325-333.

14 See Buchanan, R A 'Brunel in Bristol' in McGrath, P and Cannon, J (ed) (1976) *Essays in Bristol and Gloucestershire History* Bristol and Gloucestershire Archaeological Society: Bristol p 246: the quotation is from a letter of Brunel to J N Miles, dated 28 April 1845, in the Private Letter Books in the Brunel Collection.

15 These words come at the end of Brunel's report of 15 June 1844.

Portbury, Avonmouth and Portishead

1   Quoted in Brunel, Isambard (1971) *The Life of Isambard Kingdom Brunel, Civil Engineer (1870)* David and Charles Reprints: London p 427.

The Great Western Railway

1   MacDermot, E T (1927) *History of the Great Western Railway*, I, London pp 1-5 (revised edition printed 1964 by Ian Allan).

2   Ibid. p 5; Vaughan, A (1991) *Isambard Kingdom Brunel: engineering knight-errant* John Murray: London pp 46-7.

3   Private Diary, Brunel Collection, 5 December 1831.

4   MacDermot, op. cit. p 14; Buchanan, R A (2002) *Brunel: the life and times of Isambard Kingdom Brunel* Hambledon and London: London p 66.

5   Vaughan, op. cit. p 49.

6   Buchanan, R A 'Working for the Chief – the design team and office staff of I K Brunel', in Kentley, E, Hudson, A and Peto, J (eds) (2000) *Isambard Kingdom Brunel: recent works* Design Museum: London pp 15-16.

7   Gren, A (2003) *The Foundation of Brunel's Great Western Railway* Silver Link Publishing Ltd: Kettering pp 23-30.

8   Ibid. pp 47-57 and 68-76; MacDermot, op. cit. pp 11-25.

9   Brindle, S (2004) *Paddington Station: its history and architecture* English Heritage: London pp 13-18.

10 Bryan, T 'The Battle of the Gauges – Brunel's broad gauge', Kentley, Hudson and Peto, op. cit. pp 39-41.

11 NA RAIL 250/83, GWR London Committee, abstract of minutes, 6.6.1836.

12 Bryan in Kentley, Hudson and Peto, op. cit. p 48. The two systems had met at about ten places by 1845, but the most famous, or notorious, of these was Gloucester.

13 NA RAIL 1149/44, GWR Tenders.

14 Buchanan, in Kentley, Hudson and Peto, op. cit. p 16.

15 NA RAIL 1149/44, GWR tenders, contract 1L.

16 Vaughan, op. cit. p 74. Sir Marc's diaries show that he had been working on the Brent Viaduct designs two years previously – in March and April 1834 – so it was one of the first elements to be designed in detail. However, there is no further evidence of Sir Marc working on the GWR.

17 NA RAIL 250/82, Brunel's reports to the GWR board, 15-16, 22; RAIL 1149/2, IKB letters, 1835-7. 234,241; RAIL 1149/4, letters, 1838-40, 96, 177; NA MT 8/1, Returns of Iron Railway Bridges, 1847, ff. 44-5, lists five iron railway bridges carrying the GWR main line, but not the Uxbridge Road bridge: Brunel omitted it from his list as it had been damaged by fire in May 1847, and propped with timber. He subsequently rebuilt it in wrought-iron, and it has since been rebuilt again in steel.

18 Owen, J B B 'Arch bridges', in Pugsley, A (ed) (1976) *The Works of Isambard Kingdom Brunel: an engineering appreciation*, Institution of Civil Engineers/University of Bristol: London and Bristol pp 91-106.

19 MacDermot, op. cit. p 46, pp 51-5.

20 Brindle, op. cit. pp 13-25; Tutton, M (1999) *Paddington Station, 1833-54* Railway & Canal Historical Society: Mold. Brunel's sketch designs for a grand terminus at Paddington are in Bristol University Library, Special Collections, sketchbook 'GWR 1836', pp 47-9. No full designs have been found, but a plan apparently corresponding to this scheme is in the Network Rail Western Region Plan Room, No. 7494.

21 MacDermot, op. cit. p 55; Simmons, J (1971) (ed) *The Birth of the Great Western Railway: extracts from the diary and correspondence of George Henry Gibbs* Adams & Dart: Bath pp 37-8.

22 MacDermot, op. cit. pp 59-61; Simmons, op. cit. pp 39-43. Part of one arch of the Maidenhead bridge had settled slightly, due to the contractor having eased the centring before the Roman cement had fully set: Brunel held firm, insisting that the affected area be cut out and made good, and the bridge has successfully carried railway trains up to the present day. See Owen in Pugsley, op.cit.

23 Simmons, op. cit. pp 6-8; Gren, op. cit. pp 104-10. Mr Gren has established that in October 1835 there were just 49 Liverpool shareholders, holding just under five per cent of the share capital, but Gibbs' diary entries suggest that by 1838-9 the Liverpudlians must have controlled a considerably larger proportion of the shares than this, in order for them to have wielded so much influence. The whole subject of railway finance needs a lot more research.

24 Ibid. pp 42-3, Gibbs' diary entries for 13 and 16 July.

25 Ibid. p 43; Brunel issued a detailed defence of the broad gauge at the half-yearly meeting on 15 August, quoted in full by MacDermot, op. cit. pp 65-71.

26 The experts were Nicholas Wood, a colliery engineer; John Hawkshaw, a young northern engineer who had been working at Liverpool docks, and Dionysius Lardner, professor at University College, London, and at the time a highly regarded pundit on railway matters. Wood and Hawkshaw were northerners, probably appointed at the instance of the Liverpool investors. Ibid. pp 46-52.

27 Ibid. p 64; MacDermot, op. cit. pp 73-87.

28 MacDermot, ibid. pp 103-16.

29 Booth, L G 'Timber Works' in Pugsley, op. cit. p 121.

30 MacDermot, op. cit. pp 126-31.

31 Bristol University, Brunel Collection, Private Diaries, 26.xii.1835, quoted in Buchanan (2002), op. cit. pp 167-8, and Vaughan, op. cit. p 57.

32 Buchanan (2002), op. cit. pp 156-7.

33 MacDermot, op. cit. pp 162-75, pp 207-11. The Bristol & Gloucester began as a standard gauge line, but agreed to convert to the broad gauge in 1843.

34 Gren, op. cit. pp 17-9, pp 114-5. Mr Gren shows that, in 1842, the 36 main passenger railways in operation were making a gross operating profit of 43.2 per cent. Even when they had covered their borrowing costs, and with huge ongoing reinvestment, they were already paying dividends ranging from three per cent for the GWR, to 10 per cent for the Liverpool & Manchester: this was in an age with zero inflation.

35 Ibid. pp 211-2. The Bristol & Gloucester, which had just been converted to the broad gauge, met the standard gauge Birmingham & Gloucester.

36 MacDermot, op. cit. pp 191-2.

37 There were further rounds in the battle, relating to the GWR's struggle to control and build the Birmingham & Oxford Junction Railway, and the Birmingham, Wolverhampton & Dudley Railway, 1846-52, and to its rivalry with the London & South-Western Railway. What is given here is an extreme simplification of a ludicrously complicated story, the only proper account of which is that given by MacDermot, op. cit. pp 200-93.

38 MacDermot, ibid. pp 229-47. Bryan, op. cit.

39 MacDermot, ibid. pp 178-9.

40 Vaughan, op. cit. pp 149-51; Booth, in Pugsley, op. cit. pp 114-9.

41 Booth, op. cit. pp 125-35.

42 Brindle, S and Tucker, M 'Brunel's Lost Bridge: the rediscovery and salvage of the Bishops Road Canal Bridge, Paddington' in *Construction History* 2005 (forthcoming).

43 Sutherland, R J M 'Iron railway bridges', in Bailey, M (ed) (2003) *Robert Stephenson: the eminent engineer* Ashgate Publishing Ltd: London pp 301-336.

44 Ibid. pp 317-35.

45 Binding, J 'The final bridge – the design of the Royal Albert Bridge at Saltash', in Kentley, Hudson and Peto, op. cit. pp 91-4. Mr Binding discusses the Chepstow design as a preliminary to Saltash: oddly enough, no-one yet seems to have written a detailed study of the Chepstow bridge itself.

46 Brindle, op. cit. chapter 3.

The Bristol terminus

1 Foyle, Andrew (2004) *Bristol* Yale University Press: New Haven and London p 89.

ss *Great Britain*

1 Griffiths, D (1985) *Brunel's Great Western*. Patrick Stephens: Wellingborough pp 15-6.

2 Hyman, A (1982) *Charles Babbage: pioneer of the computer* Princeton University Press: Princeton.

3 The report to the directors of the Great Western Steam Ship Company recommending the screw propeller is printed in Appendix II of Brunel, I (1870) *Life of Isambard Kingdom Brunel, Civil Engineer* Longmans, Green and Company: London (reprinted by David & Charles Reprints 1971, pp 539-558).

4 Brunel: 'Report to the Directors of the Great Western Steamship Company on Screw Propellers' 17.10.1840' ss *Great Britain* Trust Ms.

5 Brunel to Claxton 12.10.1840: Brunel Collection Bristol University Library. Great Britain Box DM 1758.

6 See Corlett, E (1990) *The Iron Ship: the story of Brunel's Great Britain* Conway Maritime Press: London (originally published 1975 by Moonraker Press) p 96.

7 Corlett, ibid. p 113-114.

Further Reading

Allington, P. & Greenhill, B (1997) *The First Atlantic Liners: steamships in the age of paddle wheels, sail and screw* Conway Maritime Press London.

Buchanan, R A (2002) *Brunel: the life and times of Isambard Kingdom Brunel* Hambledon and London: London.

Griffiths, D (1985) *Brunel's Great Western* Patrick Stephens: Wellingborough.

Griffiths, D, Lambert, AD and Walker, F (1999) *Brunel's Ships* Chatham Press: London.

Hyde, F E (1975) *Cunard and the North Atlantic* Macmillan: London.

Lambert, A D 'The screw propeller' in Kentley, E, Hudson, A and Peto J (eds) (2000) *Isambard Kingdom Brunel: recent works* Design Museum: London pp 102-11.

Pugsley, Sir A (1976) *The Works of Isambard Kingdom Brunel* Institution of Civil Engineers/University of Bristol: London and Bristol.

Rolt, L T C (1957) *Isambard Kingdom Brunel* Longmans: London pp 282-5.

Nursing and the Crimean War

1 Quoted in Brunel, Isambard (1971) *The Life of Isambard Kingdom Brunel, Civil Engineer (1870)* David and Charles Reprints: London p 461.

ss *Great Eastern*

1 Gordon, John Steele (2002) *A Thread Across the Ocean: the heroic story of the transatlantic cable* Simon & Schuster: London p 162.

The Brunel Collection, Bristol University Library

1 Noble, C B (1938) *The Brunels, Father and Son* Cobden-Sanderson: London.

2 *University of Bristol Gazette*, 19: 3, 1975, pp 7-14.

3 Humber, W (1870) *A Complete Treatise on Cast and Wrought Iron Bridge Construction* Lockwood: London.

4 Rolt, L T C (1957) *Isambard Kingdom Brunel* Longmans: London.

5 Respective publication details as follows: David Bogue: London, 1846: Longmans, Green and Company, 1870; Day: London, 1865; J Smith Homans: New York, 1845; London, 1841; 16 January 1858.

6 This was republished by Cambridge University Press: Cambridge, 1980.

7 Private Diary, 26.12.1835.

8 Memoir of Mr Isambard Kingdom Brunel, M Inst C E: Excerpt Annual Report of the Institution of Civil Engineers, 1859-60. London: printed by William Clowes and Sons, 1862.

9 Brunel, I (1870) *Life of Isambard Kingdom Brunel, Civil Engineer* Longmans, Green and Company: London (reprinted by David & Charles Reprints, 1971).

10 Buchanan, R A (2002) *Brunel: the life and times of Isambard Kingdom Brunel* Hambledon and London: London p xvi, 271.

11 Ibid. p xvi

Professional Colleagues

1 Parnell, Sir H (1833) *A Treatise on Roads* Longman, Rees, Orme etc: London.

Bibliography

Bailey, M R (ed) (2003) *Robert Stephenson: the eminent engineer* Ashgate Publishing Ltd: Aldershot.

Brooke, D (2004) *William Mackenzie: international railway builder* The Newcomen Society: London.

Brooks, Rev Dr E C (1996) *Sir Samuel Morton Peto Bt. 1809-1889* Bury Clerical Society.

Buchanan, R A (2002) *Brunel: the life and times of Isambard Kingdom Brunel* Hambledon and London: London.

Buchanan, R A (1995) *Engineers and Engineering: papers of the Rolt Fellows* Moorland Publishing Co: Ashbourne.

Clark, E F (1983) *George Parker Bidder: the calculating boy* KSL Publications: Bedford.

Lewis, L (1994) *The Cabry Family* Railway & Canal Historical Society.

Pole, W (ed) (1877) *The Life of Sir William Fairbairn*, Bart. Reprint 1970, David & Charles (Publishers) Ltd: Newton Abbot.

Ritchie-Noakes, N (1980) *Jesse Hartley: dock engineer to the Port of Liverpool 1824-60* Merseyside County Museums.

Robbins, M 'Thomas Longridge Gooch' in *Transactions of the Newcomen Society* 56 (1984/5) pp 59-70.

Rolt, L T C (1960) *George and Robert Stephenson: the railway revolution* Longman: London (and subsequent editions).

Skempton, Prof Sir A (ed) (2002) *Biographical Dictionary of Civil Engineers* I, Thomas Telford Publishing: London. (II, 1830-1890, edited by Peter Cross-Rudkin, to contain biographies of other engineers referred to in this chapter, is due to be published in 2007).

Smith, D P 'Sir Joseph William Bazalgette (1819-1891): engineer to the Metropolitan Board of Works' in *Transactions of the Newcomen Society* 58 (1986/7) pp 89-112.

Smith, D P 'The Works of William Tierney Clark (1783-1852)' in *Transactions of the Newcomen Society* 63 (1991/2) pp181-208.

Smith, D P 'James Walker (1781-1862): civil engineer' in *Transactions of the Newcomen Society* 69 (1997/8) pp.23-56.

Vignoles, K H (1982) *Charles Blacker Vignoles: romantic engineer* Cambridge University Press: Cambridge.

Walker, C (1969) *Thomas Brassey: railway builder* F. Muller: London.

Watson, G (1988) *The Civils. The story of the Institution of Civil Engineers* Thomas Telford: London.

Webster, N W (1970) *Joseph Locke: railway revolutionary* George Allen & Unwin: London.

The Roeblings

1   Quoted in McCullough, David (1972) *The Great Bridge: the epic story of the building of the Brooklyn Bridge* Simon & Shuster: New York p 27.

Who Built Brunel?

1   Public Record Office (PRO) ref: RAIL 251/1.
2   Gren, A (2003) *The Foundation of Brunel's Great Western Railway* Silver Link Publishing Ltd: Kettering pp 102-10.
3   Ibid. p 108.

4   Other investors included Richard Fingair, a hairdresser, of Bath, who had two shares, and Arthur Guiness, brewer, of Dublin, who daringly risked his savings to put his name down for four shares. Richard, Joseph and Francis Fry, chocolate manufacturers, of Bristol, took 26, 27 and 36 shares respectively while their sisters, Anna and Caroline, had ten each. The entire family of Fox, from Wellington in Somerset, bought 45 shares between them. Henry Burslem, gentleman, of Grosvenor Square, London, took 100. George Henry Gibbs, a merchant banker of Bristol and London and a director of the GWR, held 47 shares in 1835 while I K Brunel had 50 and had therefore paid a cash deposit of £250. Sealed Register NA/PRO 251/1.
5   Simmons, J (1971) (ed) *The Birth of the Great Western Railway: extracts from the diary and correspondence of George Henry Gibbs* Adams & Dart: Bath p 56.
6   Brunel's Private Journal.
7   PLB.5.27/11/46. See also 22 and 30 November.
8   Vaughan, A (1991) *Isambard Kingdom Brunel: engineering knight-errant* John Murray: London p 131.
9   Royal Humane Society Report 1838.
10   Gravatt's obituary. Mins. Proc. ICE, 26 1867 p 566.
11   Ibid. p 571.
12   Mins.Proc. 1861-2.
13   Brunel's evidence to the Parliamentary Committee on Railway Labourers. 1846 Q.2100.
14   Ibid. Q.2120.
15   Sullivan, D (1983) *Navvyman* Coracle Books: London p 56.
16   Ibid. p 56.
17   Ibid. p 13.
18   Ibid. p 13.
19   Digging Box Tunnel. p 11. Thomas Gale's pamphlet. University of Bristol Library.
20   Brunel's evidence to the Parliamentary Committee on Railway Labourers. 1846

Battle of Mickleton Tunnel

1   Coleman, Terry (2000) *The Railway Navvies: a history of the men who made the railways* (3rd ed) Pimlico: London p 110.
2   Rolt, L T C (1957) *Isambard Kingdom Brunel* Longmans: London p 158.

Swindon: a railway town

1   For a detailed account of the phases of development of the works and village see Cattell, John and Falconer, Keith (1995) *Swindon: the legacy of a railway town* English Heritage: Swindon.
2   Quoted in Cattell and Falconer p 39.
3   Quoted in Rolt, L T C (1957) *Isambard Kingdom Brunel* Longmans: London p 140.

From slavery to industrialisation

1   See Dresser, Madge (2001) *Slavery Obscured: the social history of the slave trade in an English provincial port* Continuum Press: London and New York.

Technological and Social Change: the impact on society of the work of Brunel and his contemporaries

1   Richard Trevithick can reasonably be described as the father of the steam locomotive. A group of pioneers on Tyneside contributed to the evolution of the locomotive, including; William Hedley, William Blenkinsop, Timothy Hackworth and Edward Pease.

2   The Letters of Queen Victoria 1837-1861, 1, John Murray, London, 1908, p 369.

3   Proc. ICE, 76, 1883-4, pt. 2, p 207.

4   See: Joby, R S (1983) *The Railway Builders* David & Charles: Newton Abbot and Sullivan, Dick (1983) *Navvyman* Coracle Books: London.

5   Proc. IMechE, 1, On Railway Axles, 24 October 1849 pp 13-21.

6   Galton published three papers on railway brakes in the Proc. IMechE, in June 1878, October 1878 and April 1879.

7   Proc. Inc. Assoc. of Mun. & County Engineers, XX, 1893-4, Presidential Address, A M Fowler p 210.

8   Proc. ICE, 78, 1884, pp 414-416, Obit. of Eugenius Birch.

9   Chilcott, J (1844) *History of Bristol* Clare Street: Bristol, (6th ed.) p 113.

10 Ibid. p 295. Michael Robbins, the eminent transport historian, pointed out, however, that the clock at Christchurch College, Oxford still refuses to fall in line.

11 See Howse, D (1980) *Greenwich Time* Book IV chapter 2 OUP: Oxford pp 345-7.

12 Smiles, Samuel (ed) (1912) *James Nasmyth: Engineer, An Autobiography* John Murray: London p 230 [1st. edition, 1883].

13 Kieve, Jeffrey (1973) *The Electric Telegraph: a social and economic history* David & Charles: Newton Abbot.

14 *London and its Vicinity Exhibited* published by Weale, Holborn, 1851 p 362. The News-Rooms were in Glasgow, Edinburgh, Newcastle, Leeds, Hull, Liverpool, Manchester, and Stockport.

15 Timbs, John *The Year-Book of Facts in Science and Art: 1848* David Bogue: London p 232.

16 By 1853 five other cables were laid in Britain: Dover-Ostend, Firth of Forth, Portpatrick-Donaghdee, River Tay, and Portpatrick-Whitehead.

17 Russell, W H (1865) *The Atlantic Telegraph* Day & Son: London p 2.

18 Wilson, Roger Burdett (1972) *Sir Daniel Gooch: memoirs & diary* David & Charles: Newton Abbot p 93. See also: Cookson, Gillian (2003) *The Cable: the wire that changed the world* Tempus: Stroud.

19 *The Times* 6 August 1858.

20 *The Illustrated Exhibitor* 1, 1852 p 23.

21 Appleyard, Rollo (1939) *History of the Institution of Electrical Engineers: (1871-1931)* IEE: London p 51.

22 An interesting and accessible source is Steinberg, S H (1961) *Five Hundred Years of Printing* Penguin Books: Harmondsworth.

23 *London and its Vicinity Exhibited*, op. cit. p 6.

24 Ward, A W, and Waller, A R (1916) *The Cambridge History of English Literature* XIV, CUP: Cambridge p 174.

25 Proc. ICE, 1, 12 May 1840, 'Photography as Applicable to Engineering' Alexander Gordon, MICE, p 59. [Gordon quoting Mons. Arago].

26 See: Thompson, F M L (1968) *Chartered Surveyors: the growth of a profession* Routledge & Kegan Paul: London p 299.

27 *The Illustrated Magazine of Art* 1, Cassell, London, 1853, p 191.

28 Hawke, G R (1970) *Railways and Economic Growth in England and Wales 1840-1870* Clarendon Press: Oxford p134.

29 Butler, Samuel (1917) *The Notebooks of Samuel Butler* E P Dutton: New York p 21.

30 For a fascinating social account of working life and conditions see: Elbourne, Roger (1980) *Music and Tradition in Early Industrial Lancashire: 1780-1840* published by D S Brewer and Rowman and Littlefield for the Folklore Society.

31 Smith, Denis (2001) *Civil Engineering Heritage: London and the Thames Valley* Thomas Telford Ltd: London.

32 Binnie, G M (1981) *Early Victorian Water Engineers* Thomas Telford Ltd: London pp 80-91.

33 Frank Forster (1800-1852) is an example of a link between railways and public health as Forster was resident engineer on Robert Stephenson's Kilsby tunnel on the London & Birmingham Railway before moving into public health engineering.

34 Jephson, Henry (1907) *The Sanitary Evolution of London* T Fisher Unwin: London pp 120-121.

35 The pumping stations are at Deptford (1864), Crossness (1865), Abbey Mills (1868) and Pimlico (1875).

36 Proc. Inc. Assoc. of Mun. and San. Engineers and Surveyors, 1, 1873, Inaugural Address of Lewis Angel, President, p.20.

37 Proc. ICE, 91, 1887 p 22, Presidential Address of George B Bruce.

38 Proc. Inc. Assoc. of Mun. & County Engineers, 20, 1893-4, Presidential Address of Alfred M. Fowler, MICE, June 1894, p 220.

39 Ibid. p 212.

Emigration

1   Fogg, Nicholas (1999) ss *Great Britain*, published by ss *Great Britain* p 18.

Rain, Steam and Speed

1   Hamilton, James (1998) *Turner and the Scientists* Tate Gallery Publishing: London p 103.

2   Bailey, Anthony (1997) *Standing in the Sun: a life of JMW Turner* Sinclair-Stevenson: London p 363-364.

3   Quoted in John Rushkin's Dialecta, *Works* XXXV p 559-601 and reproduced in Simmons, Jack (1991) *Railways: an anthology* Collins: London pp 226-227.

4   Hamilton, op. cit. p 100.

Lines, Landscapes and Anti-Modernism: understanding Victorian opposition to the railways

1   Ruskin, J 'Letter 1: January 1871' *Fors Clavigera* in Cook, E T and Wedderburn, A (eds) (1903-12) *The Library Edition of the Complete Works of John Ruskin* published in 39 volumes by George Allen: London, 27 pp 11-26 (see p 15).

2   Ruskin, J 'Letter to *Birmingham Gazette* 3 March 1887' in *Works* 34 p 604.

3   Cooke Taylor, W (1842) *Notes of a Tour in the Manufacturing Districts of Lancashire* Duncan and Malcolm: London p 4.

4   Tennyson, A 'Locksley Hall' in Ricks, C (ed) (1969) *The Poems of Tennyson* Longmans, Green and Company: London, pp 688-699 (see p 699).

5   Carlyle, T 'On Heroes, Hero-Worship and the Heroic in History' *Centenary Edition of the Works of Thomas Carlyle* published in 30 volumes by Chapman and Hall: London, 1904, 5.

6   Carlyle, T *Past and Present* in *Works* 10.

7   Ruskin, J *The Cestus of Aglaia* in *Works* 19 pp 43-159.

8   Ibid. pp 60-61.

9   In this poem, the young Ruskin wrote of how 'a steamboat can / Be the most useful engine brought to man' *Works* 2, pp 254-255 (see p 254).

10 Ruskin, J *The Stones of Venice* in *Works* 10 pp 3-4.

11 More information on this subject, and on other issues treated in this essay, can be found in Richards, J 'The role of the railways' in Wheeler, M (ed) (1995) *Ruskin and Environment: the storm-cloud of the nineteenth century* Manchester University Press: Manchester pp 123-143 (see p 135).

12 Ruskin, J *The Seven Lamps of Architecture* in *Works* 8 p 159.

13 Ruskin, J 'Samuel Prout' *Reviews and Pamphlets on Art* in *Works* 12 pp 305-315 (see pp 314-315).

14 Ibid. pp 314-315.

15 Ruskin mentions the damage caused to these places in 'The Art of England' in *Works* 33 pp 255-408 (see pp 404-405).

16 Bourne, J Cooke (1839) *Drawings of the London and Birmingham Railway. With an Historical and Descriptive Account by John Britton* Ackermann: London.

17 Op. cit. pp 60-61.

18 Op. cit. p 159.

19 'Letter 35, November 1873' in *Works* 27 p. 664.

20 Ruskin, J 'The storm-cloud of the nineteenth century' in *Works* 34 pp 1-80 (see p 28).

21 Op. cit. p. 246.

22 Carlyle, T 'Hudson's Statue' *Latter-day Pamphlets* in *Works* 20 p 266.

23 'The Art of England', ibid. p 404.

24 'Letter 44: August 1874' *Fors Clavigera* in *Works* pp 125-144 (see p 129).

25 Somerville, R (1877) *A Protest Against the Extension of Railways in the Lake District, and a Preface by John Ruskin* J. Garnett: Windermere.

26 Ibid. p 6.

27 Ibid. p 6.

28 Ibid. p 34.

29 'Mr. Ruskin and Wakefield' *The Saturday Review* 4 March 1876 pp 297-298 (see p 297).

30 Letter to Henry Acland, 28 March 1874, in *Works* 20, pp. xli-xlii (see p xli).

31 Ibid. p. xli.

32 Ibid. p. xli.

33 'Letter 49: January 1875' *Fors Clavigera* in *Works*, 28, pp. 235-253 (see p 247).

34 Ibid. p 247.

35 Ruskin, J Letter to *The Daily Telegraph* (6 August 1868), 'Railways and the state' in *Works* 17 pp 530-531.

36 Ruskin, J *Munera Pulveris* in *Works* 17 pp 115-293 (p 252).

37 Wiener, M J (1981) *English Culture and the Decline of the Industrial Spirit, 1850-1980* Cambridge University Press: Cambridge.

Brunel the conservationist

1   In Wintle, J (ed) (2001) *Makers of Nineteenth Century Culture 1800-1914* Routledge Kegan Paul: London p 83.

The Brunels at Watcombe

1   Brunel, I (1971) *The Life of Isambard Kingdom Brunel, Civil Engineer (1870)* David and Charles Reprints: London p 514-515.

2   Ibid. p 515.

The Function of Ornament: the consolation of design in the Industrial Age

1   In the wake of Darwin's *The Origin of Species* 1856 and *The Descent of Man* 1871, Ruskin's and Morris' fear of moral decline had disturbing parallels in the new pseudo-scientific theories of phrenology and physiognomy which attempted to diagnose physical marks of moral decline. Phrenology claimed to devise a system for diagnosing a person's moral character by analysing the shape and surface bumps of the skull; physiognomy similarly claimed that facial features could be analysed as indicators of moral character. These troubling new theories became very popular and widespread and were paralleled in Dickens' use of detailed physical descriptions and names and William Powell Frith's social typing in paintings such as *The Railway Station*. A Victorian reader instantly recognises the moral character and personality of the schoolteachers 'Mr M'Choakumchild' and Mr Grandgrind by their names, similarly Frith identifies the moral and social status of each person in his panoramic paintings representing every strata of Victorian society from the street urchin to the nobleman. A vast array of recent Victorian Studies scholarship has persuasively argued the prevalence of the pessimistic theories of physiology, phrenology and physiognomy which dominated Victorian medicine, ethics and culture, which signalled fears of degeneration and national decline. A few particularly apposite examples of this scholarship are: Cowling, Mary C (1989) *The Artist as Anthropologist: the representation of type and character in Victorian art* Cambridge University Press: Cambridge; Da Costa Nunes, Jadviga M 'O G Rejlander's

Photographs of Ragged Children: Reflections on the idea of Urban Poverty in Mid-Victorian Society' in *Nineteenth Century Studies* vol 4 1990 pp105-136; Humphreys, A (1977) *Travels into Poor Man's Country: the work of Henry Mayhew* University of Georgia Press: Athens; Rainger, Ronald 'Race, Politics and Science: The Anthropological Society of London in the 1860s' in *Victorian Studies* vol XXII Autumn 1978 pp51-70; West, Shearer (ed) (1996) *The Victorians and Race* Ashgate: Aldershot.

2   Sussman, Herbert L (1968) *The Victorians and the Machine: the literary response to technology* Harvard University Press: Cambridge Massachusetts p 7.

3   Architecture and design during Brunel's lifetime was dominated by references to past styles. These references are usually referred to as historicist or revival styles, as S Lang explained: 'Revivals play perhaps a greater role in English eighteenth and nineteenth-century architecture and in the decorative arts of that period, than in any other country. The Gothic Revival is the most significant and widespread one among a host of others, the Chinese, the Greek, the Egyptian, the Moorish and the Indian…'. Lang, S 'The Principles of the Gothic Revival in England' *The Journal of the Society of Architectural Historians*, 25: 4 (Dec 1966) p 240.

4   Thorne, Robert 'Inventing New Design Technology' in MacKenzie, John M (2001) *The Victorian Vision: inventing New Britain* Victoria and Albert Museum Publications: London pp 176-8.

5   Carlyle, Thomas 'Signs of the Times 1829' in *The Complete Works of Thomas Carlyle* Vol III London: Chapman and Hall: London, 1858, p100.

6   Interestingly, Engineering sits two rungs further down from Mechanics and Chemistry, beside the trades of Upholstery and Masonry supporting Booksellers, suggesting the conflicted social status of the profession, although noticeably neither Architecture nor Architects are included.

7   Atterbury, Paul 'Steam and Speed Industry, Transport and Communications' in MacKenzie, op.cit. p 148.

8   Ruskin's and Morris's writings and public lectures were perhaps the most famous proponents of a widespread debate in Victorian Britain about the dehumanising effects of repetitive industrial labour practices. The social theories of Thomas Carlyle, Henry Mayhew and Matthew Arnold were other key voices in these debates, as were popular novels such as Charles Dickens *Hard Times* of 1854 or Mrs Gaskell's *North and South* of 1855 or the 'character' filled paintings of William Powell Frith and Augustus Leopold Egg. The core lines of these debates centred on the alienation of humankind in the modern industrial age, from the natural world, which then led to moral and physical decline. The horrendous working conditions of the mill or factory meant legions of urban labourers had virtually no contact with the organic world, which Ruskin and Morris felt were not only the primary source of inspiration for art and ornamental design, but also personal harmony and fulfilment. Ruskin and Morris criticised, for example, the fact that the modern workman, effectively working on an assembly line, was never allowed the satisfaction of creating a whole object, from planning stage to finished article, as a medieval craftsman did. This 'part work' divorced the craftsman from a sense of belonging and harmony within a predetermined natural and social order which images such as Cruikshank's illustrated. These fears of spiritual decline were paralleled in medical and sociological debates (see first endnote and note on physiognomy below.)

9   However, it should be noted that Ruskin is perfectly content for machinery to be used to undertake superhuman tasks such as irrigating the deserts, breaking Arctic ice, etc – after all he used iron girders in the University

Museum in Oxford. Sussman also cites similarly eloquent satirical views from Matthew Arnold: 'Your middle-class man thinks it the highest pitch of development and civilisation when his letters are carried twelve times a day from Islington to Camberwell, and from Camberwell to Islington, and if the railway runs to and fro between them every quarter of an hour. He thinks it nothing that the trains only carry him from an illiberal, dismal life at Islington to an illiberal, dismal life at Camberwell.' And by Henry D Thoreau in Walden: 'We are in great haste to construct a magnetic telegraph from Maine to Texas; but Maine and Texas it may be, have nothing important to communicate.' All three passages are cited in Sussman, op. cit. pp 94-5.

10 Opinions expressed in the *Ecclesiologist* and by Garbutt cited in Mordaunt Crook, J (1987) *The Dilemma of Style Architectural Ideas from the Picturesque to the Post-Modern* John Murray: London p 111.

11 Cited in Mordaunt Crook, ibid, p 112.

12 Ruskin, John (1907) *The Seven Lamps of Architecture* London pp 122-3 cited in Richards, Jeffrey and MacKenzie, John M (ed) (1986) *The Railway Station A Social History*: Oxford University Press: Oxford/New York pp 33-4.

13 Kaufman, Edward 'Architectural Representation in Victorian England' in *The Journal of the Society of Architectural Historians*, 46: 1 (Mar 1987) p 1.

14 Kaufman, op. cit. pp 32-3.

15 The use of 'revival' or 'historicist' styles in architecture and design also implied associations with the historical perception of the age to which they referred. This is to simplify their stances, but in many ways for Ruskin and Pugin buildings and objects decorated in the Gothic style inherently created the atmosphere of collectivity and piety which was for them the essential character of the Middle Ages in which the Gothic style originated. See Boris, Eileen (1986) *Art and Labor: Ruskin, Morris, and the craftsman ideal in America* Temple University Press: Philadelphia, 1986; Harvey, Charles and Press, Jon (1991) *William Morris: design and enterprise in Victorian Britain*, Manchester University Press: Manchester/New York; Roe, Frederick William (1969) *The Social Philosophy of Carlyle and Ruskin* Kennikat Press: London; Swenarton, Mark (1989) *Artisans and Architects: The Ruskinian tradition in architectural thought* Macmillan: Basingstoke, amongst many others. What I hope to argue here is that Brunel's inclusion of ornamental motifs which referred to historicist or exotic styles or natural motifs amidst the new stark forms of undecorated iron and glass construction, both suggested the drama and fantasy of these new spaces of modern travel experience and also minimised the potential alienation which undecorated iron and glass might inflict. As Ruskin's quotation declared, a nineteenth-century person risked becoming alienated and dehumanised, 'a living parcel', if forced to stay for any length of time in an undecorated industrial building. I use the term physiognomy here to underscore the close parallels and associations with medical and social theories of degeneration evoked in endnote 1 and 8 .

16 The notable exception being William Powell Frith's representation of Paddington in *The Railway Station* (1862), the mainstay of the Victorian survey lecture to which we shall return later.

17 The importance of Brunel's wife, Mary Elizabeth Horsley (1813-81), and the lively musical and artistic circle around her family which included Félix Mendelssohn, is well documented. Brunel's 'in-laws' included two Royal Academicians, his wife's great uncle Sir Augustus Wall Callcott (1779-1844) and her brother John Callcott Horsley (1807-1903), who painted several portraits of Brunel and was a close friend. See Faberman, Hilarie and

McEvansoneya, Philip 'Isambard Kingdom Brunel's 'Shakespeare Room' in *The Burlington Magazine*, 137: 1103 (Feb 1995) pp 108-18; Brunel Gotch, Rosamund (1934) *Mendelssohn and His Friends in Kensington: Letters from Fanny and Sophy Horsley Written 1833-36* Oxford University Press: London; and Buchanan, R A (2002) *Brunel: the life and times of Isambard Kingdom Brunel*: Hambledon and London: London/New York pp 17-18, 194-200.

18 Binding, John (2001) *Brunel's Temple Meads: a study of the design and construction of the original railway station at Bristol Temple Meads 1835-1965*: Oxford Publishing Co: Hersham and Brindle, Steven (2004) *Paddington Station: its history and architecture* English Heritage: Swindon.

19 'Much of his planting of trees, now come to maturity, together with his garden walks, has miraculously survived… He also applied himself to the design of a house, eventually settling for an Italianate villa with a belvedere and a colonnaded terrace, but when he died work had only started on the foundations, and the house was never built.' Buchanan, op. cit. p 198. Steven Pugsley includes a brief entry on Watcombe in his *Gazeteer of Devon Gardens* and in it indicates that Brunel's aesthetic links included a number of eminent Victorian garden designers whom he consulted whilst devising his projects at Watcombe, including William Nesfield, Alexander Forsyth and William Simpson: Pugsley, Steven (1994) *Devon Gardens: an historical survey* Stroud: Alan Sutton Publishing: Stroud p 182. Todd Gray identifies William Simpson as the author of the 'elaborate water system' and credits Alexander Forsyth with the design of 'the upper parkland'. He also cites the 1891 sale particulars which describes the Watcombe estate as 'selected some years ago by the late Isambard Kingdom Brunel, being seated on a well sheltered plateau, at an altitude of 500 ft above sea level, and approached by a carriage drive through a shrubbery avenue. The climate is healthy, the atmosphere pure and the scenery is truly lovely'. It refers to 300 acres of 'wilderness and woodland parks, glades, slopes, ferneries, shrubberies, dells, shady retreats and terrace walks'. The surviving garden features include extensive terraces and arboretum.' Gray, Todd (1995) *The Garden History of Devon: an illustrated guide to sources* University of Exeter Press: Exeter pp 228-9.

20 Buchanan, op. cit. p 196.

21 Ibid. p 197. I personally would advocate a less value-laden assessment of Brunel's collecting practices in light of contextualising interventions such as MacLeod, Diane Sachko 'Art Collecting and Victorian Middle-class Taste' in *Art History*, 10: 3 (September 1987) pp 328-49.

22 Ashton, G (1980) *Shakespeare's Heroines in the Nineteenth Century*: Buxton Museum and Art Gallery: Buxton p iv cited in Faberman and McEvansoneya, op. cit. p 108 who also highlight the 1843 mural decoration competition for the Houses of Parliament and the 1847 Robert Vernon gift of his modern British paintings collection to the nation as inspirations.

23 Attention is drawn to the relevant passages in Brunel, I (1870) *The Life of Isambard Kingdom Brunel* London pp 505-7 and Noble, C B (1938) *The Brunels, Father and Son* London p 187 cited by Faberman and McEvansoneya, op. cit. p 110

24 Faberman and McEvansoneya list the ten paintings as: Sir Augustus Wall Callcott and John Callcott Horsley *Launce Reproving his Dog* (1849) (whereabouts unknown) – this underlines the earlier point of the directness of Brunel's artistic sympathies with his family by marriage, painted by his wife's brother and uncle. Charles West Cope *King Lear Act IV, scene vii* (1848-50) (whereabouts unknown); Augustus Leopold Egg *Launce's Substitute for Proteus' Dog, Two Gentleman of Verona Act IV* scene iv, (1849)

(Leicestershire Museums and Art Gallery); John Callcott Horsley *Romeo and Juliet Act III* scene vii (whereabouts unknown); Sir Edwin Landseer *Scene from the Midsummer-Night's Dream. Titania and Bottom: Fairies Attending – Peas-blossom, Cob-web, Mustard-seed, Moth, etc, A Midsummer-Night's Dream* Act III, scene I (National Gallery of Victoria, Melbourne); Frederick Richard Lee *Jacques and the stag, As You Like It Act II* scene I (whereabouts unknown); Charles Robert Leslie Scene from Henry VIII, Henry VIII Act I scene iv (whereabouts unknown); Charles Robert Leslie *Scene from Henry VIII (Queen Katherine of Aragon's Interview with Capucins, Henry VIII's Ambassador at Kimbolton, Henry VIII* Act IV scene ii (Mead Art Museum, Amherst College, Amherst Massachusetts); Charles Robert Leslie *Hermione, A Winter's Tale* Act V, scene iii (Royal Shakespeare Theatre Picture Gallery and Museum, Stratford-upon-Avon); Clarkson Stanfield Macbeth Act I scene iii (Leicestershire Museums and Art Gallery).

25 Pugin, A Welby (1843) *An Apology for the Revival of Christian Architecture* cited in Binding, op. cit. p 44.

26 Bartram, Christian (1950) *An Introduction to Railway Architecture* Art and Technics: London p 10.

27 Pugin's ardent advocacy of Gothic architecture as inherently expressive of moral virtue is evident throughout his writings. *Contrasts or, A parallel between the noble edifices of the Middle Ages, and corresponding buildings of the present day: shewing the present decay of taste* of 1841 uses comparative illustrations of 'good' medieval examples vs 'bad' modern examples, comparisons polemically underlined in the accompanying commentary. Although an uncomfortable text from the multicultural viewpoint of twenty-first century Britain, a particularly clear declaration of Pugin's position appears in *An apology for the Revival of Christian Architecture in England* of 1843: 'If we worshipped Jupiter, or were votaries of Juggernaut we should raise a temple or erect a pagoda. If we believed in Mahomet, we should raise a temple or mount the crescent, and a raise a mosque. If we burnt our dead, and offered animals to the gods, we should use cinerary urns, and carve sacrificial friezes of bulls and goats. If we denied Christ, we should reject his Cross. For all these would be natural consequences: but in the name of commonsense, whilst we profess the creed of Christians, whilst we glory in being Englishmen, let us have an architecture, the arrangement and details of which will alike remind us of our faith and our country, – an architecture whose beauties we may claim as our own, whose symbols have originated in our religion and our customs…' cited in Lang, 1966, p 263. The tremendous impact of Pugin's theories has been well documented in studies such as Atterbury, Paul and Wainwright, Clive (1994) *Pugin: a Gothic passion* Yale University Press in association with the Victoria and Albert Museum: London; Banham, Joanna and Harris, Jennifer (1984) *William Morris and the Middle Ages* Manchester University Press: Manchester; Stansky, Peter (1985) *Redesigning the World: William Morris, the 1880s, and the Arts and Crafts* Princeton University Press: Princeton, NJ; Stanton, Phoebe (1971) *Pugin* London: Thames and Hudson.

28 The passages from this letter in the private collection of Nicholas Mackenzie are cited in Faberman and McEvansoneya, op. cit. pp 109-10.

29 The letter implies his hope that these works might be of 'National interest', signalling that Brunel seems to have had an ambition to provide the Victorian public with a gift like the Vernon Bequest.

30 The advice of Owen Jones, author of the manual for all Victorian artist-decorators of every hue, *The Grammar of Ornament* (1856) and meticulous

designer of the Moorish Court in the reconstructed Crystal Palace at Sydenham, was sought on the matter of polychromy, whilst the Belgian artist was asked to design the Royal Waiting Room, although this design is only known from a drawing in the RIBA collection illustrated in Brindle, op. cit. p 45, figure 3.20.

31 Cited in Brindle, op. cit. p 40.

32 Brindle indicates that the current whereabouts of this fascinating letter is unknown, but he cites it in full from a transcript in the National Archives British Rail Archive and Rolt, L T C (1957) *Isambard Kingdom Brunel* Longmans: London pp 231-1, Brindle, op. cit. p 37.

33 Hitchcock, Henry Russell (1972) *Early Victorian Architecture in Britain* Trewin and Copplestone: London pp 558-61.

34 Cowling, op. cit. pp 284-316.

35 Scott, James (1913) *Railway Romance and Other Essays* London pp 89-90 cited in Richards and MacKenzie, op. cit. pp 7-8.

36 Pierre Nora's helpful distinction between the idea of memory and history is vital to this essay's argument. He thoughtfully argued for the importance of common sites and objects as having a unique resonance, as important as "High" art and culture in the formation of nations and individuals. Nora, Pierre (1998) *Realms of Memory: The construction of the French past* London: Columbia University Press: London.

37 "This painful catalogue makes up only a small part of the inventory of the 'art' of the restaurant… I say their contemplation can give no one pleasure; they are there because their absence would be resented by the average man who regards a large amount of futile display as in some way inseparable from the conditions of the well-to-do life to which he belongs or aspires to belong. If everything were merely clean and serviceable he would proclaim the place bare and uncomfortable." Fry, R (1937) *Vision and Design* London: pp 60-4, cited in Richards and MacKenzie, op. cit. p 33.

38 1946, inspired by Noel Coward *Still Life* 1937.

39 Fred: Laura?
Laura: Yes, dear.
Fred: Whatever your dream was, it wasn't a very happy one, was it?
Laura: No.
Fred: Is there anything I can do to help?
Laura: Yes, Fred, you always have.
Fred: You've been a long way away.
Laura: Yes.
Fred: Thank you for coming back to me. (Laura weeps in Fred's arms.)
Film historians differ in their views of this destination as a place of security or claustrophobia.

Brunel and the art of engineering drawing

1 Rolt, L T C (1957) *Isambard Kingdom Brunel* Longmans: London p 16.
2 Pugh, Francis 'Design, Engineering and the Art of Drawing', Walker, Derek (ed) (1987) *The Great Engineers* Academy Editions: London p 71.
3 Walker, Derek 'Introduction', ibid. pp 15-16.
4 Rolt, op. cit. p 319.
5 Ibid. p 321.

Conclusion: Brunel: the legacy

1 Anonymous, 'Looking for Mr Brunel', *The Economist*, 2 December 1989, p 70.
2 Taken from the Science Museum's Antenna website October 2005: http://www.sciencemuseum.org.uk/antenna/building.
3 This and the other statements from modern-day engineers quoted below came from an exchange of emails during the writing of this chapter, unless otherwise stated.
4 Walker, Derek 'Today's Engineers – the legacy lives on', Walker, Derek (ed) (1987) *The Great Engineers* Academy Editions: London p 218.
5 Dyson, James, 'Conference message', The National Association of Advisers and Inspectors in Design and Technology, http://www.naaidt.org.uk/news/docs/conf2005/PatronsMessage2005.html.
6 Quoted in Rolt, L T C (1957) *Isambard Kingdom Brunel* Longmans: London p 298.

AUTHOR BIOGRAPHIES

Adrian Andrews was born in Bristol in 1950 and educated at Bristol Cathedral School. He studied Fine Art at West of England College of Art. Adrian is a freelance (architectural and graphic) designer based in Bristol. In 1998 he designed the first Clifton Suspension Bridge Visitor Centre at Bridge House and in 2002 was engaged to design the fitting-out and exhibition for the new Bridge Centre. Adrian has become a passionate 'pontist' through research for this latest project.

Dr Michael R Bailey MA DPhil is a museum consultant, author, lecturer and broadcaster on early railway history. He is past president of the Newcomen Society for the Study of the History of Engineering and Technology, an associate of the Institute of Railway Studies and Transport History (National Railway Museum/University of York) and associate trustee of the Museum of Science and Industry in Manchester.

Dr Steven Brindle is from Lancashire, was educated at Rossall and in Los Angeles, and read History at Keble College, Oxford. He now lives in London. He has worked for English Heritage since 1989, first as a historian, then as an Inspector in the Crown Buildings team, and is now the Inspector of Ancient Monuments for the London Region. He is the author of *Paddington Station, its History and Architecture*, published in 2005, and the co-author of the *Blue Guide to the Country Houses of England* (with Geoffrey Tyack, 1991) and of *Windsor Revealed: New Light on the History of the Castle* (with Brian Kerr, 1997). He has published numerous academic articles, and lectured widely, on a range of architectural subjects. He has had an interest in Isambard Kingdom Brunel for many years, and his discovery of the previously forgotten Brunel canal bridge at Paddington, and its subsequent rescue, arose out of the research for his history of Paddington Station. English Heritage is collaborating with British Waterways and Westminster City Council on a project for the bridge's reconstruction.

Professor Angus Buchanan is Emeritus Professor of the History of Technology at the University of Bath, and Honorary Director of the Centre for the History of Technology. He has written extensively on industrial archaeology, engineering history and engineering biography, especially in connection with I K Brunel. His latest book is *Brunel: the life and times of I K Brunel* Hambledon & London, 2002.

Andrew Kelly is director of Bristol Cultural Development Partnership (BCDP) and Brunel 200. He founded and led At-Bristol, the Brief Encounters and Animated Encounters film festivals, Digital Arts Development Agency and the Festival of Ideas, launched in 2005. He is the author of 11 books and a visiting professor in the Faculty of Art, Media and Design, University of the West of England.

Melanie Kelly leads the research and operations management of Brunel 200, the Bristol Great Reading Adventure, and BCDP consulting. She is a graduate of the University of Warwick and the University of Bath and has worked for the University of Bath, the University of the West of England, the Royal Photographic Society and the Waterstone's Group. She has written and edited books on the management of university collections, and co-authored with Andrew Kelly *Managing Partnerships* (2002) and *Building Legible Cities 2* (2003).

Professor Andrew Lambert is Laughton Professor of Naval History in the Department of War Studies, King's College, London. A Fellow of the Royal Historical Society, his books include *The Crimean War: British Grand Strategy against Russia 1853-1856* (1990), *The Last Sailing Battlefleet: Maintaining Naval Mastery 1815-1850* (1991), *The Foundations of Naval History; John Knox Laughton, the Royal Navy and the Historical Profession* (1998) and *War at Sea in the Age of Sail 1650-1850* (2000). His latest book, *Nelson: Britannia's God of War*, was published in October 2004. Professor Lambert has also taken naval history to the broadest audience with a number of television appearances. In 2001 he took part in the BBC series *The Ship*, a reconstruction of Captain Cook's first Pacific voyage, as a crew member and historical consultant. He wrote and presented the three-part series, *War at Sea*, for BBC2, broadcast in February 2004 and repeated in May 2004.

Nick Lee was born in Wimbledon in 1939 and went to school locally and at Kingston Grammar School. After National Service and three years in the Admiralty, he studied at Birmingham University (BA and MA) and later at London (DipLib). He came to Bristol University in 1965 and was successively lecturer in English, special collections librarian, and archivist (from 1989 to 1996 and October 2004 to October 2005). In recent years he has edited a series on Irish history.

Dr Christine MacLeod specialises in modern British economic and technological history, in particular the development of the patent system and the role of inventors in the industrial revolution. She read Modern History at Oxford, and received her PhD from Cambridge for a thesis on the English patent system and inventive activity, 1660-1753. She has taught at the University of Bristol since 1991, where she is now a senior lecturer in Economic and Social History. She is currently engaged in two major research projects: an analysis of innovation in the nineteenth-century steam-engineering industry (in collaboration with colleagues at the University of Birmingham); and *Heroes of invention: celebrating the industrial culture of nineteenth-century Britain* (a book on the cultural history of invention and its influence on the historiography of the industrial revolution). Her previous publications include *Inventing the industrial revolution: the English patent system, 1660-1800* (Cambridge University Press, 1988), which was awarded the Wadsworth Prize for 1989, and articles in journals such as *Technology & Culture*, *Economic History Review* and *Business History*. She is a member of the executive council of SHOT (Society for the History of Technology), a special adviser on the Science Museum Committee, an editorial adviser for Blackwell's History Compass, and an occasional consultant for the BBC (most recently, *What the Industrial Revolution Did for Us*); in 2002 she was Professeur Invitée at CNAM in Paris.

Dr Andrew Nahum is Senior Curator of Aeronautics at the Science Museum, London. He has a doctorate in Economic History from the London School of Economics and has written widely on the history of technology and on design for both academic and popular audiences. From 1996 to 2000 he was the Project Director and lead curator for the museum's landmark permanent gallery on the history of technology and science. He is a member of the Industrial Archaeology Panel at English Heritage and a visiting professor in the Department of Vehicle Design at the Royal College of Art.

Dr Claire I R O'Mahony specialises in the history of art, design and interior decoration of the late nineteenth and early twentieth centuries. She received her BA from the University of California at Berkeley and completed her PhD at the Courtauld Institute of Art in London, devoted to mural decoration in Third Republic French town halls. She began her professional life as a museum education officer and exhibition co-ordinator at the Courtauld Gallery and the Richard Green Gallery, London. As Director of Programmes for Lifelong Learning in the History of Art at the University of Bristol, she encourages a broad study of visual culture amongst a wide range of students of differing ages and experiences in Bristol and throughout the South West region. She has published exhibition catalogues and articles on art and design including entries in Royal Academy of Arts, *1900 Art at the Crossroads*, London, 1999, and Minneapolis Institute of Arts/High Museum, Atlanta, *Degas and America: The Early Collectors*, 2001, an introductory essay to *Essential William Morris*, 1999, and an article 'Modern Muses: representing the life model in fin de siècle France', *Art on the Line* Vol. 1, October 2003.

Michael Pascoe, a Devonian by birth, read French at the University of Bristol. After service as an infantry officer, he worked in Africa for a number of years. On returning to Bristol, he worked in public relations, marketing and education. He is the author of *The Clifton Guide*, wrote and presented a video (*Clifton - a place for all seasons*) and has published many articles on Clifton's history. Together with Adrian Andrews, he set up the Clifton Suspension Bridge Visitor Centre in 1998, which he managed for three years.

Denis Smith, PhD, MSc, DIC, CEng, is an engineering historian who has written extensively on a wide range of subjects. He is a member of the Institution of Civil Engineers Panel for Historical Engineering Works (PHEW), and of the ICE Archives Panel. He is a past president of the Newcomen Society for the Study of the History of Engineering and Technology (founded 1920). Among his recent publications are the book *Civil Engineering Heritage: London and the Thames Valley* (Thomas Telford, 2001), contributions to *Robert Stephenson - The Eminent Engineer* (Ashgate, 2003) and the new *Dictionary of National Biography* (OUP, 2004), and 'Leipzig explored' in *The Structural Engineer* (16 November 2004).

Adrian Vaughan was born in Reading in 1941 and grew up always taking a keen interest in Brunel's railway. He joined the Western Region of British Railways in 1960. He worked as a signalman but rode the steam footplate almost daily and knows Brunel's line from Bristol to London intimately. He began writing about the GWR in 1970 and has now written 25 railway history books, including a biography of Brunel, published in 1991. He has just completed a new biography of the great man.

Dr Marcus Waithe completed a PhD on the utopias of William Morris at King's College Cambridge in 2003. He is the author of several articles in prestigious academic journals, among them *Victorian Studies*, *English* and *Textual Practice*. He has contributed reviews to the *Times Literary Supplement* and to *P N Review*. In 2006, the English Association/Boydell & Brewer will publish his book on William Morris. In autumn 2004 his essay devoted to William Morris was published by the French publisher Editions du temps. He has spoken at a number of professional academic conferences and seminars, including the Ruskin Programme at Lancaster University. He took up the post of lecturer in Victorian Literature at the University of Sheffield in September 2005.

IMPERIAL-METRIC CONVERSIONS

Imperial	Metric
1 inch	2.54 centimetres
1 foot/12 inches	0.3048 metres
1 yard/3 feet	0.9144 metres
1 mile/1,760 yards	1.6093 kilometres
4 feet 8$\frac{1}{2}$ inches (standard gauge)	1.435 metres
7 feet (broad gauge)	2.1336 metres

TIMELINE OF KEY EVENTS

Year	Brunel's life and work	British events
1769	Birth of Marc Brunel (25 April).	
1793	Marc Brunel leaves France.	
1799	Marc Brunel arrives in England. Marries Sophia Kingdom (1 Nov). Promotes block-making machinery.	Political associations forbidden.
1806	Birth of Isambard Kingdom Brunel (9 April) at Portsea, Portsmouth. Family moves to Chelsea.	General election
1814	Marc Brunel's Battersea sawmills destroyed by fire.	
1818	Marc Brunel invents and patents tunnelling shield.	
1820	Brunel begins studies at Caen College.	Death of George III and accession of George IV. Failure of Cato St conspiracy to assassinate government.
1821	Marc Brunel imprisoned for debt. Released through intercession of Duke of Wellington.	Death of Queen Caroline leads to public rioting.
1822	Brunel leaves Lycée Henri-Quatre and finishes apprenticeship to the watchmaker Louis Breguet. Returns to England (21 Aug) to join father's drawing office.	
1824	Thames Tunnel Company formed. Brunel family moves to Blackfriars. Brunel given task to develop engine run on carbonic gas.	British workers allowed to unionise.

Science & engineering	Arts & letters	World events
Arkwright's waterframe spinning machine. Cugnot's steam road carriage. First lightning conductors fitted to tall buildings.		Captain Cook sails to Tahiti. Birth of Napoleon Bonaparte.
Whitney's cotton gin and Bentham's woodworking machinery. Telford begins canal work.	David paints the murder of Marat. Louvre becomes France's national gallery.	Louis XVI and Marie Antionette executed. France declares war on Britain, Holland and Russia, and Reign of Terror begins.
Rosetta Stone found in Egypt and used for deciphering hieroglyphics.	Beethoven's *Symphony No 1*. Birth of Balzac and Pushkin.	Napoleon becomes Consul. Austria, Britain, Russia and Turkey at war with France.
Humphry Davy works on electrical preparation of potassium and sodium.	Birth of John Stuart Mill and Elizabeth Barrett Browning. 'Elgin' marbles arrive in London.	Official end of the Holy Roman Empire.
Stephenson designs his first locomotive.	*The Times* becomes first newspaper printed on steam-powered presses.	Napoleon exiled to Elba. Treaty of Ghent ends war between US and Britain.
Steamship *Savannah* crosses Atlantic. Institution of Civil Engineers founded. Dover-Calais steam packet route begins.	Shelley's *Frankenstein*, Austen's *Northanger Abbey* and *Persuasion*, Byron's *Don Juan*. Birth of Emily Brontë.	Border agreed between US and Canada - the 49th parallel.
Ampère's laws of electro-dynamic action. First large-scale suspension bridge built in Britain: Brown's bridge across the Tweed.	Keats' *Ode to a Nightingale* and Scott's *Ivanhoe*. Malthus' *Principles of Political Economy*. Venus de Milo discovered.	Revolution begins in Spain.
Faraday demonstrates electro-magnetic rotation.	Hegel's *Philosophical Principles of Law* and Mill's *Elements of Political Economy*. Constable's *Haywain*. *Manchester Guardian* founded. Death of Keats.	Greek War of Liberation begins. Death of Napoleon.
First rail of Stockton & Darlington Railway laid. Babbage begins work on his calculating machine.	Death of Percy Shelley. Royal Academy of Music and *Sunday Times* founded.	Liberia founded as colony for freed American slaves.
Aspdin's Portland cement. Imperial standards of weights and measures given legal force.	Death of Byron. Beethoven's *Ninth Symphony*. Birth of Wilkie Collins. National Gallery founded.	End of Spanish influence in South America following defeat at Battle of Ayacucho.

Year	Brunel's life and work	British events
1825	Work begins on Thames Tunnel (2 Mar).	Speculation mania leads to bank failures.
1827	Formally appointed resident engineer on tunnel (3 Jan). Roof collapses (18 May) and tunnel flooded. Work resumes after draining. Banquet held in the tunnel (10 Nov).	
1828	Brunel injured in second flood at Thames Tunnel in January. Brunel learns of plans to build bridge across the Avon at Clifton while convalescing in Clifton (previously went to Brighton).	Duke of Wellington becomes Prime Minister. Non-conformists allowed to hold public office.
1829	Telford faults all submissions to first Clifton Bridge competition. Brunel commissioned to carry out drainage works at Tollesbury.	Metropolitan Police founded by Sir Robert Peel. Catholic Emancipation Act allows Roman Catholics to hold public office.
1830	Elected Fellow of the Royal Society (10 June). First full meeting of Clifton bridge fundraising committee (22 June).	Ascension of William IV. Swing Riots including machine-breaking and rick-burning.
1831	Designs for second Clifton Bridge formally accepted (16 Mar). Attends ceremony to mark start of work (21 June). Brunel completes observatory in Kensington and is commissioned to undertake new dock work at Monkwearmouth. Makes first trip by railway (5 Dec).	
1832	Brunel begins association with Bristol Dock Company. Electioneers on behalf of brother-in-law, the victorious Radical candidate for Lambeth.	First Reform Act passed, giving middle-class men over the age of 21 the right to vote.
1833	Brunel made chief engineer of the newly formed Great Western Railway (7 Mar) and starts surveying the route. Modernisation of Bristol Docks. Abandons Gaz engine.	Slave ownership abolished in British Empire. First Factory Act bans employment of children under nine. Start of railway boom.

Science & engineering	Arts & letters	World events
Stockton & Darlington Railway opened. Horse drawn buses introduced in London. Work begins on Telford's Birmingham & Liverpool Canal.		
Niepce's photographs on metal plate. Ohm formulates law of resistance. Ressel's ship's screw.	Deaths of Beethoven and Blake. Cooper's *The Last of the Mohicans*. Baedeker begins publishing travel guides.	Treaty of London declares Greek independence.
Construction of first US railway begins - Baltimore-Ohio.	Deaths of Goya and Schubert. Clausewitz's *On War* and Dumas' *The Three Musketeers*. Births of Tolstoy, Ibsen and Rossetti. Founding of *Manchester Guardian*, *Athenaeum* and *Spectator*.	
Trials of Stephenson's *Rocket*. Henry constructs early electric motor. First US steam locomotive runs on the Baltimore-Ohio. Shillibeer's horse-drawn omnibus in London. Steam coaches between London and Bath.	Birth of Millais.	
Liverpool & Manchester Railway opened (William Huskisson killed). Steam cars operating in London. Lyell's *Principles of Geology*. Royal Geographic Society founded.	Delacroix paints *Liberty Guiding the People*. Stendhal's *Rouge et Noir*.	Louis Philippe becomes King of France following revolution. First cholera epidemic in Europe.
Faraday demonstrates electro-magnetic induction. Ross determines position of magnetic North Pole. Beginning of Darwin's voyage on the Beagle. Opening of the rebuilt London Bridge. British Association for the Advancement of Science founded. William Bickford's safety fuse.	Hugo's *Notre Dame de Paris*. Constable's *Salisbury Cathedral from the Meadows*.	Foreign Legion formed. Nat Turner leads Virginia slave revolt.
Guthrie and Leibig invent chloroform. Morse's telegraph. First complete European railway - Budweis-Linz. *Royal William* crosses the Atlantic by steam. Corrugated iron made by Walker of Rotherhithe.	Tennyson's 'Lotus Eaters' and 'Lady of Shallott'. Birth of Lewis Carroll (Charles Dodgson). Deaths of Goethe and Walter Scott.	Cholera epidemic spreads from Russia to Central Europe.
Gauss and Weber devise electro-magnetic telegraph. Death of Trevithick.	Birth of Burne Jones. Ruskin's first visit to Switzerland.	

Year	Brunel's life and work	British events
1834	First Great Western Railway Bill rejected in Parliament (25 July). Underfall sluices constructed at Bristol docks. First drag boat in operation.	Tolpuddle Martyrs sentenced to transportation. National education system introduced. Fire destroys the Houses of Parliament. British workhouse system established.
1835	Second Great Western Railway Bill passed in Parliament (31 Aug). GWR accept Brunel's broad gauge proposal (29 Oct). Work restarted on Thames Tunnel. Brunel proposes building a steamship for the Bristol-New York service. Appointed engineer for Cheltenham & Great Western Union, Bristol & Exeter, Bristol & Gloucester, and Merthyr & Cardiff railways. Recommends broadening of Bristol South Lock.	London and Greenwich railway opens between London Bridge and Deptford. British Association for the Promotion of Temperance founded.
1836	Appointed engineer of Great Western Steamship Company. Brunel marries Mary Horsley (5 July) and moves to Westminster. Foundation stone for Leigh Woods abutment of Clifton Bridge laid (27 Aug). Work starts on ss *Great Western* in Bristol. Work starts on Box Tunnel. Brunel makes first crossing of the Avon in a suspended basket which became stuck 200 feet above the water.	Compulsory registration of births, marriages and deaths.
1837	Launch of hull of the *Great Western* in Bristol (19 July). Work begins on Royal Western Hotel. Daniel Gooch made GWR locomotive superintendent.	Ascension of Queen Victoria. Anti-Corn Law Association established.
1838	*Great Western* crosses the Atlantic in 15 days from Avonmouth to New York (April). Paddington to Maidenhead section of GWR opened. Work begins on Temple Meads. Brunel answers criticisms of broad gauge.	First Chartist petition demanding universal male suffrage and private ballot. Foundation of the Agricultural Society.
1839	Maidenhead bridge opened and line extended to Twyford. Great Western Steamship Company builds the new Great Western Dockyard and Engine Factory. Construction begins on *Great Britain*. Appointed engineer to South Devon Railway. Gooch's locomotive designs chosen over Brunel's. Electric telegraph installed between Paddington and Hanwell on GWR.	Chartist riots. Centralised Anti-Corn Law League formed.

Science & engineering	Arts & letters	World events
Death of Thomas Telford. Babbage develops analytic engine. Faraday's electrical self-induction. Fixed railway signals introduced on the Liverpool & Manchester.	Birth of Whistler and William Morris.	End of the Spanish Inquisition.
Halley's comet reappears. First German railroad. Fox Talbot takes negative photograph at Lacock Abbey. Geological Survey of Great Britain established.	Publication of Andersen's first collection of fairy tales.	
Ericsson patents screw propeller. The *Beagle* returns to Britain. Foundation of Edinburgh Botanical Society.	Gogol's play, *The Government Inspector*. Birth of W S Gilbert.	The siege of the Alamo. Boer farmers launch Great Trek.
Telegraph system patented by Wheatstone and Cooke. Morse exhibits electric telegraph. First Canadian railroad. Pitman's short hand.	Death of Constable. Serialisation begins of Dickens' *Oliver Twist* and *Pickwick Papers* (includes scenes set in Bristol).	
Daguerre demonstrates photography. Completion of London-Birmingham Railway Line.	Serialisation begins of Dickens' *Nicholas Nickleby*. National Gallery opens.	Boers defeat the Zulus at Battle of Blood River.
Macmillan builds first bicycle. Goodyear discovers rubber vulcanisation. Darwin's *Beagle* journals published. Fox Talbot produces photographic negative. Nasmyth's steam hammer.	Poe's *The Fall of the House of Usher*. Turner's *Fighting Temeraire*.	Start of the first Opium War and the first Afghan War.

Year	Brunel's life and work	British events
1840	*Great Britain*'s hull redesigned for screw propeller. Opening of GWR sections from London to Hay Lane, and from Bristol to Bath (31 Aug). Gooch chooses Swindon to be main base for engines.	Queen Victoria marries Prince Albert. Public postal service introduced. Creation of Railway Department of the Board of Trade. Construction begins on new Houses of Parliament.
1841	Marc Brunel knighted. Box Tunnel and Bristol Temple Meads station completed. Full London-Bristol route opened (30 June). GWR associate company Bristol & Exeter Railway formed. Appointed engineer of Oxford Railway. Eight passengers killed at Sonning cutting (24 Dec).	Birth of future King Edward VII.
1842	Opening of Bristol to Taunton section of Bristol & Exeter (1 July). Urges new measures to overcome silting of Floating Harbour.	Queen Victoria travels by train from Windsor to Paddington. Chartist riots. Mines Act prohibits women and children working underground. Chadwick's report, *Enquiry into the Sanitary Conditions of the Labouring Population of Great Britain.*
1843	Swindon locomotive works opened (1 Feb). Thames Tunnel opened to public (25 Mar). Launch of *Great Britain* at Bristol (19 July). Visits Italy to engineer two railways. Appointed engineer of South Devon Railway.	
1844	Lines opened to Exeter and Oxford. Appointed engineer to South Wales, Wilts, Somerset & Weymouth, Oxford, Worcester & Wolverhampton, Berks & Hants, Oxford & Rugby, and Monmouth & Hereford railways. Further Italian visits.	Gladstone's Railway Act makes third-class provision compulsory. Factory Act restricts women to 12 working hours a day, children to $6^1/_2$ hours. Co-operative Society Founded. Start of railway mania (share prices rise). Lord Shaftsbury organises 'ragged schools' teaching reading, writing, arithmetic and bible study to poor inner city children.
1845	Maiden voyage of *Great Britain*. Work begins on new lock at South entrance of Bristol Docks. Appointed engineer of Cornwall and West Cornwall railways. Hungerford suspension footbridge opens.	

Science & engineering	Arts & letters	World events
An atmospheric railway is demonstrated in London and seen by Brunel. Draper photographs the moon. Kew Gardens opens. Russell Wallace begins geological survey of western England and Wales. Joule begins work on heat.	Births of Thomas Hardy, Zola, Monet, Renoir, Tchaikovsky and Rodin. Turner meets Ruskin for first time.	
Bunsen's carbon-zinc battery. British Pharmaceutical Society founded. First director of Kew Gardens appointed.	First edition of *Punch*. Dickens' *The Old Curiosity Shop*, Cooper's *The Deerslayer* and Poe's *The Murders in the Rue Morgue*. Birth of Renoir.	Massacre of British officers leads to second Afghan war.
Long uses ether. Morse lays telegraph cable in New York harbour. Mayer's Law of Conservation of Energy. America's first wire cable bridge opened at Philadelphia (Ellet and Roebling).	Publication of Poe's *The Mask of the Red Death*. *The Illustrated London News* founded.	End of the Opium War leads to cession of Hong Kong to Britain (Treaty of Nanking).
Nelson's column erected in Trafalgar Square. Demonstration of suction railway system near Dublin. Paddington-Slough telegraph line in operation. Forbes publishes theory of glacial ice.	Dickens' *Christmas Carol*. Ruskin's *Modern Painters* begins. Carlyle's *Past and Present*. *The Economist* founded. Birth of Henry James. Wagner's *Flying Dutchman*.	
Reinforced concrete patented. Pratt truss for bridge construction designed. Morse sends first telegraph message from Washington to Baltimore. Fox Talbot starts publication of *The Pencil of Nature*. Hooker begins work with Darwin cataloguing Galapagos Island plant specimens. Chambers' *Vestiges of the Natural History of Creation*.	Turner's *Rain, Steam and Speed*.	
Armstrong patents hydraulic crane. McNaught's compound steam engine. Thompson's pneumatic tyre. E P Thompson's *Note-book of a Naturalist* initiates the modern study of natural history. Tarmacadam laid in Nottingham.	Engel's *Condition of the Working Class in England*. Wagner's *Tannhauser*. Turner Acting President of Royal Academy.	Start of the Irish famine.

Year	Brunel's life and work	British events
1846	Gooch's locomotive the *Great Western* arrives from the Swindon works. Line extended to Teignmouth and Newton Abbot. *Great Britain* runs aground at Dundrum Bay (23 Sept). Appointed engineer of the Birmingham & Oxford Junction Railway.	Government act favours standard over broad gauge for all new railtracks. Corn Law repealed, reducing the tax on bread.
1847	Atmospheric system used on Exeter-Teignmouth stretch. *Great Britain* refloated (27 Aug). Great Western Steamship Company bankrupted. Buys land at Watcombe in Torquay for an estate.	The Ten Hours Act limited women and children to ten hours' work a day. Poor Law Commission abolished following scandal of abuses at Andover Workhouse.
1848	Atmospheric system extended to Newton Abbot but later abandoned. Designs bridges for Slough-Windsor branch and at Chepstow, and viaducts at Landore and Newport.	Public Health Act forcing local authorities to institute sanitation measures and introduce local boards of health and medicine. Last big Chartist demonstration.
1849	Death of Marc Brunel (12 Dec). Windsor branch of GWR opened. South Devon Railway completed to Plymouth. New lock completed at Cumberland Basin, spanned by tubular swing bridge. Begins designing permanent station for Paddington.	16,000 Londoners killed by cholera.
1851	Designing timber trestle viaducts for Cornish railways.	Great Exhibition of world-wide manufacturers held in Hyde Park in Paxton's Crystal Palace, the first large pre-fabricated building (Brunel contributed to early designs and is in opening parade). Amalgamated Society of Engineers founded: the beginning of New Model Unionism.
1852	Work begins on Paddington Station. West Cornwall Railway opened. First regular steamship voyages to Australia inaugurated by *Great Britain*. Chepstow Bridge opens. Designs brick water-carrying towers for relocated Crystal Palace. Appointed engineer of Eastern Steam Navigation Company.	King's Cross Station opens, designed by Cubitt. Death of the Duke of Wellington.
1853	Work begins on the *Great Eastern*. Begins supervising work on Royal Albert Bridge over Tamar at Saltash.	Compulsory vaccination against smallpox introduced. Anaesthetic gains popularity.
1854	Paddington Station opened. Appointed engineer of unrealised Westminster Terminus Railway.	Crystal Palace reassembled at Sydenham. London Working Men's College founded. London cholera epidemic.

Science & engineering	Arts & letters	World events
Galle discovers planet Neptune. Howe's sewing machine patented in US.	Lear's *Book of Nonsense*. Second volume of *Modern Painters*.	
First operation using chloroform. Institution of Mechanical Engineers founded. Birth of Alexander Graham Bell and Edison.	Publication of *Jane Eyre*, *Agnes Grey* and *Wuthering Heights* by the Brontë sisters. Verdi's *Macbeth*.	
Death of George Stephenson. First safety matches produced. Work begins on Roebling's Niagara Railway Bridge.	Formation of Pre-Raphaelite Brotherhood. Marx and Engel's *The Communist Manifesto* and Mill's *Principles of Political Economy*. Dickens' *Dombey and Son* and Gaskell's *Mary Barton*.	Revolutionary uprisings across Europe. Start of California Gold Rush. Louis Napoleon elected President of Second Republic in France.
Hunt patents the safety pin. Fizeau measures speed of light.	Serialisation begins of Dickens' *David Copperfield*.	
Telegraph cable laid between France and Britain (Brett brothers). Singer invents practical sewing machine. Bunsen invents gas burner. Kelly's steel-making convertor. Scott Archer's wet collodion photography.	Death of Turner. Ruskin's *The Stones of Venice* and Melville's *Moby Dick*. First edition of *The New York Times*. Verdi's *Rigoletto*. Reuters formed in London - the first news agency.	
Giffard's steam airship. Gisborne's submarine cable between New Brunswick and Prince Edward Island. Death of British bridge designers Sir Samuel Brown and William Tierney Clark.	Serialisation starts of Dickens' *Bleak House*. Stowe's *Uncle Tom's Cabin*. Ford Maddox Brown's *The Last of England*.	David Livingstone explores the Zambesi. Beginning of the Second Empire in France with Louis Napoleon declared Napoleon III. Military action in Burma to protect British merchants.
Vienna-Trieste railway runs through the Alps. Photographic Society founded by Roger Fenton. First manned flight by glider.	Verdi's *La Traviata*. Birth of Van Gogh.	Haussmann begins reconstruction of Paris boulevards.
Elder's compound marine engine.	Dickens' *Hard Times* and Tennyson's 'The Charge of the Light Brigade'. Thoreau's *Walden*.	Outbreak of the Crimean War (Britain, France, Turkey and Sardinia against Russia).

Year	Brunel's life and work	British events
1855	Brunel designs pre-fabricated hospital for the Crimea and *Great Britain* carries troops to and from the war. World's first postal train on the London-Bristol route.	Friendly Societies Act protects trade union funds.
1856	Russell, partner in *Great Eastern*, is bankrupted leading to crisis in construction (4 Feb).	
1857	*Great Western* broken up at Vauxhall. Brunel honoured at Oxford University. *Great Britain* carries troops to the Indian Mutiny. Failed attempt to launch *Great Eastern* (3 Nov). Planning the East Bengal Railway.	Matrimonial Causes Act makes divorce easier.
1858	*Great Eastern* launched (31 Jan). New company (Great Ship Company) formed to undertake completion when Eastern Steam Navigation Company dissolved (25 Nov). Goes to Alps, Vichy and Egypt for health.	Abolition of property qualification for MPs. Jews are admitted to Parliament. Last Chartist Convention. Postal districts introduced in Central London.
1859	Sea trials of *Great Eastern*. Men killed in engine room when feed-water heating jacket explodes. Royal Albert Bridge completed. Death of Brunel (15 Sept).	
1863	Temporary Clifton bridge with partially laid planking first crossed on 3 July.	
1864	Opening of Clifton Suspension Bridge (8 Dec). Used chains from Hungerford bridge, purchased for £5,000.	Contagious Diseases Act. First national conference of trade union delegates.
1870	*The Life of Isambard Kingdom Brunel, Civil Engineer* published.	Education Act: all children between five and 13 must attend school. Bankruptcy Act abolishes debtors' prisons. Telegraphic service transferred to Post Office.

Science & engineering	Arts & letters	World events
Hughes' printing telegraph. Gregor Mendel begins his study of peas to see how features are passed from generation to generation. London–Balaclava electric telegraph.	Birth of William Friese-Green. Death of Charlotte Bronte. *Daily Telegraph* founded. Whitman's *Leaves of Grass*.	World Exhibition in Paris. Livingstone finds Victoria Falls. Florence Nightingale introduces hygienic standards into military hospitals. Australian Colonies become self governing.
Bessemer's converter for steel production introduced, reducing the cost of making steel. Birth of Freud. Atlantic Telegraph Co launched. Pasteur discovers airborne germs.	Serialisation begins of Dickens' *Little Dorritt*. Henry Irving's London debut. Births of Bernard Shaw and Oscar Wilde.	End of Crimean War (Treaty of Paris). Outbreak of second Opium War.
Otis installs first safety elevator. Science Museum founded.	Flaubert's *Madame Bovary*. Births of Joseph Conrad and Edward Elgar. National Portrait Gallery and Victoria & Albert Museum open.	Indian Mutiny begins.
First transatlantic telegraph cable laid (breaks). Darwin and Wallace's joint paper on variation of species.	Frith's *Derby Day*. Offenbach's *Orpheus in the Underworld*.	British Crown gains control of India from East India Company.
Publication of Darwin's *The Origin of Species*. Death of Robert Stephenson. Steam-roller invented. Merrell's electric washing machine.	Dickens' *A Tale of Two Cities* and Mill's *On Liberty*. Birth of Conan Doyle.	Work begins on Suez Canal.
Opening of GWR-operated Metropolitan Line. Thomas Huxley publishes *Evidence as to Man's Place in Nature*.	Manet's *Dejeuner sur l'Herbe*.	Lincoln's Emancipation Proclamation.
Pasteur invents pasteurisation.	Tolstoy's *War and Peace* published. Birth of Toulouse-Lautrec and Richard Strauss.	International Red Cross founded at the Geneva Convention. Confederate army defeated at Atlanta.
Rockefeller founds Standard Oil. Gramme's dynamo.	Death of Dickens. Monet and Pissarro flee to England to escape Franco-Prussian war.	Siege of Paris by Prussia. French Third Republic formed. Birth of Lenin.

Year	Brunel's life and work	British events
1877	Statue of Brunel by Marochetti unveiled at Victoria Embankment.	Society for the Protection of Rural England formed by William Morris and colleagues. Completion of St Pancras Station.
1950	Lady Celia Noble gives the University of Bristol Library the original collection of papers and other materials relating to Brunel, Marc Brunel, Isambard Brunel Jnr and Henry Marc Brunel.	Marshall Aid to Britain ends. National Service extended to two years. London dock strike.
1965	Original Temple Meads station ceases operation.	General Post Office Tower opens. Death penalty abolished. State funeral of Churchill. GLC formed.
1970	*Great Britain* towed back to Bristol from the Falkland Islands.	
1972	Clifton Suspension Bridge (CSB) Trust deposit papers relating to the Clifton Suspension Bridge at the University.	Coal strike. Direct rule imposed on Northern Ireland.
1990	University purchases additional papers still held by the Brunel family.	Margaret Thatcher resigns.
1996/97	Purchase of further Brunel papers.	Death of Princess Diana. BSE outbreak.
2002	CSB Trust deposits additional papers at University and carries out a survey of holdings with help from University staff. Opening of British Empire & Commonwealth Museum.	Queen Mother dies. British troops make ready for war with Iraq.
2005	Opening of new visitor centre at ss *Great Britain*. Digitisation of part of Brunel collection.	Historic third term Labour government. Suicide bombers in London.
2006	Brunel 200	

Science & engineering	Arts & letters	World events
Edison invents phonograph. Siemens exhibits first electric train. Momer's reinforced concrete beams. First public telephone.	Zola's *L'Assommoir*.	Victoria becomes Empress of India. Famine in India.
US Atomic Energy Commission separates plutonium and is instructed to develop hydrogen bomb.	Cocteau's *Orphee*, Kuosawa's *Rashomon*, Ophul's *La Ronde*, Wilder's *Sunset Boulevard*.	Britain recognises Israel and Communist China. Start of Korean War. UN Building, New York, completed.
Cosmonaut Leonov floats in space. US Gemini V makes 120 orbits of earth. Dungenness Atomic Power Station opens.	Death of Nat King Cole. Beatles awarded MBE.	Malcolm X assassinated. Martin Luther King leads civil rights march on Montgomery. Anti-Vietnam War protests.
Launch of Apollo 13.	Death of Rothko. *True Grit* and the film of *Woodstock*.	End of civil war in Biafra. Killing of student protestors at Kent State University. Allende elected president of Chile.
Apollo 16 lands on moon. Soviet Venus 8 lands on Venus.	Deaths of Mahalia Jackson and Maurice Chevalier. *Cabaret* and *The Godfather*.	Bangladesh established. Nixon visits China. Death of Israeli athletes at Munich Olympics.
Hubble space telescope launched. Berners-Lee writes first World Wide Web programme. Kodak introduces Photo CD player.	*Good Fellas*. Octavio Paz wins Nobel Prize for Literature.	Nelson Mandela released from prison. Iraq invades Kuwait. Lech Walesa becomes Polish President.
Mars Pathfinder probe. Scientists clone sheep. Tallest building to date built at Kuala Lumpur.	First Harry Potter book. *The Full Monty*.	Britain cedes Hong Kong back to China.
House of Lords back research on cloned embryos. Ice shelf in Antarctica melts.	Deaths of Spike Milligan and Billy Wilder.	India and Pakistan on brink of nuclear war. Enron, Worldcom and other corporate scandals. Euro notes and coins in circulation. Iraq accepts UN weapon inspectors. Zimbabwe barred from Commonwealth.
Severe weather prompts new warnings on global warming. Viaduc de Millau completed.	Harold Pinter wins Nobel Prize for Literature.	Relief efforts continue in wake of Boxing Day 2004 tsunami.

BIRTH AND DEATH DATES OF KEY INDIVIDUALS

Albert, Prince (1819-1861)

Arup, Ove (1895-1988)

Babbage, Benjamin Herschel (1815-1878)

Babbage, Charles (1791-1871)

Barlow, Peter (1776-1862)

Barlow, W H (1812-1902)

Bazalgette, Joseph William (1819-1891)

Bentham, Samuel (1757-1831)

Bidder, George Parker (1806-1878)

Birch, Eugenius (1818-1884)

Bourne, John Cooke (1814-1896)

Braithwaite, John (1797-1870)

Brereton, Robert Pearson (1819-1894)

Brindley, James (1716-1772)

Brodie, Benjamin (1783-1862)

Brown, Ford Maddox (1821-1893)

Brown, Samuel (1776-1852)

Brunel, Florence (1847-1876)

Brunel, Henry Marc (1842-1903)

Brunel, Isambard (1837-1902)

Brunel, Isambard Kingdom (1806-1859)

Brunel, Marc (1769-1849)

Burge, George (1795-1874)

Callcott, Elizabeth Hutchins (1793-1875)

Carlyle, Thomas (1795-1881)

Clark, Edwin (1814-1894)

Clark, Latimer (1822-1898)

Clark, William Tierney (1783-1852)

Clarkson, Thomas (1760-1846)

Colston, Edward (1636-1721)

Cooke, William Fothergill (1806-1879)

Cope, Charles West (1811-1890)

Crampton, T R (1816-1888)

Cruikshank, George (1792-1878)

Cubitt, William (1785-1861)

Davy, Humphry (1778-1829)

Dickens, Charles (1812-1870)

Egg, Augustus Leopold (1816-1863)

Ellet, Charles (1810-1862)

Errington, John (1806-1862)

Fairbairn, William (1789-1874)

Faraday, Michael (1791-1867)

Fenton, Roger (1819-1869)

Field, Joshua (1757-1863)

Finley, James (1756-1828)

Fox Talbot, W H (1800-1877)

Fox, Charles (1810-1874)

Frith, William Powell (1819-1909)

Froude, William (1810-1879)

Fry, Roger (1866-1934)

Galton, Douglas (1822-1899)

Gilbert, Davies MP (1767-1839)

Gooch, Daniel (1816-1889)

Gooch, Thomas Longridge (1808-1882)

Grainger, Thomas (1794-1852)

Gravatt, William (1809-1866)

Guppy, Thomas Richard (1797-1882)

Harrison, Thomas E (1808-1888)

Hartley, Jesse (1780-1860)

Hawes, Benjamin (1797-1862)

Hawkshaw, John (1811-1891)

Hawksley, Thomas (1807-1893)

Hodgkinson, Eaton (1789-1861)

Horsley, John Callcott (1817-1903)

INDEX

ACKNOWLEDGEMENTS

Brunel 200 is a partnership initiative led by Bristol Cultural Development Partnership (Arts Council England South West, Bristol City Council and Business West).

It is funded by:

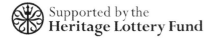

A great debt is owed to many people for their help and support in the production of *Brunel: in love with the impossible*. First, our thanks go to the authors who have provided the chapters. We would also like to thank those that have financed the production of the book. We have benefited from the support of Arup, W S Atkins, Halcrow, Hyder Consulting, Parsons Brinckerhoff, the Society of Merchant Venturers, Whitbybird and WSP. Without the authors and financial supporters this book would not have been published. We are especially grateful to Terry Hill and Alf Perry of Arup who took on the fundraising work for the book, and to the following in the companies and organisations concerned for their support: Mike Casey, Dave Darnell, Liz Gibson, Peter Kydd, Richard Morris, Marion Myers, Norman Schunter, Trudy Warrender, Mark Whitby. Alf Perry has also been of great help in much of the development of this book.

We were determined from the outset to make this a publication that would last. The authors of each chapter have made this a possibility by providing excellent overviews of the subject and revealing much new information. So have too those who have provided images for use. The copyright holders of each image are noted in the text. Our thanks to the many staff involved in each organisation for their support.

Many individuals have also been of assistance. We would like to thank Eugene Byrne, Mike Chrimes, Madge Dresser, Julia Elton, June Gadd, Sue Giles, Anna Harrison, Warwick Hulme, Andy King, Hannah Lowery, Mandy Leivers, Maggie Mayo, Carol Morgan, John Powell, Michael Richardson, Sonia Rolt, Mike Rowland, Annette Ruehlmann, John Sansom, Laura Shears, Paul Stevenson. Deserving of special thanks for service to this book, but also to the work of Bristol Cultural Development Partnership throughout its 13 years of work are two people. Without Louis Sherwood we would not have been able to get started and he was there at the end with proof reading and like all proof readers offered much more. Barry Taylor has long been a friend of our work. He copy-edited this book with the excellence and good humour that we have come to expect from him. We would also like to thank our designers, Qube Design Associates Limited, for such wonderful work on the book, Gareth Davies and Karen Walk in particular. Finally, we would like to thank all those that have played a part in Brunel 200. It has truly been a team effort.

If we have missed anyone, we hope that they will accept our apologies. Needless to say, any errors left are our responsibility.